PHYSICS LABORATORY EXPERIMENTS

Sixth Edition

JERRY D. WILSON

Lander University

CECILIA A. HERNÁNDEZ HALL

American River College

Houghton Mifflin Company

Boston   New York

Publisher: Charles Hartford
Executive Editor: Richard Stratton
Editorial Associate: Rosemary Mack
Associate Project Editor: Shelley Dickerson
Senior Marketing Manager: Katherine Greig
Marketing Associate: Alexandra Shaw
Senior Manufacturing Coordinator: Marie Barnes

Cover image credits: Photo of students and computer—Jon Hanks/PASCO scientific;
Photo of Pendulum © Loren Winters/Visuals Unlimited

Printed in the U.S.A.

Library of Congress Control Number: 2003115599

ISBN-13: 978-0-618-56856-7
ISBN-10: 0-618-56856-5
N04298
3 4 5 6 7 8 9-COS-08 07 06

"What is the meaning of it all, Mr. Holmes?"
"Ah. I have no data. I cannot tell," he said.

Arthur Conan Doyle, *The Adventures of the Copper Beeches,* 1892

Contents

Note:

In this Table of Contents, the experiments that incorporate both traditional and computer instruction are denoted with a "TI/CI" abbreviation in front of the title of the experiment. In some instances, the TI and CI components are not the same, so the title for each component is listed separately. The experiments that only involve traditional instruction do not have any abbreviations next to them.

Preface

Physics Laboratory Experiments was written for students of introductory physics—in fact, it was written at the request of my students. The main purpose of laboratory experiments is to augment and supplement the learning and understanding of basic physical principles, while introducing laboratory procedures, techniques, and equipment. This sixth edition of *Physics Laboratory Experiments* has 33 experiments, with 16 more available for customization, providing an ample number of experiments to choose from for a 2-semester or 3-quarter laboratory physics course. Those features that proved effective in previous editions have been retained, and new and exciting features have been added. Here are some of the new features of this edition that we feel are especially important.

Computerized Instruction Integrated into Selected Experiments

In the previous edition, we included a supplement on Integrating Computerized Instruction and made note of available computer equipment for particular experiments. In the last few years, the use of computerized instruction and equipment has become increasingly popular in introductory physics laboratories. Therefore, for this sixth edition, we have integrated this component into the book, providing both computerized instruction (CI) and traditional instruction (TI) for 10 of the 33 experiments.

We suggest that students first do the hands-on TI experiment to gain a basic knowledge of what is being measured and of graphical analysis. It is here that the physical parameters of an experiment are clearly associated with principles and with the desired results. Once the students have this type of acquaintance with the concept at hand, they can perform the CI experiment or can view it as a demonstration. Now the student will better understand the computer procedure and analysis of electronically recorded data, with resulting graphs immediately plotted on the screen. The computerized and traditional components treat the same principle, but from different perspectives, together giving the student a more comprehensive understanding of the underlying physical concepts.

With this sixth edition, we carefully selected ten experiments to have both TI and CI components, thereby giving the instructor the option of doing the TI experiment, the CI experiment (using PASCO equipment), or *both,* as suggested above. Four more TI and CI experiments are available for customization.

New Co-Author

To make all this possible, Professor Cecilia A. Hernández Hall at American River College in Sacramento, California, has joined the project as co-author. Having used computerized instruction with hundreds of her students, she is highly qualified to present the CI experiments. Professor Hernández received her MS in physics from the University of Puerto Rico, Mayagüez Campus, in 1993. After teaching for two years at the Mayagüez and Cayey campuses, she joined the Physics Education Research Group at the University of Nebraska, Lincoln, where research in physics education was conducted with an emphasis on computerized instruction. Professor Hernández joined the faculty of American River College in 1998 and has worked for PASCO Scientific in Roseville, California, helping develop computer-assisted lab activities for college physics, as well as writing manuals for new equipment.

Additional Experiments Available for Customized Publishing

Also new to this edition of *Physics Laboratory Experiments* is a handy customizable option—a way for instructors to build their own lab manual that fits the need of their specific course. We've pared down this sixth edition of the manual to 33 experiments from the 49 experiments that appeared in the fifth edition. To produce a shorter, more manageable text, we chose the 33 experiments most used by professors. However, the extra 16 experiments, including 4 integrated TI and CI experiments, are still available through Houghton Mifflin's Custom Courseware website, accessible via **catalog.college.hmco.com.**

Using this online book-building system, instructors can select from the complete set of 49 experiments to create text materials customized to their course objectives and can arrange the experiments in whatever sequence they prefer. Thus students pay only for the content they need. Instructors may also want to add their own materials, experiments, lab reports, or handouts to make the book even more relevant to the students.

To learn more about customizable options for *Physics Laboratory Experiments,* visit the online catalog page for this manual at **catalog.college.hmco.com** and search by author. You can also request a sample copy of the experiments by calling Houghton Mifflin's Faculty Services at 1-800-733-1717 or by requesting a copy via our online catalog at **catalog.college.hmco.com.** Because instructors can either choose this sixth-edition bound volume, with its selection of the most popular experiments, or assemble a custom package, there is an option appropriate for every introductory physics laboratory course. Here is a list of the additional experiments found on the Houghton Mifflin Custom Courseware website:

34. The Scientific Method: The Simple Pendulum
35. (TI/CI) Rotational Motion and Moment of Inertia
36. Conservation of Angular Momentum: The Ballistic Pendulum
37. Elasticity: Young's Modulus
38. Air Column Resonance: The Speed of Sound in Air
39. (TI) Latent Heats: Heats of Fusion and Vaporization of Water
    (CI) Latent Heat of Fusion for Water
40. Newton's Law of Cooling: The Time Constant of a Thermometer
41. The Potentiometer: emf and Terminal Voltage
42. The Voltmeter and Ammeter
43. Resistivity
44. Multiloop Circuits: Kirchhoff's Rules
45. Earth's Magnetic Field
46. Introduction to the Oscilloscope
47. (TI/CI) Phase Measurements and Resonance in ac Circuits
48. (TI/CI) Electromagnetic Induction
49. The Mass of an Electron: e/m Measurement

Organization of the Sixth Edition
Both the TI and the CI experiments are generally organized into the following sections:

- **Advance Study Assignment**
- **Introduction and Objectives**
- **Equipment Needed**
- **Theory**
- **Experimental Procedure**
- **Laboratory Report**
- **Post-lab Questions**

Features include:

Laboratory safety. Safety is continually stressed and highlighted in the manual. This critical issue is expanded upon in the Introduction to the manual.

Advance Study Assignments. Students often come to the laboratory unprepared, even though they should have read the experiment before the lab period to familiarize themselves with it. To address this problem, an Advance Study Assignment precedes each experiment. The assignment consists of a set of questions drawn from the Theory and Experimental Procedures sections of the experiment. To answer the questions, students must read the experiment before the lab period; consequently, they will be better prepared. We recommend that the Advance Study

Assignment be collected at the beginning of the laboratory period.

Example Calculations. In the Theory section of some experiments, sample calculations that involve the equations and mathematics used in the experiment have been included where appropriate. These demonstrate to the student how experimental data are applied.

Illustrations. Over 200 photographs and diagrams illustrate experimental procedures, equipment, and computer programs. To allow for variation in laboratory equipment, different types of equipment that can be used are often illustrated.

Laboratory Reports. Because a standardized format for laboratory reports greatly facilitates grading by the instructor, a Laboratory Report is provided for both TI and CI experiments. These reports provide a place for recording data, calculations, experimental results, and analyses. Only the Laboratory Report and post-lab Questions that follow it need to be submitted for grading. The Laboratory Report tables are organized for easy data recording and analysis. Students are reminded to include the units of measurement.

Maximum Application of Available Equipment. Laboratory equipment at many institutions is limited, and often only standard equipment, purchased from scientific suppliers, is available. The TI experimental procedures in this manual are described for different types of common laboratory apparatus, thus maximizing the application of the manual.

Instructor's Resource Manual
The Instructor's Resource Manual is a special feature and resource for the instructor. It is available online on the instructor website prepared to accompany the sixth edition of *Physics Laboratory Experiments*. To view a sampling of instructor materials, go to **http://college.hmco.com**, click on the instructor tab, and select *Physics* and then this textbook.

Professor Fred B. Otto, previously of the Maine Maritime Academy, who has over 20 years of teaching and laboratory experience, has revised this manual. He retained the general format of the previous edition. For each experiment, there are (1) Comments and Hints, (2) Answers to post-Experiment Questions, and (3) Post-lab Quiz Questions [completion and multiple-choice (with answers), and essay]. The Instructor's Resource Manual also includes laboratory safety references, lists of scientific equipment suppliers and physics software suppliers, and graph paper copy masters.

Of course, the publication of this manual would not have been possible without a great deal of help. Professor Hernández and I would like to thank the people at PASCO—in particular, Paul A. Stokstad, Dave Griffith, and Jon and Ann Hanks—for their support and help. We thank Fred B. Otto for his in-depth review of the experiments. We are grateful to Richard Stratton, executive editor, to Rosemary Mack, editorial associate, to Shelley Dickerson, associate project editor, to Katherine Greig, senior marketing manager, and to Alexandra Shaw, marketing associate—all at Houghton Mifflin—and to Merrill Peterson at Matrix Productions Inc. We both hope that you will find the sixth edition of *Physics Laboratory Experiments* helpful and educational. And we urge anyone—student or instructor—to pass on to us any suggestions that you might have for improvement.

Jerry D. Wilson
Emeritus Professor of Physics
Lander University
Greenwood, South Carolina
jwilson@greenwood.net

Cecilia A. Hernández Hall
Professor of Physics
American River College
Sacramento, California
hernanc@arc.losrios.edu

Introduction

WHY WE MAKE EXPERIMENTAL MEASUREMENTS

When you can measure what you are speaking about and express it in numbers, you know something about it; but when you cannot measure it, when you cannot express it in numbers, your knowledge is of a meager and unsatisfactory kind.

LORD KELVIN

As Lord Kelvin so aptly expressed, we measure things to know something about them—so that we can describe objects and understand phenomena. Experimental measurement is the cornerstone of the *scientific method,* which holds that no theory or model of nature is tenable unless the results it predicts are in accord with experiment.

The main purpose of an introductory physics laboratory is to provide "hands-on" experiences of various physical principles. In so doing, one becomes familiar with laboratory equipment and procedures and with the scientific method.

In general, the theory of a physical principle will be presented in an experiment, and the predicted results will be tested by experimental measurements. Of course, these well-known principles have been tested many times before, and there are accepted values for certain physical quantities. Basically you will be comparing your experimentally measured values to accepted theoretical or measured values. Even so, you will experience the excitement of the scientific method. Imagine that you are the first person to perform an experiment to test a scientific theory.

GENERAL LABORATORY PROCEDURES

Safety

The most important thing in the laboratory is your safety and that of others. Experiments are designed to be done safely, but proper caution should always be exercised.

A potential danger comes from a lack of knowledge of the equipment and procedures. Upon entering the physics lab at the beginning of the lab period, you will probably find the equipment for an experiment on the laboratory table. Restrain your curiosity and do not play with the equipment. You may hurt yourself and/or the equipment. A good general rule is:

Do not touch or turn on laboratory equipment until it has been explained and permission has been given by the instructor.

Also, certain items used in various experiments can be particularly dangerous, for example, hot objects, electricity, mercury lamps, and radioactive sources. In some instances, such as with hot objects and electricity, basic common sense and knowledge are required.

However, in other instances, such as with mercury lamps and radioactive sources, you may not be aware of the possible dangers. Mercury lamps may emit ultraviolet radiation that can be harmful to your eyes. Consequently, some sources need to be properly shielded. Some radioactive sources are solids and are encapsulated to prevent contact. Others are in liquid form and are transferred during an experiment, so there is a danger of spillage. Proper handling is therefore important.

In general, necessary precautions will be given in the experiment descriptions. *Note them well.* When you see the arrow symbol in the margin as illustrated here, you should take extra care to follow the procedure carefully and adhere to the precautions described in the text. As pointed out earlier, experiments are designed to be done safely. Yet a common kitchen match can be dangerous if used improperly. Another good rule for the laboratory is:

If you have any questions about the safety of a procedure, ask your instructor before doing it.

The physics lab is a place to learn and practice safety.

Equipment Care

The equipment provided for the laboratory experiment is often expensive and in some instances quite delicate. If used improperly, certain pieces of apparatus can be damaged. The general rules given above concerning personal safety also apply to equipment care.

Even after familiarizing oneself with the equipment, it is often advisable or required to have an experimental setup checked and approved by the instructor before putting it into operation. This is particularly true for electrical experiments. Applying power to improperly wired circuits can cause serious damage to meters and other pieces of apparatus.

If a piece of equipment is broken or does not function properly, it should be reported to the laboratory instructor. Also, after you complete an experiment, the experimental setup should be disassembled and left neatly as found, unless you are otherwise instructed,

If you accidentally break some equipment or the equipment stops working properly during an experiment, *report it to your instructor.* Otherwise, the next time the equipment is used, a great deal of time may be wasted trying to get good results.

Laboratory Reports

A Laboratory Report is provided for each experiment in which experimental data are recorded. This should be done *neatly.* Calculations of experimental results should be included. Remember, the neatness, organization, and explanations of your measurements and calculations in the Laboratory Report represent the quality of your work.

EXPERIMENT 1

Experimental Uncertainty (Error) and Data Analysis

/TI/ *Advance Study Assignment*

Read the experiment and answer the following questions.

1. Do experimental measurements give the true value of a physical quantity? Explain.

2. Distinguish between random (statistical) error and systematic error, and give an example of each.

3. What is the difference between determinate and indeterminate errors?

4. What is the difference between measurement accuracy and precision? Explain the general dependence of these properties on the various types of errors.

(continued)

5. What determines how many figures are significant in reported measurement values? What would be the effect of reporting more or fewer figures or digits than are significant?

6. In expressing experimental error or uncertainty, when does one use (a) experimental error and (b) percent difference?

7. (Optional) For a series of experimentally measured values, distinguish among (a) the average or mean value, (b) the deviation from the mean, and (c) the mean deviation.

8. (Optional) What is the statistical significance of one standard deviation? Two standard deviations?

9. How could the function $y = 3t^2 + 4$ be plotted on a Cartesian graph to produce a straight line? What would be the numerical values of the slope and intercept of the line?

Experimental Uncertainty (Error) and Data Analysis

INTRODUCTION AND OBJECTIVES

Laboratory investigations involve taking measurements of physical quantities, and the process of taking any measurement always involves some experimental uncertainty or error.* Suppose you and another person independently took several measurements of the length of an object. It is highly unlikely that you both would come up with exactly the same results. Or you may be verifying the value of a known quantity and want to express uncertainty, perhaps on a graph. Therefore, questions such as the following arise:

- Whose data are better, or how does one express the degree of uncertainty or error in experimental measurements?
- How do you compare your experimental result with an accepted value (if known)?
- How does one graphically analyze and report experimental data?

In this introductory study experiment, we examine the types of experimental uncertainties and some methods of error and data analysis that may be used in subsequent experiments.

After performing the experiment and analyzing the data, you should be able to do the following:

1. Categorize the types of experimental uncertainty (error), and explain how they may be reduced.
2. Distinguish between measurement accuracy and precision, and understand how they may be improved experimentally.
3. Define the term *least count* and explain the meaning and importance of significant figures (or digits) in reporting measurement values.
4. Express experimental results and uncertainty in appropriate numerical values so that someone reading your report will have an estimate of the reliability of your data.
5. Represent measurement data in graphical form so as to illustrate experimental data and uncertainty visually.

---

* Although *experimental uncertainty* is more descriptive, the term *error* is commonly used synonymously.

EQUIPMENT NEEDED

- Pencil and ruler
- Hand calculator
- 3 sheets of Cartesian graph paper
- French curve (optional)

THEORY

A. Types of Experimental Uncertainty

Experimental uncertainty (error) generally can be classified as being of two types: (1) random or statistical error and (2) systematic error: These are also referred to as (1) indeterminate error and (2) determinate error, respectively. Let's take a closer look at each type of experimental uncertainty.

RANDOM (INDETERMINATE) OR STATISTICAL ERROR

Random errors result from unknown and unpredictable variations that arise in all experimental measurement situations. The term *indeterminate* refers to the fact that there is no way to determine the magnitude or sign (+, too large; −, too small) of the error in any individual measurement. Conditions in which random errors can result include

1. Unpredictable fluctuations in temperature or line voltage
2. Mechanical vibrations of an experimental setup
3. Unbiased estimates of measurement readings by the observer

Repeated measurements with random errors give slightly different values each time. The effect of random errors may be reduced and minimized by improving and refining experimental techniques.

SYSTEMATIC (DETERMINATE) ERRORS

Systematic errors are associated with particular measurement instruments or techniques, such as an improperly calibrated instrument or bias on the part of the observer. The term *systematic* implies that the same magnitude and sign of experimental uncertainty are obtained when the

3

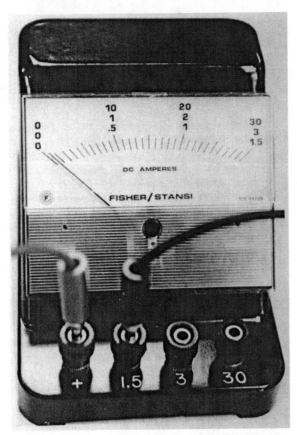

Figure 1 Systematic error. An improperly zeroed instrument gives rise to systematic error. In this case the ammeter, which has no current through it, would systematically give an incorrect reading larger that the true value. (After correcting the error by zeroing the meter, which scale would you read when using the ammeter?)

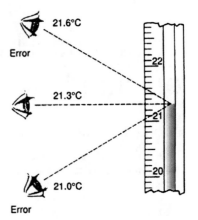

(a) Temperature measurement

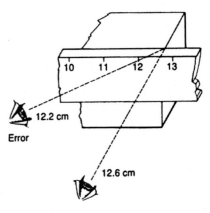

(b) Length measurement

Figure 2 Personal error. Examples of personal error due to parallax in reading (a) a thermometer and (b) a meterstick. Readings may systematically be made either too high or two low.

measurement is repeated several times. *Determinate* means that the magnitude and sign of the uncertainty can be determined if the error is identified. Conditions from which systematic errors can result include

1. An improperly "zeroed" instrument, for example, an ammeter as shown in ● Fig. 1.
2. A faulty instrument, such as a thermometer that reads 101°C when immersed in boiling water at standard atmospheric pressure. This thermometer is faulty because the reading should be 100°C.
3. Personal error, such as using a wrong constant in calculation or always taking a high or low reading of a scale division. Other examples of personal systematic error are shown in ● Fig. 2. Reading a value from a scale generally involves lining up something, such as a mark on the scale. The alignment—and hence the

value of the reading—can depend of the position of the eye (parallax).

Avoiding systematic errors depends on the skill of the observer to recognize the sources of such errors and to prevent or correct them.

B. Accuracy and Precision

Accuracy and *precision* are commonly used synonymously, but in experimental measurements there is an important distinction. The **accuracy** of a measurement signifies how close it comes to the true (or accepted) value—that is, how nearly correct it is.

Example 1 Two independent measurement results using the diameter *d* and circumference *c* of a circle in the determination of the value of π are 3.140

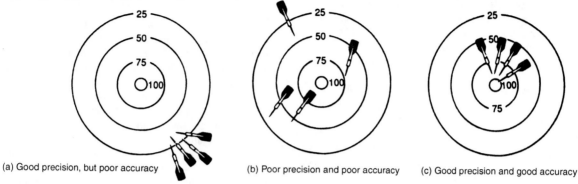

(a) Good precision, but poor accuracy   (b) Poor precision and poor accuracy   (c) Good precision and good accuracy

Figure 3   Accuracy and precision.   The true value in this analogy is the bull's eye. The degree of scattering is an indication of precision—the closer together a dart grouping, the greater the precision. A group (or symmetric grouping with an average) close to the true value represents accuracy.

and 3.143. (Recall that $\pi = c/d$.) The second result is more accurate than the first because the true value of π, to four figures, is 3.142.

Precision refers to the agreement among repeated measurements—that is, the "spread" of the measurements or how close they are together. The more precise a group of measurements, the closer together they are. However, a large degree of precision does not necessarily imply accuracy, as illustrated in ● Fig. 3.

Example   2   Two independent experiments give two sets of data with the expressed results and uncertainties of 2.5 ± 0.1 cm and 2.5 ± 0.2 cm, respectively.

The first result is more precise than the second because the spread in the first measurements is between 2.4 and 2.6 cm, whereas the spread in the second measurements is between 2.3 and 2.7 cm. That is, the measurements of the first experiment are less uncertain than those of the second.

Obtaining *greater accuracy* for an experimental value depends in general on *minimizing systematic errors*. Obtaining *greater precision* for an experimental value depends on *minimizing random errors*.

C. Least Count and Significant Figures

In general, there are *exact* numbers and *measured* numbers (or quantities). Factors such as the 100 used in calculating percentage and the 2 in $2\pi r$ are exact numbers. Measured numbers, as the name implies, are those obtained from measurement instruments and generally involve some error or uncertainty.

In reporting experimentally measured values, it is important to read one's instruments correctly. The degree of uncertainty of a number read from a measurement instrument depends on the quality of instrument and the fineness of its measuring scale. When one is reading the value from a calibrated scale, only a certain number of figures or digits can properly be obtained or read. That is, only a certain number of figures are *significant*. This depends on the **least count** of the instrument scale, which is the smallest subdivision on the measurement scale. This is the unit of the smallest reading that can be made without estimating. For example, the least count of a meterstick is usually the millimeter (mm). We commonly say "the instrument is calibrated in centimeters (numbered major divisions) with a millimeter least count." (See ● Fig. 4.)

The **significant figures** (sometimes called **significant digits**) of a measured value include all the numbers that can be read directly from the instrument scale, *plus* one doubtful or estimated number—the fractional part of the least

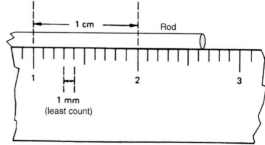

Figure 4   Least count.   Metersticks are commonly calibrated in centimeters (cm), the numbered major divisions, with a least count, or smallest subdivision, of millimeters (mm).

count smallest division. For example, the length of the rod in Fig. 4 (as measured from the zero end) is 2.64 cm. The rod's length is known to be between 2.6 cm and 2.7 cm. The estimated fraction is taken to be 4/10 of the least count (1 mm), so the doubtful figure is 4, giving 2.64 cm with three significant figures.

Thus, measured values contain inherent uncertainty or doubtfulness because of the estimated figure. However, the greater the number of significant figures, the greater the reliability of the measurement the number represents. For example, the length of an object may be read as 2.54 cm (three significant figures) on one instrument scale and as 2.5405 cm (five significant figures) on another. The latter reading from an instrument with a finer scale (why?) gives more information and reliability.

Zeros and the decimal point must be properly dealt with in determining the number of significant figures in a result. For example, how many significant figures does 0.0543 m have? What about 209.4 m? 2705.0 m? In such cases, we will use the following rules:

1. Zeros at the beginning of a number are not significant. They merely locate the decimal point. For example,

   0.0543 m has three significant figures (5, 4, and 3).

2. Zeros within a number are significant. For example,

   209.4 m has four significant figures (2, 0, 9, and 4).

3. Zeros at the end of a number after the decimal point are significant. For example,

   2705.0 has five significant figures (2, 7, 0, 5, and 0).

Some confusion may arise with whole numbers that have one or more zeros at the end without a decimal point. Consider, for example, 300 kg, where the zeros (called trailing zeros) may or may not be significant. In such cases, it is not clear which zeros serve only to locate the decimal point and which are actually part of the measurement (and hence are significant). That is, if the first zero from the left (3$\underline{0}$0 kg) is the estimated digit in the measurement, then only two digits are reliably known, and there are only two significant figures.

Similarly, if the last zero is the estimated digit (30$\underline{0}$ kg), then there are three significant figures. This ambiguity may be removed by using *scientific* (powers-of-10) *notation:*

$$3.0 \times 10^2 \text{ kg has two significant figures.}$$

$$3.00 \times 10^2 \text{ kg has three significant figures.}$$

This procedure is also helpful in expressing the significant figures in large numbers. For example, suppose that the average distance from Earth to the Sun, 93,000,000 miles, is known to only four significant figures. This is easily expressed in powers-of-10 notation: 9.300×10^7 mi.

D. Computations with Measured Values

Calculations are often performed with measured values, and error and uncertainty are "propagated" by the mathematical operations—for example, multiplication or division. (That is, error is carried through to the results by the mathematical operations.)

The error can be better expressed by statistical methods; however, a widely used procedure for *estimating* the uncertainty of a mathematical result involves the use of significant figures.

The number of significant figures in a measured value gives an indication of the uncertainty or reliability of a measurement. Hence, you might expect that the result of a mathematical operation can be no more reliable than the quantity with the least reliability, or smallest number of significant figures, used in the calculation. (That is, you can't *gain* reliability through a mathematical operation.)

It is important to report the results of mathematical operations with the proper number of significant figures. This is accomplished by using rules for (1) multiplication and division, and (2) addition and subtraction. To obtain the proper number of significant figures, one rounds the results off. Here are some general rules that will be used for mathematical operations and rounding.

SIGNIFICANT FIGURES IN CALCULATIONS
1. When multiplying and dividing quantities, leave as many significant figures in the answer as there are in the quantity with the least number of significant figures.
2. When adding or subtracting quantities, leave the same number of decimal places (rounded) in the answer as there are in the quantity with the least number of decimal places.

RULES FOR ROUNDING*
1. If the first digit to be dropped is less than 5, leave the preceding digit as is.
2. If the first digit to be dropped is 5 or greater, increase the preceding digit by one.

Notice that in this method, five digits (0, 1, 2, 3, and 4) are rounded down and five digits (5, 6, 7, 8, and 9) are rounded up.

What the rules for significant figures mean is that the result of a calculation can be no more accurate than the least accurate quantity used. That is, **you cannot gain accuracy in performing mathematical operations.**

These rules come into play frequently when one is doing mathematical operations with a hand calculator that may give a string of digits. ● Fig. 5 shows the result of the division of 374 by 29. The result must be rounded off to two significant figures—that is, to 13. (Why?)

---

* It should be noted that these rounding rules give an approximation of accuracy, as opposed to the results provided by more advanced statistical methods.

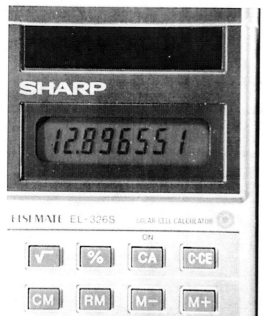

Figure 5 Insignificant figures. The calculator shows the result of the division operation 374/29. Because there are only two significant figures in the 29, a reported result should have no more than two significant figures, and the calculator display value should be rounded off to 13.

Example 3   Applying the rules.

Multiplication:

$$2.5 \text{ m} \times 1.308 \text{ m} = 3.3 \text{ m}$$
$$(2 \, sf) \qquad (4 \, sf) \qquad (2 \, sf)$$

Division:

$$\frac{882.0 \text{ s}}{0.245 \text{ s}} = 3600 \text{ s} = 3.60 \times 10^3 \text{ s}$$

(4 sf)

(3 sf)   *(represented to three significant figures; why?)*

Addition:

$$
\begin{array}{r}
46.4 \\
1.37 \\
0.505 \\
\hline
48.275 \longrightarrow 48.3
\end{array}
$$

(rounding off)
(46.4 has the least number of decimal places)

Subtraction:

$$
\begin{array}{r}
163 \\
-4.5 \\
\hline
158.5 \longrightarrow 159
\end{array}
$$

(rounding off)
(163 has the least number of decimal places, none)

E. Expressing Experimental Error and Uncertainty

PERCENT ERROR

The object of some experiments is to determine the value of a well-known physical quantity—for example, the value of π.

The **accepted or "true" value** of such a quantity found in textbooks and physics handbooks is the most accurate value (usually rounded off to a certain number of significant figures) obtained through sophisticated experiments or mathematical methods.

The **absolute difference** between the experimental value E and the accepted value A, written $|E - A|$, is the positive difference in the values, e.g., $|2 - 4| = |-2| = 2$ and $|4 - 2| = 2$. Simply subtract the smaller value from the larger, even though the symbols may be written in reverse order. For a set of measurements, E is taken as the average value of the experimental measurements.

The **fractional error** is the ratio of the absolute difference to the accepted value:

$$\text{Fractional error} = \frac{\text{absolute difference}}{\text{accepted value}}$$

or

$$\boxed{\text{Fractional error} = \frac{|E - A|}{A}} \qquad (1)$$

The fractional error is commonly expressed as a percentage to give the **percent error** of an experimental value.*

$$\text{Percent error} = \frac{\text{absolute difference}}{\text{accepted value}} \times 100\%$$

or

$$\boxed{\text{Percent error} = \frac{|E - A|}{A} \times 100\%} \qquad (2)$$

Example   4 A cylindrical object is measured to have a diameter d of 5.25 cm and a circumference c of 16.38 cm. What are the experimental value of π and the percent error of the experimental value if the accepted value of π is 3.14?

Solution   with $d = 5.25$ cm and $c = 16.38$ cm,

$$c = \pi d \quad \text{or} \quad \pi = \frac{c}{d} = \frac{16.38}{5.25} = 3.12$$

* It should be noted that percent error only gives a measure of experimental error or uncertainty when the accepted or standard value is highly accurate. If an accepted value itself has a large degree of uncertainty, then the percent error does not give a measure of experimental uncertainty.

Then $E = 3.12$ and $A = 3.14$, so

$$\text{Percent error} = \frac{|E - A|}{A} \times 100\%$$

$$= \frac{|3.12 - 3.14|}{3.14} \times 100\%$$

$$= \frac{0.02}{3.14} \times 100\% = 0.6\%$$

Note: To avoid rounding errors, the preferred order of operations is addition and subtraction before multiplication and division.*

If the uncertainty in experimentally measured values as expressed by the percent error is large, you should suspect and check for possible sources of error. Additional measurements should then be made to reduce the uncertainty. Your instructor may wish to set a maximum percent error for experimental results.

PERCENT DIFFERENCE

It is sometimes instructive to compare the results of two measurements when there is no known or accepted value. The comparison is expressed as a **percent difference,** which is the ratio of the absolute difference between the experimental values E_2 and E_1 to the average or mean value of the two results, expressed as a percent.

$$\text{Percent difference} = \frac{\text{absolute difference}}{\text{average}} \times 100\%$$

or

$$\boxed{\text{Percent difference} = \frac{|E_2 - E_1|}{(E_2 + E_1)/2} \times 100\%} \qquad (3)$$

Dividing by the average or mean value of the experimental values is logical, because there is no way of deciding which of the two results is better.

Example 5 What is the percent difference between two measured values of 4.6 cm and 5.0 cm?

Solution With $E_1 = 4.6$ cm and $E_2 = 5.0$ cm,

$$\text{Percent difference} = \frac{|E_2 - E_1|}{(E_2 + E_1)/2} \times 100\%$$

$$\text{Percent difference} = \frac{|5.0 - 4.6|}{(5.0 + 4.6)/2} \times 100\%$$

$$= \frac{0.4}{4.8} \times 100\% = 8\%$$

As in the case of percent error, when the percent difference is large, it is advisable to check the experiment and make more measurements.

In many instances there will be more than two measurement values.

When there are three or more measurements, the percent difference is found by dividing the absolute value of the difference of the extreme values (i.e., the values with greatest difference) by the average or mean value of all the measurements.

AVERAGE (MEAN) VALUE

Most experimental measurements are repeated several times, and it is very unlikely that identical results will be obtained for all trials. For a set of measurements with predominantly random errors (i.e., the measurements are all equally trustworthy or probable), it can be shown mathematically that the true value is most probably given by the average or mean value.

The **average** or **mean value** $\bar{x}$ of a set of N measurements is

$$\boxed{\bar{x} = \frac{x_1 + x_2 + x_3 + \cdots + x_N}{N} = \frac{1}{N}\sum_{i=1}^{N} x_i} \qquad (4)$$

where the summation sign Σ is a shorthand notation indicating the sum of N measurements from x_1 to x_N. ($\bar{x}$ is commonly referred to simply as the *mean.*)

Example 6 What is the average or mean value of the set of numbers 5.42, 6.18, 5.70, 6.01, and 6.32?

$$\bar{x} = \frac{1}{N}\sum_{i=1}^{N} x_i$$

$$= \frac{5.42 + 6.18 + 5.70 + 6.01 + 6.32}{5}$$

$$= 5.93$$

DEVIATION FROM THE MEAN (OPTIONAL)[†]

Having obtained a set of measurements and determined the mean value, it is helpful to report how widely the individual measurements are scattered from the mean. A quantitative description of this scatter, or dispersion, of measurements will give an idea of the precision of the experiment.

---

* Although percent error is generally defined using the absolute difference $|E - A|$, some instructors prefer to use $(E - A)$, which results in positive $(+)$ or negative $(-)$ percent errors, e.g., -0.6% in Example 4. In the case of a series of measurements and computed percent errors, this gives an indication of systematic error.

[†] A discussion of standard deviation and the method of least squares may be found in Appendix C.

The **deviation** d_i from the mean of any measurement with a mean value $\bar{x}$ is

$$d_i = x_i - \bar{x} \qquad (5)$$

(d_i is sometimes referred to as the **residual** rather than the deviation.)

As defined, the deviation may be positive or negative, since some measurements are larger than the mean and some are smaller. The average of the deviations of a set of measurements is always zero, so the mean of the deviations is not a useful way of characterizing the dispersion.

MEAN DEVIATION (OPTIONAL)

To obtain what is called the **mean** or **average deviation** of a set of N measurements, the absolute deviations $|d_i|$ are determined; that is,

$$|d_i| = |x_i - \bar{x}| \qquad (6)$$

The *mean deviation* $\bar{d}$ is then

$$\bar{d} = \frac{|d_1| + |d_2| + |d_3| + \cdots + |d_N|}{N}$$

$$= \frac{1}{N}\sum_{i=1}^{N}|d_i| \qquad (7)$$

(Although $\bar{d}$ is commonly called the mean deviation, a more appropriate term would be the **mean absolute deviation.**)

Example 7 What is the mean deviation of the set of numbers given in Example 6?

Solution First find the absolute deviation of each of the numbers, using the determined mean of 5.93.

$$|d_1| = |5.42 - 5.93| = 0.51$$

$$|d_2| = |6.18 - 5.93| = 0.25$$

$$|d_3| = |5.70 - 5.93| = 0.23$$

$$|d_4| = |6.01 - 5.93| = 0.08$$

$$|d_5| = |6.32 - 5.93| = 0.39$$

Then

$$\bar{d} = \frac{1}{N}\sum_{i=1}^{N}|d_i|$$

$$= \frac{0.51 + 0.25 + 0.23 + 0.08 + 0.39}{5} = 0.29$$

The mean deviation is a measure of the dispersion of experimental measurements about the mean (i.e., a measure of precision). It is common practice to report the experimental value E of a quantity in the form

$$E = \bar{x} \pm \bar{d}$$

In Example 7, this would be $E = 5.93 \pm 0.29$. The $\pm$ term gives a measure of the *precision* of the experimental value. The *accuracy* of the mean value of a set of experimental measurements (5.93 in the above example) may be expressed in terms of percent error or percent difference.

The dispersion of an experimental measurement may be expressed by other means (such as standard deviation; see Appendix C), so the method should be specified when reporting.

F. Graphical Representation of Data

It is often convenient to represent experimental data in graphical form, not only for reporting, but also to obtain information.

GRAPHING PROCEDURES

Quantities are commonly plotted using rectangular Cartesian axes (X and Y). The horizontal axis (X) is called the *abscissa*, and the vertical axis (Y), the *ordinate*. The location of a point on the graph is defined by its coordinates x and y, written (x, y), referenced to the origin O, the intersection of the X and Y axes.

When plotting data, choose axis scales that are easy to plot and read. The graph in ● Fig. 6A shows an example of scales that are too small. This "bunches up" the data, making the graph too small, and the major horizontal scale values make it difficult to read intermediate values. Choose scales so that most of the graph paper is used. The graph in ● Fig. 6B shows data plotted with more appropriate scales.*

Also note in Fig. 6.A that scale units on the axes are not given. For example, you don't know whether the units of displacement are feet, meters, kilometers, or whatever. *Scale units should always be included,* as in Fig. 6B. It is also acceptable, and saves time, to use standard unit abbreviations, such as N for newton and m for meter. This will be done on subsequent graphs.

When the data points are plotted, draw a smooth line described by the points. *Smooth* means that the line does not have to pass exactly through each point but connects the general areas of significance of the data points (*not* connecting the data points as in Fig. 6A). The graph in Fig. 6B with an approximately equal number of points on each side of the line gives a "curve of best fit."

---

* As a general rule, it is convenient to choose the unit of the first major scale division to the right or above the origin or zero point as 1, 2, or 5 (or multiples or submultiples thereof, e. g., 10 or 0.1) so that the minor (intermediate) scale divisions can be easily interpolated and read.

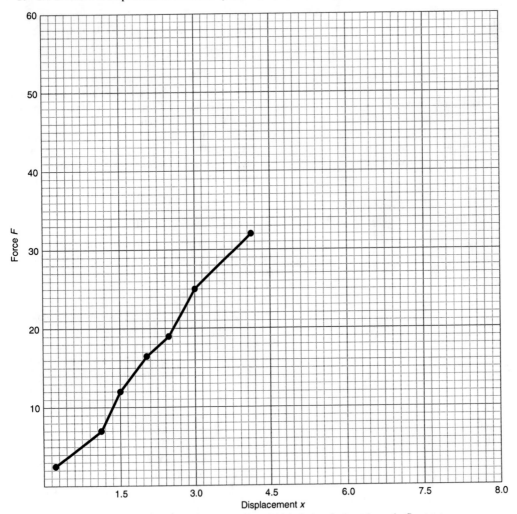

Figure 6A   Poor graphing.   An example of an improperly labeled and plotted graph. See text for description.

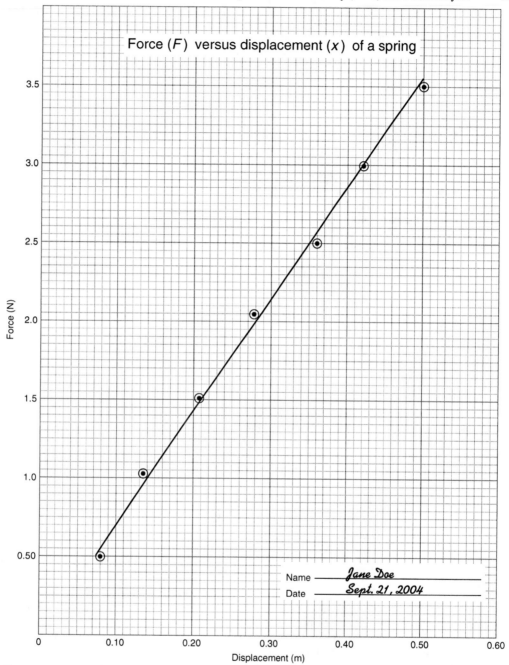

Figure 6B   Proper graphing.   An example of a properly labeled and plotted graph. See text for description.

TABLE 1   Data for Figure 7

Mass (kg)	Period	±	$\bar{d}$ (s)
0.025	0.38	±	0.17
0.050	0.53	±	0.21
0.10	0.75	±	0.25
0.15	0.92	±	0.20
0.20	1.1	±	0.31
0.25	1.2	±	0.13
0.30	1.3	±	0.20

In cases where several determinations of each experimental quantity are made, the average value is plotted and the mean deviation or the standard deviation may be plotted as *error bars*. For example, the data in Table 1 are plotted in ● Fig. 7. A smooth line is drawn so as to pass within the error bars. (Your instructor may want to explain the use of a French curve at this point).

Graphs should have the following elements (see Fig. 7):

1. Each axis labeled with the quantity plotted.
2. The units of the quantities plotted.
3. The title of the graph on the graph paper (commonly listed as the *y* coordinate versus the *x* coordinate).
4. Your name and the date.

STRAIGHT-LINE GRAPHS

Two quantities (*x* and *y*) are often linearly related; that is, they have an algebraic relationship of the form $y = mx + b$, where *m* and *b* are constants. When the values of such quantities are plotted, the graph is a straight line, as shown in ● Fig. 8.

The *m* in the algebraic relationship is called the **slope** of the line and is equal to the ratio of the intervals $\Delta y/\Delta x$. Any set of intervals may be used to determine the slope of a straight-line graph; for example, in Fig. 8,

$$m = \frac{\Delta y_1}{\Delta x_1} = \frac{15 \text{ cm}}{2.0 \text{ s}} = 7.5 \text{ cm/s}$$

$$m = \frac{\Delta y_2}{\Delta x_2} = \frac{45 \text{ cm}}{6.0 \text{ s}} = 7.5 \text{ cm/s}$$

Points should be chosen relatively far apart on the line. For best results, points corresponding to data points should not be chosen, even if they appear to lie on the line.

The *b* in the algebraic relationship is called the **y-intercept** and is equal to the value of the *y* coordinate where the graph line intercepts the *Y* axis. In Fig. 8, $b = 3$ cm. Notice from the relationship that $y = mx + b$, so that when $x = 0$, then $y = b$. If the intercept is at the origin (0, 0), then $b = 0$.

The equation of the line in the graph in Fig. 8 is $d = 7.5t + 3$. The general equation for uniform motion has the form $d = vt + d_0$. Hence, the initial displacement $d_0 = 3$ cm and the speed $v = 7.5$ cm/s.

Some forms of nonlinear functions that are common in physics can be represented as straight lines on a Cartesian

graph. This is done by plotting nonlinear values. For example, if you plotted

$$y = ax^2 + b$$

on a regular *y*-versus-*x* graph, you would get a parabola. But if you let $x^2 = x'$, the equation becomes

$$y = ax' + b$$

which has the form of a straight line.

This means that if you plotted *y* versus *x'*, you would get a straight line. Since $x' = x^2$, you must plot the squared values of *x*. That is, square all the values of *x* in your data table, and plot these numbers with the corresponding *y* values.

Other functions can be "straightened out" by this procedure, including

$$y = Ae^{ax}$$

and, in this case,

$$\ln y = \ln A + \ln e^{ax}$$

or

$$\ln y = ax + \ln A$$

Plotting the values of the natural (base *e*) logarithm versus *x* gives a straight line with slope *a* and an intercept ln *A*.

Similarly, for

$$y = ax^n$$

using the common (base 10) logarithm,

$$\log y = \log a + \log x^n$$

and

$$\log y = n \log x + \log a$$

Plotting the values of log *y* versus log *x* gives a straight line with slope *n* and intercept log *a*.

LINEAR REGRESSION AND METHOD OF LEAST SQUARES (OPTIONAL)

The straight line of "best fit" for a set of data points on a graph can be determined by a statistical procedure called **linear regression,** using what is known as the **method of least squares.** This method determines the best-fitting straight line by means of differential calculus, which is beyond the scope of this manual.

Your instructor may wish to explain the resulting equations (Appendix C) and the procedure for determining the slope and intercept of the best-fitting straight line, particularly since they can be computed automatically by some hand calculators and by computers with the appropriate software.

EXPERIMENTAL PROCEDURE

Complete the exercises in the Laboratory Report, showing calculations and attaching graphs as required. (*Note:* In this experiment and throughout, attach an additional sheet for calculations if necessary.)

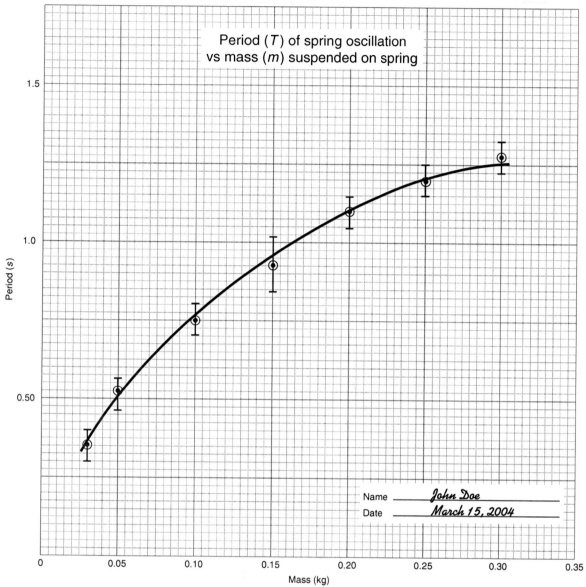

Figure 7   Error bars.   An example of graphically presented data with error bars. An error bar indicates the precision of a measurement. In this case, the error bars represent mean deviations.

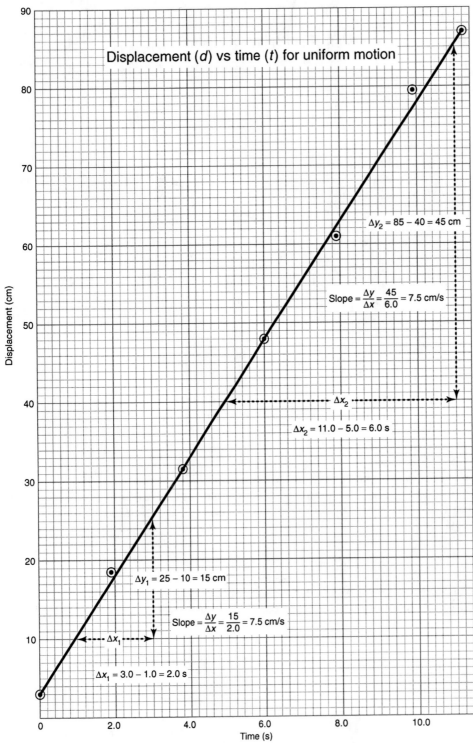

Figure 8 Straight-line slope.   Examples of intervals for determining the slope of a straight line. The slope is the ratio of $\Delta y/\Delta x$ (or $\Delta d/\Delta t$). Any set of intervals may be used, but the endpoints of an interval should be relatively far apart, as for $\Delta y_2/\Delta x_2$.

Name _____ Section _____ Date _____

Lab Partner(s) _____

EXPERIMENT 1

Experimental Uncertainty (Error) and Data Analysis

/TI/ *Laboratory Report*

1. Significant Figures
    (a) Express the numbers listed in Data Table 1 to three significant figures, writing the numbers in the first column in normal notation and the numbers in the second column in powers-of-10 (scientific) notation.

DATA TABLE 1
Purpose: To practice expressing significant figures.

0.524	_____	5280	_____
15.08	_____	0.060	_____
1444	_____	82.453	_____
0.0254	_____	0.00010	_____
83,909	_____	2,700,000,000	_____

(b) A rectangular block of wood is measured to have the dimensions 11.2 cm × 3.4 cm × 4.10 cm. Compute the volume of the block, showing explicitly (by underlining) how doubtful figures are carried through the calculation, and report the final answer with the correct number of significant figures.

Calculations
(show work)

Computed volume
(in powers-of-10 notation) _____

(units)

Don't forget units

(continued)

(c) In an experiment to determine the value of π, a cylinder is measured to have an average value of 4.25 cm for its diameter and an average value of 13.39 cm for its circumference. What is the experimental value of π to the correct number of significant figures?

Calculations
(show work)

Experimental value of π _____
(units)

2. Expressing Experimental Error
   (a) If the accepted value of π is 3.1416, what are the fractional error and the percent error of the experimental value found in 1(c)?

Calculations
(show work)

Fractional error _____

Percent error _____

(b) In an experiment to measure the acceleration g due to gravity, two values, 9.96 m/s^2 and 9.72 m/s^2, are determined. Find (1) the percent difference of the measurements, (2) the percent error of each measurement, and (3) the percent error of their mean. (Accepted value: $g = 9.80$ m/s^2.)

Calculations
(show work)

Percent difference _____

Percent error of E_1 _____

Percent error of E_2 _____

Percent error of mean _____

EXPERIMENT 1

(c) Data Table 2 shows data taken in a free-fall experiment. Measurements were made of the distance of fall (y) at each of four precisely measured times. Complete the table. Use only the proper number of significant figures in your table entries, even if you carry extra digits during your intermediate calculations.

DATA TABLE 2
Purpose: To practice analyzing data.

Time t (s)	Distance (m)					$\bar{y}$	(Optional) $\bar{d}$	t^2 ( )
	y_1	y_2	y_3	y_4	y_5			
0	0	0	0	0	0			
0.50	1.0	1.4	1.1	1.4	1.5			
0.75	2.6	3.2	2.8	2.5	3.1			
1.00	4.8	4.4	5.1	4.7	4.8			
1.25	8.2	7.9	7.5	8.1	7.4			

(d) Plot a graph of $\bar{y}$ versus t (optional: with $2\bar{d}$) error bars for the free-fall data in part (c). Remember that $t = 0$ is a known point.

(e) The equation of motion for an object in free fall starting from rest is $y = \frac{1}{2}gt^2$, where g is the acceleration due to gravity. This is the equation of a parabola, which has the general form $y = ax^2$.

Convert the curve into a straight line by plotting $\bar{y}$ versus t^2. That is, plot the square of the time on the abscissa. (Optional: Use regional circles rather than error bars: Around each data point, place a circle with a diameter the size of the error bar dispersion.) Determine the slope of the line and compute the experimental value of g from the slope value.

Calculations
(show work)

Experimental value of g from graph _____

(units)

(continued)

(f) Compute the percent error of the experimental value of g determined from the graph in part (e). (Accepted value: $g = 9.8$ m/s^2.)

Calculations
(show work)

Percent error _____

(g) The relationship of the applied force F and the displacement x of a spring has the general form $F = kx$, where the constant k is called the *spring constant* and is a measure of the "stiffness" of the spring. Notice that this equation has the form of a straight line. Find the value of the spring constant k of the spring used in determining the experimental data plotted in the Fig. 6B graph. (*Note:* Because $k = F/x$, the units of k in the graph are N/m.)

Calculations
(show work)

Value of spring constant of
spring in Fig. 6B graph _____
(units)

(h) The general relationship of the period of oscillation T of a mass m suspended on a spring is $T = 2\pi\sqrt{m/k}$, where k is the spring constant. Replot the data in Fig. 7 so as to obtain a straight-line graph, and determine the value of the spring constant used in the experiment. [*Hint:* Square both sides of the equation, and plot in a manner similar to that used in part (e).] Show the final form of the equation and calculations.

Calculations
(show work)

Value of spring constant of
spring in Fig. 7 _____
(units)

(i) The data in sections (g) and (h) above were for the same spring. Compute the percent difference for the values of the spring constants obtained in each section.

Name_____ Section _____ Date _____
Lab Partner(s)_____

Laboratory Report

TI/ QUESTIONS

1. Read the measurements on the rulers in ● Fig. 9, and comment on the results.

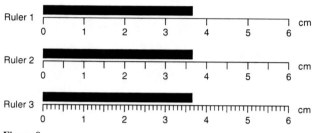

Figure 9

2. Were the measurements on the block in part (b) of Procedure 1 of this laboratory report all done with the same instrument? Explain.

3. Referring to the dart analogy in Fig. 3, draw a dart grouping that would represent poor precision, but good accuracy with an average value.

4. Do percent error and percent difference give indications of accuracy or precision? Discuss each.

(continued)

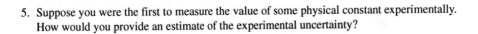

5. Suppose you were the first to measure the value of some physical constant experimentally. How would you provide an estimate of the experimental uncertainty?

6. (Optional) Why is the average of the deviations of a set of measurement values always zero?

7. In Table 1, the periods of oscillation were measured (timed) down to 0.38 s. How could this be done with common laboratory timers? (*Hint:* Are the times *exact* values?)

Name _____ Section _____ Date _____

Lab Partner(s) _____

Measurement Instruments (Mass, Volume, and Density)

/TI/ *Advance Study Assignment*

Read the experiment and answer the following questions.

1. What is the least count of a measurement instrument, and how is it related to the number of significant figures of a measurement reading?

2. Does a laboratory balance measure weight or mass? Explain.

3. What is the function of the vernier scale on the vernier caliper? Does it extend accuracy or precision? Explain.

4. Distinguish between positive and negative zero errors and how corrections are made for such errors. For what kind of error does a zero correction correct?

(continued)

5. What is the purpose of the ratchet mechanism on a micrometer caliper?

6. Explain how readings from 0.00 through 1.00 mm are obtained from the micrometer thimble scale when it is calibrated only from 0.00 through 0.50 mm.

7. If the density of one object is greater than that of another, what does this indicate? Do the sizes of the objects affect their densities? Explain.

8. Explain how the volume of a heavy, irregularly shaped object may be determined experimentally. Are there any limitations?

Measurement Instruments (Mass, Volume, and Density)

INTRODUCTION AND OBJECTIVES

Common laboratory measurements involve the determination of the fundamental properties of mass and length. Most people are familiar with the use of scales and rulers or metersticks. However, for more accurate and precise measurements, laboratory balances and vernier calipers or micrometer calipers are often used, particularly in measurements involving small objects.

In this initial experiment on measurement, you will learn how to use these instruments and what advantages they offer. Density, the ratio of mass to volume, will also be considered, and the densities of several materials will be determined experimentally.

After performing this experiment and analyzing the data, you should be able to do the following:

1. Use the vernier caliper and read the vernier scale.
2. Use the micrometer caliper and read its scale.
3. Distinguish between mass and density, and know how to determine experimentally the density of an object or substance.

EQUIPMENT NEEDED

- Laboratory balance
- Vernier caliper
- Micrometer caliper (metric)
- Meterstick
- Graduated cylinder

- Cylindrical metal rod (e.g., aluminum, brass, or copper)
- Sphere (metal or glass, e.g., a ball bearing or marble)
- Short piece of solid copper wire
- Rectangular piece of metal sheet (e.g., aluminum)
- Irregularly shaped metal object

THEORY

A. Laboratory Balances

Some common types of laboratory balances are shown in
● Fig. 1. Mechanical balances or "scales" are used to balance the weight of an unknown mass m against that of a known mass m_1 (i.e., $mg = m_1g$ or $m = m_1$), and the mass of the unknown is read directly in mass units, usually grams. The weight w of an object is its mass m times a constant g, the acceleration due to gravity; $g = 9.80$ m/s^2 = 980 cm/s^2 (i.e., $w = mg$ or $m = w/g$). Some scales, such as bathroom scales, are calibrated in weight (force) units (such as pounds) rather than in mass units.

A set of known masses is used to balance an unknown mass on a pan balance (Fig. 1a). On a beam balance, the riders on the beams are used to balance the unknown mass on the platform (Fig. 1b). The common laboratory beam balance is calibrated in grams. In this case, the least count is 0.1 g and a reading can be estimated to 0.01 g.* (See Experiment 1 for a review of least count.)

Before making a mass determination, check whether the balance is zeroed. Adjustments can be made by various means on different scales.

Electronic balances with digital readouts are becoming increasingly common (Fig. 1c). They have the advantages of accuracy and ease of operation. However, electronic balances are much more delicate. The mass value is displayed automatically, and the accuracy or number of significant figures depends on the particular balance. Some electronic balances have autocalibration, and other have a keypad for calibration by the user.† Most electronic balances are zeroed by pressing a "tare" button. This has the advantage that one can place an empty dish on the balance before pressing the "tare" button, and then, when the material is added to the dish, the balance displays the mass of the contents alone.

---

* The official abbreviation of the gram unit is g (in roman type). The standard symbol for acceleration due to gravity is g (in italic type), where weight is given by mg, which is not to be confused with mg for miligram. Look closely so as to avoid confusion with these symbols.

† In general, an electronic balance has a suspended beam, and the balancing force on the end of the beam opposite the weighing pan is electromagnetic. The force is supplied by a current-carrying coil of wire in the field of a permanent magnet. The force is directly proportional to the current, which is controlled automatically by a photosensitive diode whose resistance is a function of the light incident on it.

Any tilting of the beam increases the light from a source on the diode, and a feedback circuit calls for more current in the coil. The increase in current (and hence in force) is adjusted so as to keep the beam in horizontal equilibrium. The current that balances the beam is read out on a digital ammeter calibrated in grams or milligrams.

(a)

(b)

(c)

(d)

Figure 1   Laboratory balances.   (a) A double-beam, double platform Harvard trip balance, which is also called an *equal-arm balance*. (b) A single-platform, triple-beam balance. (c) High-form beam balances. The balance on the left has a dial mechanism that replaces the lower-mass beams. (d) A digital electronic balance. (Courtesy of Sargent-Welch.)

Because of the wide variety of electronic balances available, if you are using one in this experiment you should first familiarize yourself with its operation. Your instructor may brief you, or an operation manual should be available. (When first using an electronic instrument, it is always advisable to read the operation manual supplied by the manufacturer.)

B. The Vernier Caliper

In 1631, a Frenchman, Pierre Vernier, devised a way to improve the precision of length measurements. The **vernier caliper** (● Fig. 2), commonly called a **vernier,** consists of a rule with a main engraved scale and a movable jaw with an engraved vernier scale. The span of the lower jaw is used to measure length and is particularly convenient for measuring the diameter of a cylindrical object. The span of the upper jaw is used to measure distances between two surfaces (such as, the inside diameter of a hollow cylindrical object).

The main scale is calibrated in centimeters with a millimeter least count, and the movable vernier scale has 10 divisions that cover 9 divisions on the main scale. When making a measurement with a meterstick, it is necessary to estimate, or "eyeball," the fractional part of the smallest scale division (the tenth of a millimeter). The function of the vernier scale is to assist in the accurate reading of the fractional part of the scale division, thus increasing the precision.

The leftmost mark on the vernier scale is the zero mark (lower scale for metric reading and upper scale for inches). The zero mark is often unlabeled. A measurement is made by closing the jaws on the object to be measured and read-

Figure 2   A vernier caliper.   A good instrument for measuring rectangular dimensions and circular diameters. This caliper has scales for both metric and English measurements. See text for description. (Courtesy of Sargent-Welch.)

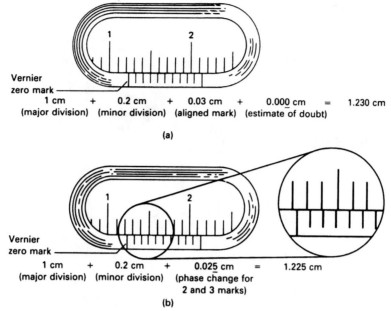

Figure 3   The vernier scale.   An example of reading the vernier scale on a caliper. See text for description.

ing where the zero mark on the vernier scale falls on the main scale (See ● Fig. 3).   Some calipers, as the one here, have vernier scales for both metric and British units.

The first two significant figures are read directly from the main scale in Fig. 3.   The vernier zero mark is past the 2-mm line after the 1-cm major division mark, so we have a reading of 1.2 cm for both (a) and (b). The next significant figure is the fractional part of the smallest subdivision on the main scale. This is obtained by referring to the vernier scale markings below the main scale.

If a vernier mark coincides with a mark on the main scale, then the vernier mark number is the fractional part of the main-scale division (see Fig. 3a).   In the figure, this is the third mark to the right of the vernier zero, so the third significant figure is 3 (0.03 cm). Finally, since the 0.03-cm reading is known exactly, a zero is added as the doubtful figure, for a reading of 1.230 cm or 12.30 mm. Note how the vernier scale gives more significant figures or extends the precision.

However, a mark on the vernier scale may not always line up exactly with one on the main scale (Fig. 3b).   In this case, there is more uncertainty in the 0.001-cm or 0.01-mm figure, and we say there is a change of "phase" between two successive vernier markings.

Notice how in Fig. 3b   the second vernier mark after the zero is to the right of the closest main-scale mark, and the third vernier mark is to the left of the next main-scale mark. Hence, the marks change "phase" between the 2 and 3 marks, which means the reading is between 1.22 cm and 1.23 cm. Most vernier scales are not fine enough for us to make an estimate of the doubtful figure, so a suggested

method is to take the middle of the range. Thus we would put a 5 in the thousandth-of-a-centimeter digit, for a reading of 1.225 cm.*

ZEROING

Before making a measurement, one should check the zero of the vernier caliper with the jaws completely closed. It is possible that through misuse the caliper is no longer zeroed and thus gives erroneous readings (systematic error). If this is the case, a zero correction must be made for each reading.

In zeroing, if the vernier zero lies to the right of the main-scale zero, measurements will be too large and the error is said to be *positive*. In this case, the zero correction is made by subtracting the zero reading from the measurement reading. For example, the "zero" reading in ● Fig. 4 is +0.05 cm, and this   amount must   be subtracted from each measurement reading for more accurate results.

Similarly, if the error is *negative,* or the vernier zero lies to the left of the main-scale zero, measurements will be too small, and the zero correction must be added to the measurement readings.

Summarizing these corrections in equation form,

Corrected reading = actual reading − zero reading

For example, for a *positive* error of +0.05 cm as in Fig. 4,

Corrected reading = actual reading − 0.05 cm

If there is a *negative* correction of −0.05 cm, then

Corrected reading = actual reading − (−0.05) cm
= actual reading + 0.05 cm

* E. S. Oberhofer, "The Vernier Caliper and Significant Figures," *The Physics Teacher,* Vol. 23 (November 1985), 493.

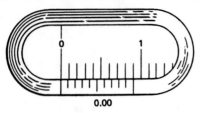

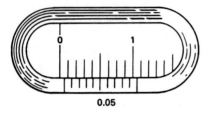

Figure 4   Zeroing and error. The zero of the vernier caliper is checked with the jaws closed. (a) Zero error. (b) Positive error, +0.05 cm.

C. The Micrometer Caliper

The **micrometer caliper** (● Fig. 5), commonly called a **mike,** provides for accurate measurements of small lengths and is particularly convenient in measuring the diameters of thin wires and the thicknesses of thin sheets. It consists of a movable spindle (jaw) that is advanced toward another, parallel-faced jaw (called an anvil) by rotating the thimble. The thimble rotates over an engraved sleeve (or "barrel") mounted on a solid frame.

Most micrometers are equipped with a ratchet (ratchet handle is to the far right in the figure) that allows slippage of the screw mechanism when a small and constant force is exerted on the jaw. This permits the jaw to be tightened on an object with the same amount of force each time. Care should be taken not to force the screw (particularly if the micrometer does not have a ratchet mechanism), so as not to damage the measured object and/or the micrometer.

The axial main scale on the sleeve is calibrated in millimeters, and the thimble scale is calibrated in 0.01 mm (hundredths of a millimeter). The movement mechanism of the micrometer is a carefully machined screw with a pitch of 0.5 mm. The pitch of a screw, or the distance between screw threads, is the lateral linear distance the screw moves when turned through one rotation.

The axial line on the sleeve main scale serves as a reading line. Since the pitch of the screw is 0.5 mm and there are 50 divisions on the thimble, when the thimble is turned through one of its divisions, the thimble moves (and the jaws open or close) $\frac{1}{50}$ of 0.5 mm, or 0.01 mm ($\frac{1}{50} \times 0.5$ mm = 0.01 mm).

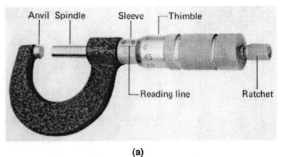

(a)

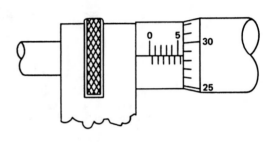

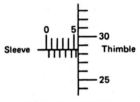

(b) Reading of 5.785 mm

Figure 5   A micrometer caliper and an example of a micrometer reading.   (a) This particular mike has the 1.0-mm and 0.5-mm scale divisions below the reading line. (b) In this diagram, as on some mikes, the 1.0-mm divisions are above the reading line and the 0.5-mm divisions are below it. The thimble in the diagram is in the second rotation of millimeter movement, as indicated by its being past the 0.5-mm mark. The reading is 5.500 + 0.285 mm, or 5.785 mm, where the last 5 is the estimated figure. (Photo courtesy of Sargent-Welch.)

One complete rotation of the thimble (50 divisions) moves it through 0.5 mm, and a second rotation moves it through another 0.5 mm, for a total of 1.0 mm, or one scale division along the main scale. That is, the first rotation moves the thimble from 0.00 through 0.50 mm, and the second rotation moves the thimble from 0.50 through 1.00 mm.

It is sometimes instructive to think of the 1-mm main-scale divisions as analogous to dollar ($) divisions and of the thimble scale divisions as cents ($0.01). The first rotation of the thimble corresponds to going from $0.00 to $0.50 (50 cents), and the second rotation corresponds to going from $0.50 to $1.00, so that two complete rotations go through 100 cents, or $1.00, of the main scale.

Some micrometers have a scale that indicates the 0.5-mm marks of the main-scale divisions and hence tells which rotation the thimble is in (see Fig. 5). Cheaper mikes do not have this extra graduation, and the main scale must be closely examined to determine which rotation the thimble is in.

If a mike does not have the 0.5-mm scale, you must determine whether the thimble is in its first rotation, in which case the thimble reading is between 0.00 and 0.50 mm (corresponding to the actual engraved numbers on the thimble), or in the second rotation, in which case the reading is between 0.50 and 1.00 mm (the actual thimble scale reading plus 0.50).

This can be determined by judging whether the edge of the thimble is in the first or the second half of the main-scale division. Notice that the zero mark on the thimble is used to indicate both 0.00 mm (beginning of the first rotation) and 0.50 mm (beginning of the second rotation).

Measurements are taken by noting the position of the edge of the thimble on the main scale and the position of the reading line on the thimble scale. For example, for the drawing in Fig. 5, the mike has a reading of 5.785 mm. On the main scale is a reading of 5.000 mm plus one 0.500-mm division (scale below reading line), giving 5.500 mm. That is, in the figure, the thimble is in the second rotation of a main-scale division.

The reading on the thimble scale is 0.285 mm, where the 5 is the estimated or doubtful figure; that is, the reading line is estimated to be midway between the 28 and the 29 marks. (Some mikes have vernier scales on the sleeves to help the user read this last significant figure and further extend the precision.)

As with all instruments, a zero check should be made and a zero correction applied to each reading if necessary, as described in Section B. A zero reading is made by rotating the screw until the jaw is closed or the spindle comes into contact with the anvil. The contacting surfaces of the spindle and anvil should be clean and free of dust. (Micrometers can be adjusted to zero readings by means of a spanner wrench. *Do not attempt to do this* without your instructor's permission or supervision.)

D. Density

The **density** ρ of a substance is defined as the mass m per unit volume V (i.e., $\rho = m/V$). Thus the densities of substances or materials provide comparative measures of the amounts of matter in a particular (unit) space. Note that there are two variables in density—mass and volume. Hence, densities can be affected by the masses of atoms and/or by their compactness.

As can be seen from the defining equation ($\rho = m/V$), the SI units of density are kilogram per cubic meter (kg/m^3). However, measurements are commonly made in

Figure 6   Density, mass, and volume. The marble and the Styrofoam ball have equal masses but different densities ($\rho = m/V$). Because the volume of the ball is greater than that of the marble, its density is less.

the smaller metric units of grams per cubic centimeter (g/cm^3), which can easily be converted to standard units.*

Density may be determined experimentally by measuring the mass and volume of a sample of a substance and calculating the ratio m/V. The volume of regularly shaped objects may be calculated from length measurements. For example;

Rectangle
$$V = l \times w \times h$$
(length $\times$ width $\times$ height)

Cylinder
$$V = Al = (\pi r^2)l$$
(circular cross-sectional area $A = \pi r^2$, where r is the radius, times the length l of the cylinder)

Sphere
$$V = \tfrac{4}{3}\pi r^3$$
(where r is the radius of the sphere)

To illustrate how density provides a measure of compactness of matter, consider the marble and Styrofoam ball in ● Fig. 6. Both have the same mass (5.0 g), but the marble has greater density. (Why?) With measured radii of $r_m = 0.75$ cm and $r_b = 6.0$ cm for the marble and ball, respectively, the calculated densities are

$$\rho_m = \frac{m_m}{V_m} = \frac{m_m}{\tfrac{4}{3}\pi r_m^3} = \frac{5.0 \text{ g}}{\tfrac{4}{3}\pi(0.75 \text{ cm})^3} = 2.8 \text{ g/cm}^3$$

$$\rho_b = \frac{m_b}{V_b} = \frac{m_b}{\tfrac{4}{3}\pi r_b^3} = \frac{5.0 \text{ g}}{\tfrac{4}{3}\pi(6.0 \text{ cm})^3} = 0.0055 \text{ g/cm}^3$$

---

* In the British fps (foot–pound–second) system, density is expressed in terms of weight rather than mass. For example, the weight density of water is 62.4 lb/ft^3.

(Notice that the calculated results have only two significant figures. Why?) In standard SI units, these results are 2.8×10^3 kg/m^3 and 5.5 kg/m^3, respectively.

But how does one find the volume of an irregularly shaped object? This may be done by immersing it in water (or some other liquid) in a graduated container. Since the object will displace a volume of water equal to its own volume, the difference in the container readings before and after immersion is the volume of the object. Cylinders commonly have scale divisions of milliliters (mL) and 1 mL = 1 cm^3.* [cm^3 (cubic centimeter) is sometimes written on glassware as cc.]

The physical property of density can be used to identify substances in some cases. If a substance is not pure or is not homogeneous (that is, its mass is not evenly distributed), an average density is obtained, which is generally different from that of a pure or homogeneous substance.

EXPERIMENTAL PROCEDURE

A. *Least Count of an Instrument Scale*

1. List the least count and the estimated fraction of the least count for each of the measuring instruments in Data Table 1 of the laboratory report. For example, for a meterstick, these would be 1 mm and 0.1 mm, respectively. (Review Experiment 1 if necessary.)

B. *Thickness Measurements*

2. Using the micrometer caliper, take a zero reading and record it in Data Table 3. Then take several measurements of a single page of this manual, incorporating the zero correction if necessary, to determine the average thickness per page. Record the data and result in Data Table 2.

3. With the micrometer, take thickness measurements of a group of several pages together [e.g., 10 pages (sheets of paper)], and record the data in Data Table 2. Calculate the average thickness per page.

4. With the vernier caliper, take several measurements of the total thickness of the manual (*excluding* covers).† Record the data in Data Table 2, and compute the average overall thickness of the manual. (Did you remember to take a zero reading and record in Data Table 3?)

5. Using the values of the average thickness per page determined in procedures 2 and 3 and the overall average thickness of the manual from procedure 4, compute the number of pages (sheets of paper) in your manual. For example, if the average thickness per page is 0.150 mm and the average overall thickness is 35.5 mm (3.55 cm), the calculated number of papers is

$$\frac{35.5 \text{ mm}}{0.150 \text{ mm/page}} = 236.6666 = 237 \text{ pages}$$

6. Determine the actual number of pages (sheets of paper) in the manual. (Remember to subtract any pages handed in from Experiment 1, the Advance Study Assignment for this experiment, and any others that might be missing.) Compute the percent error for each of the two experimentally determined values.

C. *Density Determinations*

7. The densities of the materials of the various objects are to be determined from mass and volume (length) measurements. Taking the mass and length measurements will give you experience in using the laboratory balance and the vernier and micrometer calipers.

8. Using the appropriate measuring instrument(s), take several measurements to determine the average dimensions of the regularly shaped objects so that their volumes can be calculated. Record the data in Data Table 3. Remember to make a zero correction for each reading if necessary.

9. Calculate the volume of each of the objects, and record in Data Table 4.

10. Determine the volume of the irregularly shaped metal object by the method described in Theory section D. Record the volume in Data Table 4.

11. Using a laboratory balance, determine the mass of each object, and record the results in Data Table 4.

12. Calculate the density of the material of each object, and find the percent error of each experimental result. (Accepted density values are given in Appendix A, Table A1.)

---

* *Milliliter* is abbreviated both ml and mL. The mL abbreviation is generally preferred in order to avoid confusion of a lowercase l ("ell") with the number 1.

† Be sure the pages are compacted as much as possible before you take the measurements.

Name _____ Section _____ Date _____

Lab Partner(s) _____

Measurement Instruments (Mass, Volume, and Density)
$\boxed{TI}$ *Laboratory Report*

A. *Least Count of an Instrument Scale*

DATA TABLE 1

Purpose: To practice determining least count and estimated fraction of least count.

Instrument	Least count	Estimated fraction
Meterstick		
Vernier caliper		
Micrometer caliper		
Balance		
Graduated cylinder		

Calculations
(show work)

Don't forget units

(continued)

B. Thickness Measurements

DATA TABLE 2

Purpose: To practice using calipers. (Indicate units in the parentheses.)

Reading	Thickness of single page ( )	Thickness of _____ pages ( )	Average page thickness ( )	Thickness of manual, excluding covers ( )
1				
2				
3				
4				
Average				

Actual number of pages (sheets)
in manual   _____   Percent error

Computed number of pages
(from single-page measurement)   _____   _____

(from multiple-page measurement)   _____   _____

Calculations
(show work)

Name _____ Section _____ Date _____

Lab Partner(s) _____

Laboratory Report

C. Density Determination

DATA TABLE 3

Purpose: To record dimensional measurements.

Zero reading: Vernier caliper _____ Micrometer caliper _____

	Rod		Wire		Sphere	Rectangular sheet		
Instrument used								
Reading	Diameter ( )	Length ( )	Diameter ( )	Length ( )	Diameter ( )	Length ( )	Width ( )	Thickness ( )
1								
2								
3								
4								
Average								

Calculations
(show work)

(continued)

31

DATA TABLE 4

Purpose: To compare experimental and accepted density values.

Object	Mass ( )	Volume ( )	Experimental density ( )	Accepted density (from Table A1)	Percent error
Rod Type of material: _____					
Wire Type of material: _____					
Sphere Type of material: _____					
Rectangular sheet Type of material: _____					
Irregularly shaped object Type of material: _____					

Calculations
(attach additional sheet if necessary)

EXPERIMENT 2 | *Laboratory Report*

/TI/ QUESTIONS

1. Explain the probable source of error(s) in the experimental determination of the number of manual pages.

2. In the first four density determinations in Data Table 4, what major factors might account for the experimental errors that were obtained?

3. In determination of the volume of the irregularly shaped object, any air bubbles sticking to the surface of the object when it is submerged cause systematic errors. Will this error give an experimental density that is too high or too low? Explain.

4. Suppose that you were given an irregularly shaped object that floats. Describe how you would experimentally determine its volume.

(continued)

5. A thin circular sheet of aluminum has a radius of 20 cm and a thickness of 0.50 mm. Find the mass of the sheet.

6. Archimedes, a famous Greek scientist, was given a problem by King Hieron II of Syracuse (Sicily). The king suspected that his crown, which was supposed to be made of pure gold, contained some silver alloy, and he asked Archimedes to prove or disprove his suspicion. (It turned out that the crown did contain silver.) How would you have experimentally determined whether or not the crown was pure gold?

Name _____ Section _____ Date _____

Lab Partner(s) _____

EXPERIMENT 3

Uniformly Accelerated Motion

/TI/ *Advance Study Assignment*

Read the experiment and answer the following questions.

A. *Object in Free Fall*

1. What effect might the distance of fall have on your experimental results? (*Hint:* Assume that you dropped an object from a height of 1.0 m. How long would it take the object to reach the floor? Show your work.)

2. Objects of different masses are dropped. Does this make a difference in the time of fall? Explain.

3. Suppose that the initial height of the object were measured from the top of the object at the release point to the floor. How would this affect your experimental result? Is this a random or a systematic error?

(continued)

B. Linear Air Track

4. How is the acceleration of a car traveling on an elevated air track related to
   (a) the angle of elevation; (b) the height of elevation?

5. What is the equation describing the instantaneous velocity of a car on an elevated air track,
   and what is the shape of the graph of the instantaneous velocity versus time?

6. Will the graph of instantaneous velocity versus time have a Y-axis intercept of zero?
   Explain.

7. Describe how the instantaneous velocity of a car traveling on an elevated air track can be
   calculated from displacement and time data.

CI *Advance Study Assignment*

1. What precautions need to be taken when working with a fan-propelled car?

2. For an object moving with constant acceleration, what will be the shape of a graph of
   position versus time? What will be the shape of a graph of velocity versus time?

Uniformly Accelerated Motion

OVERVIEW

Experiment   examines uniformly accelerated motion using complementary TI and CI approaches. The TI procedures investigate the accelerations of (1) an object in free fall, and (2) a car on a linear air track, for both horizontal and inclined motions.

The CI procedures extend the investigation by considering not only the linear relationship for uniformly accelerated motion, $v = at$, but also the parabolic relationship $x = \frac{1}{2}at^2$. This is done using a fan car and a rotary motion sensor.

INTRODUCTION AND OBJECTIVES

An important case in kinematics is that of an object in uniformly accelerated motion—one having a uniform or *constant* acceleration. Probably the most common example is a falling object at the surface of the Earth. An object falling solely under the influence of gravity is said to be in *free fall*, and that object falls with an acceleration g (the acceleration due to gravity). Near the Earth's surface, the acceleration due to gravity is approximately constant, with a common value of

$$g = 9.80 \text{ m/s}^2 = 980 \text{ cm/s}^2 = 32.2 \text{ ft/s}^2$$

Of course, air resistance affects the acceleration of a falling object. But for relatively dense objects over short distances of fall, the effects of air resistance are negligible, and objects fall with an acceleration of g.

In this experiment we make use of the acceleration due to gravity to investigate an object undergoing uniformly accelerated motion to see how its velocity and displacement change with time. Conversely, with displacement and time measurements, the value of g can be determined. The experimental data and their analyses will yield a better understanding of the kinetic equations describing the motion.

After performing this experiment and analyzing the data, you should be able to do the following:

TI OBJECTIVES

1. Clearly distinguish between average and instantaneous velocity.
2. Express how the velocity of a uniformly accelerated object changes with time.
3. Express how the distance traveled by a uniformly accelerated object changes with time.
4. Explain how the uniform acceleration of an object may be determined from distance and time measurements.

CI OBJECTIVES

1. Analyze the motion of an object that moves with constant acceleration.
2. Understand what it means to say that the position varies with the square of the time.

Uniformly Accelerated Motion

EQUIPMENT NEEDED

A. Object in Free Fall

- 3 objects of different masses (steel balls or lead weights)
- Meterstick
- Laboratory timer or stopwatch

B. Linear Air Track

- Linear air track
- Several laboratory timers or stopwatches
- Wooden blocks of two different heights
- 1 sheet of Cartesian graph paper

(*Optional* A TI experiment for the free-fall spark timer is given in the supplement at the end of this experiment.)

TI THEORY

A. Object in Free Fall

An object in free fall (air resistance negligible) falls under the influence of gravity with an acceleration g. The distance y an object falls in a time t is given by

$$y = y_0 + v_0 t + \tfrac{1}{2} g t^2 \qquad \textbf{(TI1)}$$

where v_0 is the initial velocity.

If an object is dropped from rest ($v_0 = 0$, $y_0 = 0$), then

$$y = \tfrac{1}{2} g t^2 \qquad \textbf{(TI2)}$$

(downward taken as positive to avoid minus signs). Hence, by measuring the time t it takes for an object to fall a distance y, the acceleration g due to gravity can be easily calculated.

The time it takes for a released object to fall a distance y to the floor may be measured manually or electronically (● TI Fig. 1).

B. Linear Air Track

Types of linear air tracks are shown in ● TI Fig. 2. Air is supplied to the interior of the hollow track and emerges through a series of small holes in the track. This provides a cushion of air on which a car or glider travels along the track with very little friction (an example of the use of a gaseous lubricant).

To have the car move under the influence of gravity, one end of the air track is elevated on a block. The acceleration of the car along the air track is then due to a component of the force due to gravity, $F = ma = mg \sin\theta$ (● TI Fig. 3). The acceleration a of the glider along the air track is

$$a = g \sin\theta \qquad \textbf{(TI3)}$$

and from the geometry, $\sin\theta = h/L$ (side opposite the angle over the hypotenuse). Hence,

$$a = \frac{gh}{L} \qquad \textbf{(TI4)}$$

The instantaneous velocity v of the uniformly accelerating glider at a time t is given theoretically by

$$v = v_0 + at \qquad \textbf{(TI 5)}$$

Hence, a graph of v versus t is a straight line ($y = mx + b$) with a slope $m = a = \Delta v/\Delta t$ and an intercept $b = v_0$. If the car starts from rest, the initial velocity v_0 is zero, and

$$v = at \qquad \textbf{(TI 6)}$$

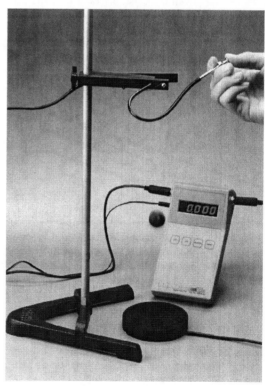

TI Figure 1   Distance–time measurements. A digital free-fall apparatus. The time of fall is recorded electronically. (Courtesy of Sargent-Welch.)

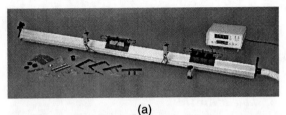

(a)

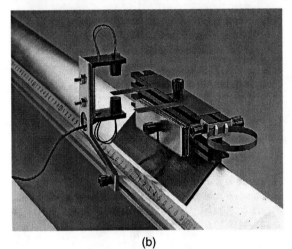

(b)

TI Figure 2   An air track system.   (a) A blower supplies air to the track through the hose on the right. The cars or gliders travel on a thin cushion of air, which greatly reduces friction. (b) The system is equipped with photogates for automatic timing. (Courtesy of Sargent-Welch.)

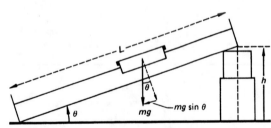

TI Figure 3   Accelerating car on air track.   When one end of an air track is elevated, the acceleration of the car is due to the component of the (weight) force mg, and $a = g \sin \theta$.

It can be shown that the instantaneous velocities of the car can be found from the experimental data of the measured displacements y_i of the glider along the air track at times t_i (with $v_0 = 0$) by

$$v_i = \frac{2y_i}{t_i}$$ **(TI 7)**

TI EXPERIMENTAL PROCEDURE

A. *Object in Free Fall*

1. One person should drop the object and do the timing. Lab partners should alternate.

2. Distinguish the objects as m_1, m_2, and m_3. Drop one of them from a fixed height y above the floor, and measure its time of fall. Drop it with the arm held horizontally or held upward. (Depending on your height, it may be advantageous to stand on a small step stool. Why?) Do a couple of practice runs to become familiar with the procedure. Record the data for four trials in Data Table 1. Repeat this procedure for the other two objects.

3. Compute the acceleration g due to gravity from TI Eq. 2, using the corrected times of fall. Find the average or mean value and the mean deviation of your results from the average value (see Experiment 1). *Note:* The results obtained by this procedure may have very poor accuracy and precision. (Why?)

B. *Linear Air Track*

4. The air track should be set up and leveled by the instructor or laboratory assistant. *Do not* attempt to make any adjustments of the air track. Ask your instructor for assistance if you need it.

5. Turn on the air supply, and place the car in motion by applying a small force on the car in a direction parallel to the air track. *Do not* attempt to move the car on the air track if the air supply is not turned on. Use the same small force for each trial—for example, by compressing a spring attached to the car.

6. Using laboratory timers or stopwatches, determine the times required for the car to travel several convenient distances, such as 0.20 m, 0.40 m, 0.50 m, 0.75 m, and so on. Record the times and distances in TI Data Table 2.*

    Several students should work together, each with a timer, taking a time reading as the car passes his or her assigned distance mark. Make several practice trials before taking actual data. (Remember that the distances are length intervals and need not be measured from the end of the air track. Make use of as much of the air track as is conveniently possible.)

7. After completing procedure 6, ask the instructor to elevate one end of the air track on a block, or obtain per-

---

* If electronic photogate timers are available, your instructor will give you instruction in their use. Electronic timing greatly improves the accuracy and precision of the results. (Why?)

mission to do so. Measure h and L (see TI Fig. 3) and enter your results in TI Data Table 2.

8. Start the car from rest near the elevated end of the air track. (To minimize error, it is better to put a block or pencil in front of the car and pull this away smoothly rather than releasing the car by hand.) Measure and record the times required for the car to travel the distances listed, and record your results in TI Data Table 2. Use the experimental method described in procedure 6.

9. Have the end of the air track elevated to a different height, and repeat the time measurements for this height.

10. Using Eq. TI 7, compute the instantaneous velocity of the car for each of the times in the three experimental sets of data in TI Data Table 2.

11. Plot v versus t for each case on the same graph, and determine the slope of each line.

12. Using Eq. TI 4, compute an experimental value of the acceleration due to gravity g for each of the elevated-air-track cases. Compute the percent error for each experimental result.

Uniformly Accelerated Motion

/TI/ *Laboratory Report*

A. *Object in Free Fall*

/TI/ **DATA TABLE 1**

Purpose: To determine *g* experimentally (and check mass dependence).

y _____

m_1 Trial	Time of fall, *t* ( )	Calculated *g* ( )	Deviation ( )
1			
2			
3			
4			
	Mean (average) value		

Result (average value ± mean deviation)

_____

m_2 Trial	Time of fall, *t* ( )	Calculated *g* ( )	Deviation ( )
1			
2			
3			
4			
	Mean (average) value		

Result (average value ± mean deviation)

_____

m_3 Trial	Time of fall, *t* ( )	Calculated *g* ( )	Deviation ( )
1			
2			
3			
4			
	Mean (average) value		

Result (average value ± mean deviation)

_____

Calculations
(show work, attach page to report)

B. Linear Air Track

/T1/ **DATA TABLE 2**

Purpose: To determine *g* experimentally.

Distances ( )					
1. Time t_i ( )　　　1					
Level air track　　　2					
3					
Average					
Computed v_i ( )					
2. Time t_i ( )　　　1					
Elevated air track　　2					
h_1 —————　3					
Average					
Computed v_i ( )					
3. Time t_i ( )　　　1					
Elevated air track　　2					
h_2 —————　3					
Average					
Computed v_i ( )					

Calculations
(show work)

Length of air track　*L* _____

Slopes of graphs　1. _____

2. _____

3. _____

Experimental values of *g*
(computed from data)　1. _____

2. _____

3. _____

Percent error　1. _____

2. _____

3. _____

EXPERIMENT 3   *Laboratory Report*

/TI/ QUESTIONS

A. *Object in Free Fall*

1. Objects of different mass were used to see whether the acceleration due to gravity depends on the mass of a falling object. In other words, does a heavier object fall faster than a lighter object? What do your experimental results show? (If you haven't already studied this, check your textbook for the correct answer.)

2. What is probably the greatest source of error in the experimental procedure? Is it random error or systematic error?

B. *Linear Air Track*

1. What are the major sources of error in this procedure?

2. What would be the shapes of the curves for a graph of y versus t of the data in each experimental case? How would you determine the value of the car's acceleration from a graph using only y and t values (i.e., not computing v_i)?

(continued)

3. What is the physical significance of the slope of the graph for the case of the level air track?

4. What is the maximum possible value of the slope of a v-versus-t curve for a car released from rest on an air track elevated at one end? Describe the experimental setup in this case.

Uniformly Accelerated Motion

CI EQUIPMENT NEEDED

- 1 collision (or plunger) cart

- 1 fan accessory
- 1 dynamics track
- 1 rotary motion sensor (RMS)
- Brackets and pulley mounts:
    1 cart-string bracket
    1 dynamics track mount accessory
    1 RMS/IDS adapter
- String
- Optional:
    Track end-stop

Pasco Collision Cart ME-9454 (or ME-9430) (Any of the classic carts or the Pascars will work fine.)
Pasco ME-9491

CI-6538

CI-6569
CI-6692 (to mount the RMS to the track)
ME-6569 (track pulley bracket)

CI THEORY

When an object moves with a uniform or constant acceleration, the position of the object at a time t is given by

$$x = x_0 + v_0 t + \tfrac{1}{2} at^2 \qquad \textbf{(CI 1)}$$

where v_0 is the initial velocity and a is the constant acceleration. For an initial position arbitrarily chosen to be $x_0 = 0$, and for an object starting from rest ($v_0 = 0$), the position at any time reduces to

$$x = \tfrac{1}{2} at^2 \qquad \textbf{(CI 2)}$$

(Compare this to the free-fall equation $y = \tfrac{1}{2} gt^2$, (TI Eq. 2.) Note that for any case where the acceleration is constant, the relationship between position and time is not linear: The position is proportional to the square of the time, not just to the time. A graph of x versus t will be a parabola, not a straight line.

On the other hand, if the object has constant acceleration, then the velocity is changing at a steady rate. The velocity of the object at any time after it starts from rest is given by

$$v = at \qquad \textbf{(CI 3)}$$

which is a linear function of time. A graph of v versus t will be a straight line.

The purpose of this experiment is to analyze the motion of an object undergoing constant acceleration, to better understand what it means to say that the position varies with the square of the time, and to compare this to the velocity function, which is directly proportional to the time. The results apply to any type of uniformly accelerated motion.

EQUIPMENT SETUP

1. The cart-string bracket and the fan accessory are mounted on top of the cart.
2. The rotary motion sensor (RMS) is mounted to one side of the track, with the small pulley of the RMS/IDS adapter mounted on the opposite end of the same side of the track. ● CI Fig. 1 is a diagram of the setup.
3. The string makes a full loop connecting the cart-string bracket with the large pulley of the RMS sensor and the small pulley on the opposite bracket. That string should be tense, but not tight.
4. Adjust the height of the string so that the fan blade clears the string as it spins. The RMS and the small pulley can be moved as far down as needed for the blades to clear the string.

SETTING UP DATA STUDIO

1. Open Data Studio and choose Create Experiment.
2. From the sensor list, choose a Rotary Motion Sensor (RMS).
3. Connect the sensor to the interface as shown on the computer screen.
4. Double-click on the RMS icon. The Sensor Properties window will open. The Sensor Properties window is illustrated in ● CI Fig. 2. Go to Measurement and choose for the sensor to read Position [m] and Velocity [m/s]. Deselect any other type of measurement. When done, press OK. The Data list on the left of the screen should now have two icons: one for the position data, the other for the velocity data.

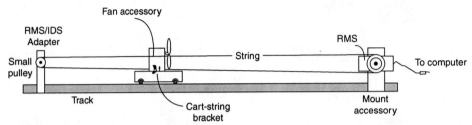

CI Figure 1   Rotary motion sensor and cart setup.   A string makes a full loop from the cart-string bracket, to the RMS pulley, to the small pulley on the opposite end of the track, and back to the cart. The height of the RMS and the pulley must be adjusted so that the fan blades do not touch the string as they spin.

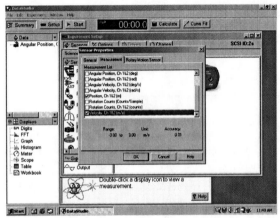

CI Figure 2   The sensor properties window. Under "Measurements," the sensor properties window enables you to choose what type of data you want the particular sensor to collect. In this case, choose Position [m] and Velocity [m/s], and deselect all others. (Data displayed using DataStudio Software. Reprinted courtesy of PASCO scientific.)

5. Create a graph by dragging the position data icon from the data list and dropping it on top of the graph icon of the displays list. A graph of position versus time will open in a window called "Graph 1."
6. Now drag the velocity data icon and drop it somewhere in the middle of the graph. The graph display will split into two graphs: one of position, the other of velocity, as shown in ● CI Fig. 3.

CI EXPERIMENTAL PROCEDURE

WARNING! Be careful not to touch the fan blades while they are spinning.

1. Turn the fan on, but hold on to the car so that it does not yet move.
2. Have a partner press the START button. Let go of the car. Your partner must press the STOP button before the car reaches the end of the track, to prevent data

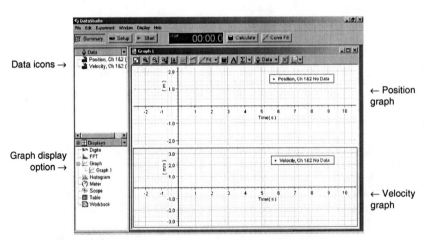

CI Figure 3   Graph displays.   The graph display in this picture has been maximized to occupy most of the screen. (Data displayed using DataStudio Software. Reprinted courtesy of PASCO scientific.)

being taken of collisions and rebounds with pulleys or end-stops. You may need to do a few practice runs to become familiar with the procedure.

Note: If the graphs show negative values, reverse the fan so that it is facing in the opposite direction, and start again from the opposite side of the track.

3. Carefully turn off the fan.
4. You should have two graphs on the screen. Use the Scale-to-fit button of the graph toolbar to display the data clearly. Notice that the graph of position versus time is a smooth parabola. The graph of velocity versus time is a straight line.
5. Click anywhere on the position-versus-time graph to make it active. Use the Smart Tool (a button of the graph toolbar, labeled "xy") to choose a data point that is close to the beginning of the motion but for which the position is not zero. Record the position and the time of this point in CI Data Table 1, as the first data points, x_1 and t_1.

6. Find the position at a time $t_2 = 2t_1$. That is, where was the car when the previous time doubled? Record x_2.
7. Repeat for times $t_3 = 3t_1, t_4 = 4t_1, \ldots$ as many multiples of t_1 as you can get from the graph. (The longer the track you use, the more you can get.)
8. Determine by what factor the distance traveled at time t_2 is greater than the distance at time t_1. Then determine by what factor the distance traveled at time t_3 is greater than the distance at time t_1. Continue until CI Data Table 1 is complete.
9. Now click anywhere on the velocity-versus-time graph to activate it. Use the Smart Tool to find the velocity of the cart at each of the times $t_1, t_2, \ldots, t_n$. Record the velocities in CI Data Table 2, and calculate by how much the velocity increases as the times double, triple, etc.
10. Use the Fit Tool to determine the slope of the velocity graph. (The fit tool is on the graph toolbar; it is a drop menu called "Fit.") Choose a "Linear Fit" for your graph. Report the slope in CI Data Table 2. What is the slope of a velocity-versus-time plot measuring? (*Hint:* Think of the units!)

Name _____ Section _____ Date _____

Lab Partner(s) _____

Uniformly Accelerated Motion

CI *Laboratory Report*

CI   DATA TABLE 1

Purpose: To investigate a position function that is proportional to the square of the time.

	Time		Position	How many times larger is x_n than x_1? x_n/x_1
t_1		x_1		
t_2		x_2		
t_3		x_3		
t_4		x_4		

CI   DATA TABLE 2

Purpose: To investigate a velocity function that is proportional to the time.

	Time		Velocity	How many times larger is v_n than v_1? v_n/v_1
t_1		v_1		
t_2		v_2		
t_3		v_3		
t_4		v_4		

Slope of the graph: _____

(units)

Don't forget units

(continued)

CI QUESTIONS

1. In CI Data Table 1 you measured the position of the car at different times. When the time doubled, did the distance from the origin double also? When the time tripled, did the distance from the origin triple also? Can you see the pattern?

2. Discuss what it means to say that the position function is proportional not to the time but to the time squared.

3. Judging on the basis of the observed pattern, and without using theoretical equations, predict the position of the car when time is $10t_1$. What will the position be at $20t_1$?

4. In CI Data Table 2 you repeated the procedure for the velocities. What is the pattern now?

5. On the basis of the observed pattern, predict the velocity of the car for times $10t_1$ and $20t_1$.

6. A graph of x versus t is a parabola, because $x \propto t^2$. But if you plot x versus t^2, the resulting graph will be a straight line, with slope $\frac{1}{2} a$, as shown below. Make a graph with your values of x_n on the vertical axis and your *times squared* on the horizontal. Determine the slope, and use it to find the acceleration of the car. (Attach a graph to Lab Report.)

$$
\begin{array}{ccc}
x & = & \frac{1}{2}a \quad t^2 \\
\downarrow & & \downarrow \quad \downarrow \\
y & = & m \quad x
\end{array}
$$

7. By determining a percent difference, compare the acceleration of the car determined from your graph to that measured as the slope of the velocity graph.

 **T I**   **E X P E R I M E N T 3**   **(OPTIONAL)**

Uniformly Accelerated Motion: Free-Fall Spark-Timer Apparatus

/TI/ *Advance Study Assignment*

Read the experiment and answer the following questions.

1. How are data recorded on the tape strip, and what information does the data tape give?

2. What precautions should be taken in using the apparatus? What could happen if this is not done?

3. What equation describes the instantaneous velocity of an object in free fall, and what is the shape of the graph of the instantaneous velocity versus time?

(continued)

4. Should the graph of instantaneous velocity versus time have a *Y*-axis intercept of zero? Explain.

5. Describe how the instantaneous velocity of an object in free fall can be calculated from displacement and time data.

Uniformly Accelerated Motion

See page for the Introduction and Objectives

 EQUIPMENT NEEDED

- Free-fall apparatus
- Meterstick

/TI/ THEORY

Some free-fall timer apparatuses are shown in ● TI Fig. 1A. The free-fall spark-timer assembly consists of a metal object that falls freely between two wires with a tape strip of specially treated paper between the object and one of the wires. The spark timer is a fast timing device that supplies a high voltage across the wires periodically for preset time intervals (e.g., a frequency of 60 Hz, or time interval of $\frac{1}{60}$ s, since $t = 1/f$). The free-fall apparatus is equipped with an electromagnet that releases the metal object when the spark timer is activated.

A high voltage causes a spark to jump between two electrical conductors in close proximity. The wires are too far apart for a spark to jump directly from one wire to the other. However, as the metal object falls between the wires, the spark (electrical current) jumps from one wire to the metal object, travels through the object, and jumps to the other wire. In so doing, the spark burns a spot on the paper tape strip.

The spots on the tape are then a certain time interval apart, as selected and usually preset on the spark timer. The series of spots on the tape gives the vertical distance of fall as a function of time, from which can be measured the distance y_i that the object falls in a time t_i.

The instantaneous velocity v of a free-falling object (neglecting air resistance) at a time t is given theoretically by

$$v = v_o + gt \qquad \textbf{(TI 1A)}$$

(where downward is taken as the positive direction). Hence, a graph of v versus t is a straight line ($y = mx + b$) with a slope $m = \Delta v/\Delta t = g$ and an intercept $b = v_o$, the initial

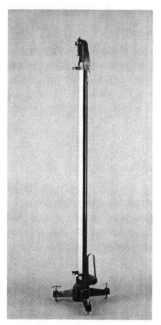

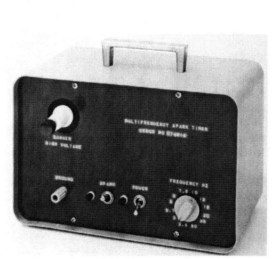

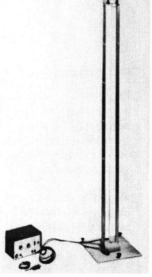

TI Figure 1A   Spark timers.   Types of free-fall spark timer apparatuses. [Photos (left and center) courtesy of Sargent-Welch. (right) Harvard Apparatus, Inc., Holliston, MA 01746.]

velocity. Recall that t in TI Eq. 1A is really a time *interval* measured from an arbitrary starting time $t_o = 0$. At this time, the velocity of the object is v_o, which may or may not be zero.

The motion of the falling object as recorded on the experimental data tape is analyzed as follows. The average velocity $\bar{v}$ of an object traveling a distance y_i in a time t_i is defined as

$$\bar{v} = \frac{y_i}{t_i} \qquad \textbf{(TI 2A)}$$

Keep in mind that y_i and t_i are really length and time *intervals*, or the differences between corresponding instantaneous lengths and times. Referenced to an initial position and time (y_o and t_o), $\Delta y_i = y_i - y_o$ and $\Delta t_i = t_i - t_o$. Arbitrarily taking $y_o = 0$ and $t_o = 0$, we have $\Delta y_i = y_i$ and $\Delta t_i = t_i$. (It is these intervals that will be measured from the data tape.)

For a uniformly accelerated object (moving with a constant acceleration), as in the case of free fall, the average velocity is given by

$$\bar{v} = \frac{v_i + v_o}{2} \qquad \textbf{(TI 3A)}$$

where v_i and v_o are the instantaneous velocities at times t_i and t_o, respectively. (Why is this? Consult your textbook.) Then, equating the expressions for $\bar{v}$, given by TI Eq. 2A and TI Eq. 3A and solving for v_i, we have

$$\frac{v_i + v_o}{2} = \frac{y_i}{t_i}$$

and

$$v_i = \frac{2y_i}{t_i} - v_o \qquad \textbf{(TI 4A)}$$

If $v_o = 0$ (that is, the object falls from rest), then

$$v_i = \frac{2y_i}{t_i} \qquad \textbf{(TI 5A)}$$

◢TI◣ EXPERIMENTAL PROCEDURE

1. Your laboratory instructor will make a data tape for you or assist and direct you in obtaining one. Care must be taken in aligning the apparatus.

   Caution: When working with high voltages, one must be careful not to receive an electrical shock. Do not touch metal parts when the spark timer is on.

2. Record the time interval of the spark timer used on the data tape, and draw small circles around the burn spots so that their locations can be easily seen. Occasionally, a spot of the sequence may be missing (e.g., due to local misalignment of the wires). However, it is usually easy to tell that a spot is missing by observation of the tape. Do not try to guess where the spot should be. Simply make a mark on the tape to indicate that a spot is missing.

3. Through each spot, draw a straight line perpendicular to the length of the tape. Using the line through the beginning spot as a reference ($y_o = 0$), measure the distance of each spot line from the reference line (y_1, y_2, y_3, etc.). Write the measured value of the distance on the tape by each respective spot line.

   Making use of the known spark-timer interval, write the time taken for the object to fall a given distance on the tape by each spot line, taking $t_o = 0$ at $y_o = 0$. For example, if the timer interval is $\frac{1}{60}$ s, the time interval between the reference line ($y_o = 0$) and the first spot line (y_1) is $t_1 = \frac{1}{60}$ s, and the time taken to fall to the second spot line (y_2) is $t_2 = \frac{1}{60} + \frac{1}{60} = \frac{2}{60} = \frac{1}{30}$ s. (Do not forget to account for the time intervals associated with missing spots, if any.)

4. Record the data measured from the tape in TI Data Table 1. Using TI Eq. 5A, compute the instantaneous velocity of the falling object at each spot line from the experimental data, and record.

5. At this point, you should realize that the instantaneous velocities given by TI Eq. 5A ($v_i = 2y_i/t_i$) are *not* the actual instantaneous velocities of the falling object, since it had a nonzero initial velocity or was in motion at the first spot line (y_o). TI Eq. 4A really applies to the situation, and $2y_i/t_i = v_i + v_o$. Note that the instantaneous velocities you computed ($2y_i/t_i$) included v_o.

   Even so, plot the computed v_i's on a v-versus-t graph and determine the slope. This will still be an experimental value of g. Compute the percent error of your experimental result. (Accepted value, $g = 9.80$ m/s² = 980 cm/s².)

6. You will notice on your graph that the line does not intercept the Y axis at the origin ($t = 0$). This is because $t = 0$ usually was measured not at the actual time of release, but at some time later. From TI Eqs. 4A and 1A we see that at $t = 0$ in the measurement time frame

$$\frac{2y_o}{t_o} = 2v_o \left[= v_{i_{(t=0)}} \right]$$

where y_o and t_o are, respectively, the distance and time measured by the zero values *from* the point of release.

The initial velocity at the first line spot is then $v_o = y_o/t_o$. This gives you the extra bonus of being able to determine v_o from your graph, since

$$v_o = \frac{v_{i_{(t=0)}}}{2}$$

where $v_{i_{(t=0)}}$ is the intercept value. Compute the initial velocity that the falling object had at your first line spot, and record in TI Data Table 1.

Name _____ Section _____ Date _____

Lab Partner(s) _____

T I   EXPERIMENT 3

Uniformly Accelerated Motion

/TI/ *Laboratory Report*

Free-Fall Timer Apparatus

/TI/ **DATA TABLE 1**

Purpose: To determine g experimentally.

Spark-timer interval _____

Distance y_i ( )		Time t_i ( )		Computed velocity $v_i = 2y_i/t_i$ ( )	
y_1		t_1		v_1	
y_2		t_2		v_2	
y_3		t_3		v_3	
y_4		t_4		v_4	
y_5		t_5		v_5	
y_6		t_6		v_6	
y_7		t_7		v_7	
y_8		t_8		v_8	
y_9		t_9		v_9	
y_{10}		t_{10}		v_{10}	
y_{11}		t_{11}		v_{11}	
y_{12}		t_{12}		v_{12}	
y_{13}		t_{13}		v_{13}	
y_{14}		t_{14}		v_{14}	
y_{15}		t_{15}		v_{15}	

Calculations
(show work)

Value of g from graph
(attach graph to lab report) _____

(units)

Percent error _____

Initial velocity at y_0 _____

(continued)

/TI/ QUESTIONS

1. Suppose that a different spark-timer interval were used. How would this affect the slope of the graph of v versus t?

2. What would be the shape of the curve of a y-versus-t graph of the experimental data?

3. If $t = 0$ were taken to be associated with some line spot other than y_0 (e.g., y_3 instead), how would this affect the v-versus-t graph?

4. Calculate v_0 directly from the first two measurement entries in TI Data Table 1, using the equation $v_0 = 2y_1/t_1 - y_2/t_2$. (Your instructor can derive this for you.) How does this compare with the value determined from your graph?

Name _____ Section _____ Date _____

Lab Partner(s) _____

The Addition and Resolution of Vectors: The Force Table

TI/ *Advance Study Assignment*

Read the experiment and answer the following questions.

1. Distinguish between scalar and vector quantities, and give an example of each.

2. How are vectors represented graphically, and how are scalars and vector quantities distinguished when written as symbols?

3. What is meant by drawing a vector to scale? Give a numerical example.

4. Are the parallelogram and triangle methods of vector addition related? Explain.

(continued)

5. How may the resultant of two vectors be computed analytically from a vector parallelogram?

6. How many vectors may be added by the polygon method? Are other methods of vector addition limited to the number of vectors that can be added? Explain.

7. What is meant by resolving a vector into components? Give an example.

8. Briefly describe the steps in the component method of vector addition.

9. On a force table, what is the difference between the equilibrant and the resultant? Why is only one of these actually determined experimentally?

The Addition and Resolution of Vectors: The Force Table

INTRODUCTION AND OBJECTIVES

Physical quantities are generally classified as either scalar or vector quantities. The distinction is simple. A **scalar** quantity (or *scalar*) is one with magnitude only (including units)—for example, speed (15 m/s) and temperature (20°C). A **vector** quantity (or *vector*), on the other hand, has both magnitude *and* direction. Such quantities include displacement, velocity, acceleration, and force—for example, a velocity of 15 m/s north or a force of 10 N along the $+x$ axis.

Because vectors have the property of direction, the common method of addition, scalar addition, is not applicable to vector quantities. To find the **resultant** or **vector sum** of two or more vectors, we use special methods of vector addition, which may be graphical and/or analytical.

Several of these methods will be described, and we shall investigate the addition of force vectors. The results of graphical and analytical methods will be compared with the experimental results obtained from a force table. The experimental arrangements of forces (vectors) will physically illustrate the principles of the methods of vector addition.

After performing this experiment and analyzing the data, you should be able to do the following:

1. Add a set of vectors graphically to find the resultant.
2. Add a set of vectors analytically to find the resultant.
3. Appreciate the difference in convenience between using graphical and using analytical methods of vector addition.

EQUIPMENT NEEDED

- Force table with four pulleys
- Four weight hangers
- Set of slotted weights (masses), including three of 50 g and three of 100 g
- String
- Protractor
- Ruler
- Level
- 3 sheets of Cartesian graph paper

THEORY

A. *Methods of Vector Addition: Graphical*

PARALLELOGRAM METHOD

Vectors are represented graphically by arrows (● Fig. 1). The length of a vector arrow (drawn to scale on graph paper) is proportional to the magnitude of the vector, and the arrow points in the direction of the vector.

The length scale is arbitrary and is usually selected for convenience and so that the vector graph fits nicely on the graph paper. A typical scale for a force vector might be 1 cm:10 N. That is, each centimeter of vector length represents 10 newtons. The *scaling factor* in this case in terms of force per unit length is 10 N/cm. (Note the similarity with the common food cost factor of price/lb—for example, 10 ¢/lb.)

When two vectors **A** + **B** (Fig. 1a) are added, the parallelogram of which **A** and **B** are adjacent sides is formed. The arrow diagonal of the parallelogram **R** is the resultant or vector sum of **A** + **B**; in other words, by vector addition, **R** = **A** + **B**. The magnitude R of the resultant vector is proportional to the length of the diagonal arrow, and the direction of the resultant vector is that of the diagonal

arrow **R**. The direction of **R** may be specified as being at an angle θ relative to **A**.

TRIANGLE METHOD

An equivalent method of finding **R** is to place the vectors to be added "head to tail" (head of **A** to tail of **B**, Fig. 1b). Vector arrows may be moved as long as they remain pointed in the same direction. The head-to-tail method gives the same resultant as the parallelogram method.

POLYGON METHOD

If more than two vectors are added, the head-to-tail method forms a polygon (Fig. 1c). For three vectors, the resultant **R** = **A** + **B** + **C** is the vector arrow from the tail of the **A** arrow to the head of the **C** vector. The length (magnitude) and the angle of orientation of **R** can be measured from the vector diagram. Note that this is equivalent to applying the head-to-tail method (two vectors) twice (**A** and **B** are added to give **R**$_1$, and then **C** is added to **R**$_1$ to give **R**).

The magnitude (length) R and the orientation angle θ of the resultant vector **R** in a graphical method can be measured directly from the vector diagram using a ruler and a protractor.

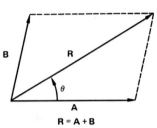

(a) Parallelogram Method

$R = A + B$

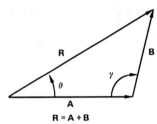

(b) Triangle "Head-To-Tail" Method

$R = A + B$

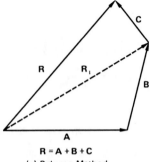

(c) Polygon Method

$R = A + B + C$

Figure 1   Vector addition.   Different methods of vector addition. Vectors are represented graphically by arrows. See text for description.

Example 1   To illustrate scaling and the graphical triangle method, let **A** and **B** represent forces at angles of 0° and 60°, respectively, with magnitudes of $A = 2.45$ N and $B = 1.47$ N.

Then, choosing a scaling factor (say, 0.50 N/cm), we find a vector length by dividing its magnitude by the scaling factor (magnitude/scaling factor). Note the unit cancellation:

A: 2.45 N/(0.50 N/cm) = 4.9 cm

B: 1.47 N/(0.50 N/cm) = 2.9 cm

Here, the 0.50-N/cm scaling factor was chosen so as to keep ● Fig. 2 an appropriate size.   In drawing your vector diagrams, you should choose a scaling factor that will use most of the allotted space on the graph paper—much as in plotting a graph in Experiment 1. Also, a factor with two significant figures was chosen because graph paper grids are usually not fine enough to plot more digits accurately.

The triangle has been drawn in Fig. 2, where $R = A + B$. The **R** vector is measured (with ruler and protractor) to have a length of 6.8 cm and a direction

angle of $\theta = 22°$ relative to the **A** vector. The magnitude of **R** in newtons is found using the scaling factor:

$$R = \text{(scaling factor)(measured length)}$$
$$= (0.50 \text{ N/cm})(6.8 \text{ cm}) = 3.4 \text{ N}$$

B. Methods of Vector Addition: Analytical

TRIANGLE METHOD

The magnitude of **R** in Fig. 2 can be computed from the law of cosines if the angle γ (angle opposite **R**) is known. We know γ, and with the given magnitudes of **A** and **B**,

$$R^2 = A^2 + B^2 - 2AB \cos \gamma \qquad (1)$$

The angle θ (between **R** and **A**) can then be computed using the law of sines because the magnitudes of sides **B** and **R** are known, and

$$\frac{B}{\sin \theta} = \frac{R}{\sin \gamma} \qquad (2)$$

From Example 1,   the magnitudes of **A** and **B** are 2.45 N and 1.47 N, respectively, and, as can be seen directly from Fig. 2, $\gamma = 120°$. (Why?) Then, using the law of cosines (Eq. 1):

$$R^2 = A^2 + B^2 - 2\ AB \cos \gamma$$
$$= (2.45)^2 + (1.47)^2 - 2(2.45)(1.47) \cos 120°$$
$$= 6.00 + 2.16 - 2(3.60)(-0.500)*$$
$$= 11.76$$

and

$$R = 3.43 \text{ N}$$

(Units were neglected in the initial calculation for convenience.)

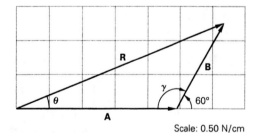

Scale: 0.50 N/cm

Figure 2   Drawing to scale. Figures are often scaled down so as to maintain a convenient size. Here the vector triangle is shown to scale, with a scaling factor of 0.50 N/cm. See text for description.

---

* Value obtained by calculator or from trig table with cos 120° = cos (180° − 120°) = −cos 60° using the trigonometric identity cos $(A − B) = \cos A \cos B + \sin A \sin B$.

The directional angle θ may be found using the law of sines (Eq. 2):

$$\theta = \sin^{-1}\left(\frac{B \sin \gamma}{R}\right)$$

$$= \sin^{-1}\left(\frac{1.47g \sin 120°}{3.43g}\right) = 21.8°$$

Remember that this is the angle between vectors **R** and **A**.

COMPONENT METHOD

If two vectors **A** and **B** are at right (90°) angles (● Fig. 3a), then the magnitude of their resultant is given by the **Pythagorean theorem**, $R = \sqrt{A^2 + B^2}$ (the hypotenuse of a right triangle is equal to the square root of the sum of the squares of the legs of the triangle). Notice that the law of cosines reduces to this formula with $\gamma = 90°$ (because $\cos 90° = 0$). The angle of orientation is given by $\tan \theta = B/A$, or $\theta = \tan^{-1}(B/A)$.

By the inverse process, we may resolve a vector into x and y components (Fig. 3b). That is, the vector **R** is the resultant of $\mathbf{R}_x$ and $\mathbf{R}_y$, and $\mathbf{R} = \mathbf{R}_x + \mathbf{R}_y$, where $R_x = R \cos \theta$ and $R_y = R \sin \theta$. The magnitude of R is given by

$$R = \sqrt{R_x^2 + R_y^2} \qquad (3)$$

and

$$\tan \theta = \frac{R_y}{R_x} \qquad (4)$$

or

$$\theta = \tan^{-1}\left(\frac{R_y}{R_x}\right)$$

(resultant, magnitude, and angle)

The vector sum of any number of vectors can be obtained by using the component method. This is conveniently done by having all the vectors originate from the origin and resolving each into x and y components, as shown in ● Fig. 4 for $\mathbf{R} = \mathbf{A} + \mathbf{B} + \mathbf{C}$.

The procedure is to add vectorially all of the x components together and all of the y components together. The $\mathbf{R}_x$ and $\mathbf{R}_y$ resultants are then added together to get the total resultant **R**. Doing this for the vectors in Fig. 4, we have

$$\mathbf{R}_x = \mathbf{A}_x + \mathbf{B}_x + \mathbf{C}_x$$

$$= 6.0 \cos 60° + 0 - 10 \cos 30°$$

$$= -5.7 \text{ N}$$

$$\mathbf{R}_y = \mathbf{A}_y + \mathbf{B}_y + \mathbf{C}_y$$

$$= 6.0 \sin 60° + 5.0 - 10 \sin 30°$$

$$= 5.2 \text{ N}$$

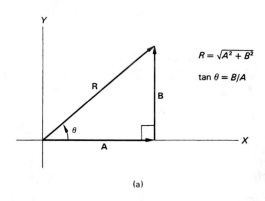

$$R = \sqrt{A^2 + B^2}$$
$$\tan \theta = B/A$$

(a)

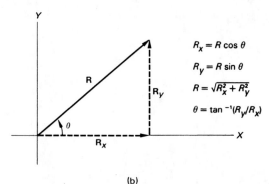

$$R_x = R \cos \theta$$
$$R_y = R \sin \theta$$
$$R = \sqrt{R_x^2 + R_y^2}$$
$$\theta = \tan^{-1}(R_y/R_x)$$

(b)

Figure 3   Vector resultant and components.   (a) The vector addition of **A** and **B** gives the resultant **R**. (b) A vector, such as **R**, can be resolved into x and y (rectangular) components $\mathbf{R}_x$ and $\mathbf{R}_y$, respectively.

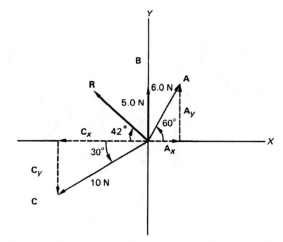

Figure 4   Component method.   Rather than using the head-to-tail method of vector addition, it is generally more convenient to use the component method, in which all vectors are drawn originating from the origin and resolved into components.

where the component directions are indicated by the positive and negative signs (arbitrary units). Note that **B** has no x component and that $\mathbf{C}_x$ and $\mathbf{C}_y$ are in the negative x and y directions, as indicated by the minus signs. Then the magnitude of **R** is (Eq. 3):

$$R = \sqrt{R_x^2 + R_y^2} = \sqrt{(5.7)^2 + (5.2)^2} = 7.7 \text{ N}$$

and, by Eq. 4,

$$\theta = \tan^{-1}\left|\frac{R_y}{R_x}\right| = \tan^{-1}\left(\frac{5.2}{5.7}\right) = 42°$$

relative to the $-X$ axis (or $180° - 42° = 138°$ relative to the $+X$ axis). It is convenient to measure all component angles as acute angles from the X axis. The minus R_x and positive R_y indicate that the resultant is in the second quadrant.*

C. Methods of Vector Addition: Experimental

THE FORCE TABLE

The **force table** is an apparatus that makes possible the experimental determination of the resultant of force vectors (● Fig. 5). The rim of the circular table is calibrated in degrees. Weight forces are applied to a central ring by means of strings running over pulleys and attached to weight hangers. The magnitude (mg) of a force (vector) is varied by adding or removing slotted weights, and the direction is varied by moving the pulley.

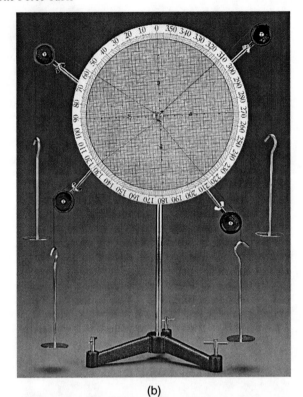

(b)

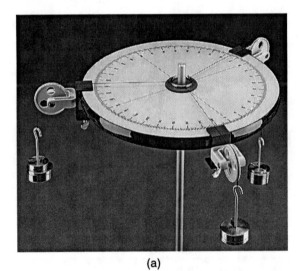

(a)

(c)

Figure 5   Force tables.   Various types of force tables. The table in (c) may be used vertically for demonstration (b), or horizontally in the laboratory. [Courtesy of Sargent-Welch.]

---

* Although it is customary to measure angles counterclockwise from the positive X axis, this procedure is convenient in eliminating the need for double-angle equations.

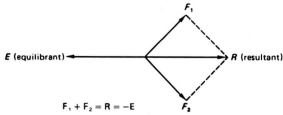

Figure 6 Resultant and equilibrant. On a force table, the magnitude and direction of the equilibrant **E** are measured, rather than those of the resultant **R**, and **R** = −**E**.

The resultant of two or more forces (vectors) is found by balancing the forces with another force (weights on a hanger) so that the ring is centered around the central pin. The balancing force is *not* the resultant **R** but rather the *equilibrant* **E**, or the force that balances the other forces and holds the ring in equilibrium.

The equilibrant is the vector force of equal magnitude, but in the *opposite direction*, to the resultant (i.e., **R** = −**E**). See ● Fig. 6. For example, if an equilibrant has a magnitude of $(0.30)g$ N in a direction of 225° on the circular scale, the resultant of the forces has a magnitude of $(0.30)g$ N in the opposite direction, 225° − 180° = 45°. It should be evident that the resultant cannot be determined directly from the force table. (Why?)*

EXPERIMENTAL PROCEDURE

1. Set up the force table with strings and suspended weights and perform the following cases of vector addition.

2. *Vector addition 1.* Given two vectors with magnitudes $F_1 = (0.200)g$ N and $F_2 = (0.200)g$ N at 30° and 120°, respectively, find their vector sum or resultant $\mathbf{F} = \mathbf{F}_1 + \mathbf{F}_2$ by each of the following procedures. (*Note:* Orientation angles of vectors are given relative to the 0° reference line or positive X axis.)

   (a) *Graphical.* Using the parallelogram method of vector addition, draw a vector diagram to scale. Use a scale such that the finished vector diagram fills about half a sheet of graph paper. Measure the magnitude and direction of the resultant (with ruler and protractor) and record the results in the data table. Save your graphical sheets to attach to the Laboratory Report.

(b) *Analytical.* Using the law of cosines, compute the magnitude of the resultant force. Compute the angle of orientation from the relationship $\tan \theta = F_2/F_1$. (Why can you use $\tan \theta$? Remember that θ is the angle between **F** and $\mathbf{F}_1$.) Record the results in the data table.

(c) *Experimental.* On the force table, clamp pulleys at 30° and 120° and add enough weights to each weight hanger to total 0.200 kg, so as to give weight forces of $F_1 = F_2 = (0.200)g$ N in these directions. (The weight hangers usually have masses of 50 g, or 0.050 kg.)

Using a third pulley and weights, determine the magnitude and direction of the equilibrant force that maintains the central ring centered in equilibrium around the center pin. Record the magnitude and direction of the resultant of the two forces in the data table. Remember, the resultant has the same magnitude as the equilibrant but is in the opposite direction.

(*Note:* The string knots on the central ring should be of a nontightening variety so that the strings will slip freely on the ring and allow the strings to pull directly away from the center. Pulling the center ring straight up a short distance and releasing it helps adjust the friction in the pulleys as the ring vibrates up and down so that it can settle into an equilibrium position involving only the applied forces. When the forces are balanced, the pin may be carefully removed to see whether the ring is centered around the central hole.)

3. *Vector addition 2.* Repeat procedure 2 for $F_1 = (0.200)g$ N at 20° and $F_2 = (0.150)g$ N at 80°. Use the other half of the sheet of graph paper used in procedure 2(a) for the graphical analysis. Be careful in the analytical analysis. Can you use $\tan \theta = F_2/F_1$ in this case?

4. *Vector addition 3.* Repeat procedure 2 with $F_1 = F_x = (0.200)g$ N (at 0°) and $F_2 = F_y = (0.150)g$ N (at 90°). In this case, $\mathbf{F} = \mathbf{F}_x + \mathbf{F}_y$, where $\mathbf{F}_x$ and $\mathbf{F}_y$ are the x and y components of **F**, respectively. That is, the resultant can be resolved into these components. Use half of another sheet of graph paper for the graphical method.

5. *Vector resolution.* Given a force vector of $F = (0.300)g$ N at 60°, resolve the vector into its x and y components and find the magnitudes of $\mathbf{F}_x$ and $\mathbf{F}_y$ by the following procedures:

   (a) *Graphical.* Draw a vector diagram to scale (on the other half of the sheet of graph paper used in procedure 4) with the component vectors (see Fig. 3b) and measure the magnitudes of $\mathbf{F}_x$ and $\mathbf{F}_y$. Record the results in the data table.

---

* The magnitude of the (weight) force vectors are in general given in the form $R = mg = (0.150)g$ N, for example, where it is understood that the mass is in kilograms and g is the acceleration due to gravity. It is convenient to leave g in symbolic form so as to avoid numerical calculations until necessary. This is similar to carrying along π in symbolic form in equations. Also, note that the masses of the laboratory "weights" usually have values stamped in grams. Don't forget to change grams to kilograms when working in the SI: for example, 150 g = 0.150 kg.

(b) *Analytical.* Compute the magnitudes of $\mathbf{F}_x$ and $\mathbf{F}_y$ (see the Theory section). Record the results in the data table.

(c) *Experimental.* Clamp pulleys at 240°, 90°, and 0° on the force table. Place a *total* of 0.300 kg on the 240° pulley string using a weight hanger. This force is then the equilibrant of $F = (0.300)g$ N at 60° (since 60° + 180° = 240°), which must be used on the force table rather than the force itself. Add weights to the 0° and 90° hangers until the system is in equilibrium. The 0° and 90° forces are then the $\mathbf{F}_x$ and $\mathbf{F}_y$ components, respectively, of $\mathbf{F}$. Record their magnitudes in the data table.

6. *Vector addition 4.* Given the force vectors $F_1 = (0.100)g$ N at 30°, $F_2 = (0.200)g$ N at 90°, and $F_3 =$

(0.30)g N at 225°, find the magnitude and direction of their resultant $\mathbf{F} = \mathbf{F}_1 + \mathbf{F}_2 + \mathbf{F}_3$ by the following procedures:

(a) *Graphical.* Use the polygon method.
(b) *Analytical.* Use the component method.
(c) *Experimental.* Use the force table. Record the results in the data table.

7. *Vector addition 5.* Instructor's choice (optional). Your instructor will give you a set of vectors to add. Record the results in the data table as you did for previous procedures.

Name _____ Section _____ Date _____

Lab Partner(s) _____

The Addition and Resolution of Vectors: The Force Table

$\boxed{TI}$ *Laboratory Report*

Note: Attach graphical analyses to Laboratory Report.

DATA TABLE

Purpose: To analyze results of different methods of vector addition.

	Forces ( )	Resultant R (magnitude and direction)		
		Graphical	Analytical*	Experimental
Vector addition 1	$F_1 = (0.200)g$ N, $\theta_1 = 30°$ $F_2 = (0.200)g$ N, $\theta_2 = 120°$			
Vector addition 2	$F_1 = (0.200)g$ N, $\theta_1 = 20°$ $F_2 = (0.150)g$ N, $\theta_2 = 80°$			
Vector addition 3	$F_1 = F_x = (0.200)g$ N, $\theta_1 = 0°$ $F_2 = F_y = (0.150)g$ N, $\theta_2 = 90°$			
Vector resolution	$F = (0.300)g$ N, $\theta = 60°$	F_x F_y	F_x F_y	F_x F_y
Vector addition 4	$F_1 = (0.100)g$ N, $\theta_1 = 30°$ $F_2 = (0.200)g$ N, $\theta_2 = 90°$ $F_3 = (0.300)g$ N, $\theta_3 = 225°$			
Vector addition 5				

*Show analytical calculations below.

Calculations
(attach additional sheet if necessary)

Don't forget units

(continued)

/TI/ QUESTIONS

1. Considering the graphical and analytical methods for obtaining the resultant, which method is more accurate? Give the probable sources of error for each method.

2. Vector subtraction $(\mathbf{A} - \mathbf{B})$ is a special case of vector addition, since $\mathbf{A} - \mathbf{B} = \mathbf{A} + (-\mathbf{B})$. Suppose that the cases of vector addition 1, 2, and 3 in this experiment were vector subtraction $(\mathbf{F}_1 - \mathbf{F}_2)$.
   (a) What effect would this have on the directions of the resultants? (Do not calculate explicitly. Simply state in which quadrant the resultant would be in each case.)

   (b) Would the magnitude of the resultant be different for vector subtraction than for vector addition in each case? If so, state whether the subtractive resultant would be greater or less than the additive resultant.

3. A picture hangs on a nail as shown in ● Fig. 7. The tension T in each string segment is 3.5 N.
   (a) What is the equilibrant or the upward reaction force of the nail?
   (b) What is the weight of the picture?

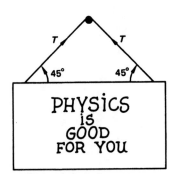

Figure 7   See Question 3.

Name _____ Section _____ Date _____

Lab Partner(s) _____

EXPERIMENT 5

Conservation of Linear Momentum

/TI/ *Advance Study Assignment*

Read the experiment and answer the following questions.

1. What do we mean when we say that a quantity, such as linear momentum, is conserved?

2. What is the condition for the conservation of linear momentum of a system?

3. Is linear momentum conserved in common applications? Explain.

4. Is the conservation of linear momentum consistent with Newton's first and third laws of motion? Explain.

(continued)

5. In a system of particles for which the total linear momentum is conserved, is the linear momentum of the individual particles constant? Explain.

6. Suppose that a particle of mass m_1 approaches a stationary mass m_2 and that $m_2 >> m_1$.
   (a) Describe the velocity of m_2 after an elastic collision—that is, one in which both momentum and kinetic energy are conserved. Justify your answer mathematically.

   (b) What is the approximate momentum of m_1 after collision?

CI *Advance Study Assignment*

Read the experiment and answer the following questions.

1. What mechanism will be used to make the collision between the cars an elastic collision?

2. What mechanism will be used to make the collision between the cars an inelastic collision?

Conservation of Linear Momentum

OVERVIEW

Experiment examines the conservation of linear momentum by TI procedures and/or CI procedures. The TI procedure uses distance-time measurements to determine the velocities of air track cars before and after collisions to investigate the conservation of linear momentum.

The CI procedure measures the velocities electronically and graphs the data. The velocities, total momentum, and total kinetic energy are obtained from the graphs.

INTRODUCTION AND OBJECTIVES

The conservation of linear momentum ($\mathbf{p} = m\mathbf{v}$) is an important physical concept. However, the experimental investigation of this concept in an introductory physics laboratory is hampered by ever-present frictional forces.

An air track provides one of the best methods to investigate linear momentum (see Fig. 3.3). Aluminum cars or gliders riding on a cushion of air on the track approach frictionless motion—a necessary condition for the conservation of linear momentum.

In the absence of friction (and other external forces), the total linear momentum of a system of two cars will be conserved during a collision of the cars. That is, the total linear momentum of the system should be the same after collision as before collision. By measuring the velocities of cars of the same and different masses before and after collision, we can determine the total momentum of a system and investigate the conservation of linear momentum.

After performing this experiment and analyzing the data, you should be able to do the following:

TI OBJECTIVES

1. Explain when linear momentum is conserved and what this means in terms of force and motion.
2. Apply the conservation of linear momentum to a system.
3. Describe two-body collisions in terms of the conservation of linear momentum.

CI OBJECTIVES

1. Understand that momentum is conserved for both elastic and inelastic collisions.
2. Distinguish between elastic and inelastic collisions in terms of the conservation of kinetic energy.

Conservation of Linear Momentum

EQUIPMENT NEEDED

- Air track
- Three cars (two of similar mass)
- Four laboratory timers or stopwatches*

- Laboratory balance
- Masking tape
- Meterstick (if no length scale on air track)
- Velcro (optional)

TI THEORY

The linear momentum **p** of a particle or object is defined as

$$\mathbf{p} = m\mathbf{v} \qquad \text{(TI 1)}$$

where m is the mass of the object and **v** its velocity.[†] Since velocity is a vector quantity, so is linear momentum.

Newton's second law of motion, commonly expressed in the form $\mathbf{F} = m\mathbf{a}$, can also be written in terms of momentum:

$$\mathbf{F} = \frac{\Delta \mathbf{p}}{\Delta t} \qquad \text{(TI 2)}$$

If there is no net or unbalanced external force acting on the object ($F = 0$), then

$$\mathbf{F} = \frac{\Delta \mathbf{p}}{\Delta t} = 0$$

or

$$\Delta \mathbf{p} = 0$$

That is, the change in the momentum is zero, or the momentum is conserved. By *conserved* we mean that the momentum remains constant (in time).

$$\Delta \mathbf{p} = \mathbf{p}_f - \mathbf{p}_i = 0$$

or

$$\mathbf{p}_f = \mathbf{p}_i \qquad \text{(TI 3)}$$

and the "final" momentum $\mathbf{p}_f$ at any time t_f is the same as the initial momentum $\mathbf{p}_i$ at time t_i.

Notice that this is consistent with Newton's first law of motion, since

$$\mathbf{p}_f = \mathbf{p}_i \qquad \text{or} \qquad m\mathbf{v}_f = m\mathbf{v}_i$$

and

$$\mathbf{v}_f = \mathbf{v}_i$$

That is, an object remains at rest ($\mathbf{v}_i = 0$) or in uniform motion ($\mathbf{v}_i = \mathbf{v}_f$) unless acted on by some external force.

The previous development also applies to the total momentum of a system of particles or objects. For example, the total linear momentum (**P**) of a system of two objects m_1 and m_2 is $\mathbf{P} = \mathbf{p}_1 + \mathbf{p}_2$, and if there is no net external force acting on the system, then

$$\Delta \mathbf{P} = 0$$

In the case of a collision between two objects of a system (with only internal forces acting), the initial total momentum before the collision is the same as the final total momentum after the collision. That is,

$$\begin{array}{cc} \text{(before)} & \text{(after)} \\ \mathbf{p}_{1_i} + \mathbf{p}_{2_i} = \mathbf{p}_{1_f} + \mathbf{p}_{2_f} & \text{(TI 4)} \end{array}$$

or

$$m_1\mathbf{v}_{1_i} + m_2\mathbf{v}_{2_i} = m_1\mathbf{v}_{1_f} + m_2\mathbf{v}_{2_f}$$

In one dimension, the directions of the velocity and momentum vectors are commonly indicated by plus and minus signs, i.e., $+v$ and $-v$.**

The internal forces do not change the total momentum because, according to Newton's third law, $\mathbf{F}_{12} = -\mathbf{F}_{21}$ (the force on object 1 due to object 2 is equal to and opposite in direction [minus] to the force on object 2 due to object 1). Thus the change in momentum for one object will be equal in magnitude and opposite in direction to the change in momentum for the other object, and the total momentum will be unchanged.

---

* If electronic photogates/timers and computer-assisted data analysis are available, your instructor will give you instruction on their use.* Boldfaced symbols indicate vectors (see Expt. 4).

[†] Boldfaced symbols indicate vectors (see Expt. 4).

** In two (or three) dimensions, the momentum is conserved in both (or all) directions. That is, $\mathbf{P} = \mathbf{P}_x + \mathbf{P}_y = 0$, and $\mathbf{P}_x = 0$ and $\mathbf{P}_y = 0$. Why? *Note:* $\mathbf{P}_x = \Sigma \mathbf{p}_x$ and $\mathbf{P}_y = \Sigma \mathbf{p}_y$.

TI EXPERIMENTAL PROCEDURES

1. Determine the mass of each car and record it in the TI Trial Data Table. Let the masses of the two cars of nearly equal mass be m_1 and m_2 and the mass of the third car be m_3.

2. Mark off two equal and convenient lengths (for example, $\frac{1}{2}$ or 1 m) on both sides of the center position of the air track. Make full use of the length of the track, but leave some space near the ends of the track. Place the four tape reference marks at the lower edges of the track so as not to interfere with the car motion. *Do not mark the air track surface itself with tape or anything else.*

3. *Time trials.* By measuring the time interval Δt it takes a car to move the reference mark length d, one can determine the magnitude of the velocity $v = d/\Delta t$ of the car, where $\Delta t = t_2 - t_1$. The actual timing of the motion of a car moving between the two sets of reference marks is done by either method (A), involving four observers, each with a timer and assigned to an individual reference mark, or method (B), involving two observers, each with a timer and assigned to one set of reference marks, as described below. Time trials will be done to determine the better method.*

   In addition to giving timing practice and determining the better method of timing, the time trials check out the experimental setup for possible systematic errors. The time intervals for the individual cars to travel the equal distances between the reference marks should be very similar for any one trial. If not, the air track may need leveling and/or there may be some frictional problem with part of the track. Should this be the case, call your instructor. *Do not* attempt to level the air track on your own.

   Experimentally carry out each of the following timing methods to determine which is better.

   Method A—Four Timers. Set one of the cars in motion with a *slight* push so that it moves with moderate speed up and down the track. (A few practice starts help.) As the car hits the bumper at one end of the track, all four observers should start their timers. As the leading edge of the car passes the reference marks, each respective observer stops his or her timer. (Making a dry run or two to become familiar with the timing sequence is helpful.) Carry out this procedure twice for each of the three cars, and record the data in the TI Trial Data Table.

   Method B—Two Timers. Set the car in motion. The two observers should start and stop their individual timers as the leading edge of the car passes their respective reference marks. Carry out this procedure twice for each of the three cars, and record the data in the TI Trial Data Table.

4. Compute the Δt's for each trial and calculate the percent difference for each trial set. From the data, decide which timing method should be used on the basis of consistency or precision.

CASE 1: COLLISION BETWEEN TWO CARS OF (NEARLY) EQUAL MASS, WITH ONE INITIALLY AT REST

5. With one of the cars (m_2) of nearly equal mass stationary at the center position of the air track, start the other car (m_1) moving toward the stationary car. See ● TI Fig. 1. (It may be more convenient to start m_1 moving away from m_2 and take measurements as m_1 returns from rebounding from the end of the track.) A trial run should show that m_1 remains at rest, or nearly at rest, after collision and that m_2 is in motion.

   Determine the time it takes for m_1 to travel between the reference marks as it approaches m_2 and the time it takes for m_2 to travel between the other set of reference marks after collision. Carry out this procedure three times and record the data in TI Data Table 1.

   Compute the velocities and the total momentum before and after collision and the percent difference in these values for each trial.

CASE 2: COLLISION BETWEEN TWO CARS OF UNEQUAL MASS, WITH THE MORE MASSIVE CAR INITIALLY AT REST

6. Repeat procedure 5 with m_2 replaced by m_3 (more massive than m_1 and m_2). See TI Fig. 1. In this case, m_1 will travel in the opposite direction after collision, as a trial run will show. Make appropriate adjustments

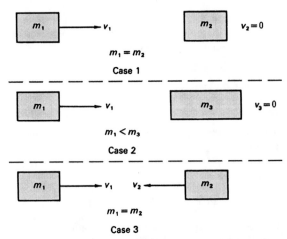

TI Figure 1 Experimental collision cases. See text for descriptions.

* If electronic photogate timers are available, your instructor will give you instruction in their use. Electronic timing greatly improves the accuracy and precision of the results. Why?

in the timing procedure to measure the velocity of m_1 before *and* after collision. Record the data and the required calculations in TI Data Table 2. Be careful with the directional signs of the velocities and momenta.

CASE 3: COLLISION BETWEEN TWO CARS OF (NEARLY) EQUAL MASS INITIALLY TRAVELING IN OPPOSITE DIRECTIONS

7. With m_1 and m_2 initially moving toward each other (TI Fig. 1), determine the total momentum before and after collision. (*Note:* Speeds do not have to be, and probably won't be, equal.)

   Make appropriate adjustments in the timing procedure to measure the velocities of m_1 and m_2 before and after collision. Carry out the procedure three times and record the data in TI Data Table 3.

   Compute the percent difference for the total momentum before and after collision for each trial.

(OPTIONAL PROCEDURE)

Another procedure, which may be done at the instructor's option, is as follows:

8. Attach pieces of Velcro to the collision bumpers of both cars, and repeat one or more of the preceding cases as directed by your instructor. Make up a data table, and analyze your results as done previously. (*Hint:* Read in your textbook about elastic and inelastic collisions—in particular, completely inelastic collisions.)

T I EXPERIMENT 5

Conservation of Linear Momentum

$\boxed{TI}$ *Laboratory Report*

Distance between marks _____

$\boxed{TI}$ **TRIAL DATA TABLE**

Purpose: To determine the better method of timing.

Car mass ( )	METHOD A							METHOD B		
	t_1 ( )	t_2 ( )	Δt_{12} ( )	t_3 ( )	t_4 ( )	Δt_{34} ( )	Percent diff.	Δt_{12} ( )	Δt_{34} ( )	Percent diff.
m_1										
m_2										
m_3										

$\boxed{TI}$ **DATA TABLE 1**

Purpose: To analyze $m_1 = m_2$ case, with $v_{2_i} = 0$.

Trial	Before collision			After collision			
	m_1			m_2			
	Δt_1 ( )	v_{1_i} ( )	p_{1_i} ( )	Δt_2 ( )	v_{2_f} ( )	p_{2_f} ( )	Percent diff.
1							
2							
3							

Don't forget units

(continued)

/T1/ **DATA TABLE 2**

Purpose: To analyze $m_3 > m_1$ case, with $v_{3_i} = 0$.

Trial	Before collision			After collision								Percent diff.
	m_1			m_1			m_3					
	Δt_{1_i} ( )	v_{1_i} ( )	Total momentum ( )	Δt_{1_f} ( )	v_{1_f} ( )	p_{1_f} ( )	Δt_{3_f} ( )	v_{3_f} ( )	p_{3_f} ( )	Total momentum ( )		
1												
2												
3												

/T1/ **DATA TABLE 3**

Purpose: To analyze $m_1 = m_2$ case, initial motions in opposite directions.

Trial	Before collision						Total momentum ( )
	m_1			m_2			
	Δt_{1_i} ( )	v_{1_i} ( )	p_{1_i} ( )	Δt_{2_i} ( )	v_{2_i} ( )	p_{2_i} ( )	
1							
2							
3							

Trial	After collision						Total momentum ( )	Percent diff. ( )
	m_1			m_2				
	Δt_{1_f} ( )	v_{1_f} ( )	p_{1_f} ( )	Δt_{2_f} ( )	v_{2_f} ( )	p_{2_f} ( )		
1								
2								
3								

Name _____ Section _____ Date _____

Lab Partner(s) _____

EXPERIMENT 5   *Laboratory Report*

/TI/ QUESTIONS

1. Do the results of the experiment support the conservation of linear momentum? Consider possible sources of error.

2. Was it necessary to have equal length intervals in the experiment to investigate properly the conservation of momentum? Explain.

3. In Cases 1 and 2, one of the cars was initially at rest, so it must have received an acceleration. Is the car accelerating as it passes between the reference marks? Explain.

4. In each of the three cases, was kinetic energy conserved? Justify your answers with a sample calculation for a trial from each case. If the kinetic energy is not conserved, where did it go?

EXPERIMENT 5

Conservation of Linear Momentum

CI EQUIPMENT NEEDED

- 2 rotary motion sensors
- Brackets and pulley mounts:
    2 cart-string brackets
    2 dynamics track mount accessories
    2 RMS/IDS adapters
- 2 collision carts
- 1 track
- Clay or Velcro strips
- String
- Optional: track end stop

PASCO CI-6538

CI-6569
CI-6692 (to mount the RMS to the track)
ME-6569 (track pulley bracket)
PASCO Classic Cars, ME-9454

CI THEORY

The purpose of this experiment is to investigate the momentum and kinetic energy for elastic and inelastic collisions. The momentum and kinetic energy before the collision of two cars are compared with the momentum and kinetic energy after the collision by looking at a plot of these quantities versus time.

In a collision between two objects, the total momentum at any time is found by adding the momentum of one of the objects to that of the other:

$$\vec{P}_{Total} = \vec{p}_1 + \vec{p}_2 = m_1\vec{v}_1 + m_2\vec{v}_2 \qquad \textbf{(CI 1)}$$

This is vector addition, which means the directions of motion of both objects must be taken into account. The sensor used to measure the speeds of the objects will also assign a positive or negative sign, depending on direction. In general, an object moving toward the sensor is assigned a positive velocity, and an object moving away from the sensor is assigned a negative velocity.

An object in motion also has kinetic energy. We can determine the total kinetic energy in a system by adding the kinetic energies of all objects in the system.

$$K_{Total} = K_1 + K_2 = \tfrac{1}{2}m_1v_1^2 + \tfrac{1}{2}m_2v_2^2 \qquad \textbf{(CI 2)}$$

The total momentum and the total kinetic energy just before and just after a collision are determined and compared. First, an elastic collision between two cars is considered. The cars have magnets that make them repel each other when they get close enough. The effect is that the cars bounce off each other (collide) without touching. Next, an inelastic collision is considered. The magnets are replaced by a piece of clay (or Velcro) that will make the cars stick to each other after the collision.

BEFORE YOU BEGIN

1. Install a cart-string bracket on each of the collision carts. The cart-string bracket is mounted on the side of the cart, as shown in ● CI Fig. 1.
2. Choose one cart to be Car 1 and measure its mass, in kilograms, including the cart-string bracket. Report the mass of Car 1 in the laboratory report.
3. The other cart will be Car 2. Measure its mass and also record that mass in the laboratory report.
4. Do not lose track of which is Car 1 and which is Car 2. If needed, put a small tape label on the cars so that you will not confuse them later.

This information will be needed during the setup of Data Studio.

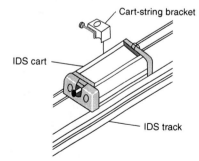

CI Figure 1   Installing cart-string brackets.   The cart-string brackets are installed on top of the collision carts, secured with a side screw. The top screw is used to tie a string. When measuring the mass of the car, include the cart-string bracket.

SETTING UP DATA STUDIO

1. Open Data Studio and choose "Create Experiment."
2. From the sensor list, choose a Rotary Motion Sensor. Connect the sensor to the interface, as shown in the computer screen, to channels 1 and 2. This will be the sensor to track the motion of Car 1.
3. Choose a second rotary motion sensor from the sensor list. This one will be connected to channels 3 and 4. This will be the sensor to track the motion of Car 2.
4. Double-click on one of the Rotary Motion Sensor icons. The Sensor Properties window will open. Select the following:

    Under General: set the sample rate to 100 Hz, Fast.
    Under Measurement: choose Velocity (m/s) and deselect any others.
    Under Rotary Motion Sensor: set the Divisions/Rotations to 1440, and set the Linear Calibration to Large Pulley (Groove).

    Click OK to accept the choices and to close the Sensor Properties window.
5. Repeat this process with the other sensor.
6. Open the program's calculator by clicking on the Calculate button, on the top main menu. Usually a small version of the calculator opens, as shown in ● CI Fig. 2. Expand the calculator window by clicking on the button marked "Experiment Constants."
7. The expanded window (shown in ● CI Fig. 3) is used to establish values of parameters that will remain constant throughout the experiment. In this case, these are the masses m_1 and m_2 of the carts, which have already been measured. This is how to do it:
    a. Click on the lower New button (within the "Experiment Constants" section of the calculator window) and enter the name of the constant as m1, the value as the mass of Car 1 measured before, and the units as kg.

    b. Click the lower Accept button.
    c. Click on the New button again and enter the name of the constant as m2, the value as the mass of Car 2 measured before, and the units as kg.
    d. Click the lower Accept button.
    e. Close the experiment constants portion of the calculator window by pressing the button marked "Experiment Constants" again.
8. **Calculation of the total momentum of the system:**
    a. In the same calculator window, clear the definition box and enter the following equation: TotalP = m1 * smooth(10,v1) + m2 * smooth(10,v2)

    This is the calculation of the total momentum, $\vec{P}_{\text{Total}} = m_1\vec{v}_1 + m_2\vec{v}_2$, that we will call TotalP. The smooth function will help produce a cleaner graph.
    b. Press the top Accept button after entering the formula. Notice that the variables m1, m2, v1, and v2 will appear in a list. The masses were already assigned values, but v1 and v2 are waiting to be defined.
    c. To define variables v1 and v2, do them one at a time by clicking on the drop menu button on the left side of each variable. A list of options appears, asking what type of variable this is.
      • Define v1 as a Data Measurement and, when prompted, choose Velocity(Ch1&2).
      • Define v2 as a Data Measurement and, when prompted, choose Velocity(Ch3&4).
    d. Press the Accept button again.

    Please notice that channels 1&2 will keep track of Car 1 and that channels 3&4 will track Car 2. Make sure the equipment is set up accordingly.
9. **Calculation of the total kinetic energy of the system:**
    a. Still in the same calculator window, press the top New button again to enter a new equation.

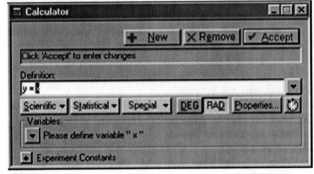

CI Figure 2   The calculator window.   This small version of the calculator window opens when the Calculate button is pressed. The calculator will be used to enter formulas that handle the values measured by the sensor. The computer will perform the calculations automatically as the sensor takes data. (Data displayed using DataStudio Software. Reprinted courtesy of PASCO scientific.)

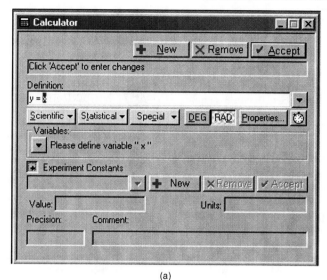

(a)

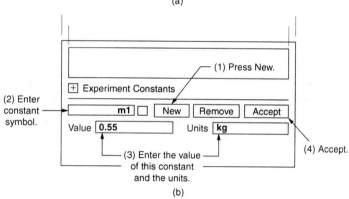

(b)

CI Figure 3   The expanded calculator window.   (a) After the Experiment Constants button is pressed, the calculator window expands to full size. (b) The "Experiment Constants" section is the lower part of the expanded calculator. This section is used to define parameters that are to remain constant during the experiment. The diagram shows the steps needed to enter experimental constants into the calculator. (Data displayed using DataStudio Software. Reprinted courtesy of PASCO scientific.)

b. Clear the definition box and enter the following equation: TotalKE = 0.5* m1 * smooth(10,v1)^2 + 0.5* m2 * smooth(10, v2)^2

This is the calculation of the total kinetic energy, $K_{\text{Total}} = \frac{1}{2}m_1 v_1^2 + \frac{1}{2}m_2 v_2^2$, that we will call TotalKE.

c. Press the <u>Accept</u> button after entering the formula. Notice that the variables will again appear in a list. Define them exactly as before.

d. Press the <u>Accept</u> button again.

10. Close the calculator window.

11. The data list at the top left of the screen should now have four items: Velocity from Ch1&2, Velocity from Ch3&4, TotalP, and TotalKE. A small calculator icon identifies the quantities that are calculated.

12. Create a graph by dragging the "Velocity Ch1&2" icon from the data list and dropping it on the "Graph" icon on the displays list. A graph of velocity versus time will open, in a window titled Graph 1.

13. Double-click anywhere on the graph. The Graph Settings window will open. Make the following changes and selections:

Under the tab Appearance:
Data:
Connect data points in bold
Deselect the buttons marked "Show Data Points" and "Show Legend Symbols"
Under the tab Layout:
Multiple graphs:
Vertical

Layering:
Do not layer
Measurement adding:
Replace matching measurement
Group measurement:
Do not group
Click OK to accept the changes and to exit the graph settings window.

14. Drag the "Velocity Ch3&4" data icon and drop it in the middle of Graph 1. The graph will split in two. At the top you will see the Velocity Ch1&2 and at the bottom the Velocity Ch3&4, on separate Y axes.

15. Drag the "TotalP" icon and drop it on the split graph. The graph will split again, this time into three sections.

16. Drag the "TotalKE" icon and also drop it on the graph. The result should be a graph split into four sections, one section for each of the quantities.

17. Press the "Align Matching X Scales" button on the graph's toolbar. (It is a button with a picture of a padlock.) This will make all graphs aligned to a common $t = 0$ on the X axis.

18. ● CI Fig. 4 shows what the screen should look like after all setup is complete. The size of the graph window can be maximized so that you can observe the plots better.

CI EXPERIMENTAL PROCEDURE

The complete experimental setup is shown in ● CI Fig. 5. Each car is connected to its own sensor and pulley system, one on each side of the track. Here are the instructions for setting up the carts:

1. Place Cars 1 and 2 (with the cart-string brackets attached) on the track with the magnetic sides facing each other. The cart-string brackets may need repositioning so that they face the outside of the track, as shown in CI Fig. 5.

2. Install a rotary motion sensor (RMS) on each side of the track, with the pulleys facing the inside of the track.

3. On the opposite side of the track, install the RMS/IDS adapters (small pulleys). See ● CI Fig. 6 for reference.

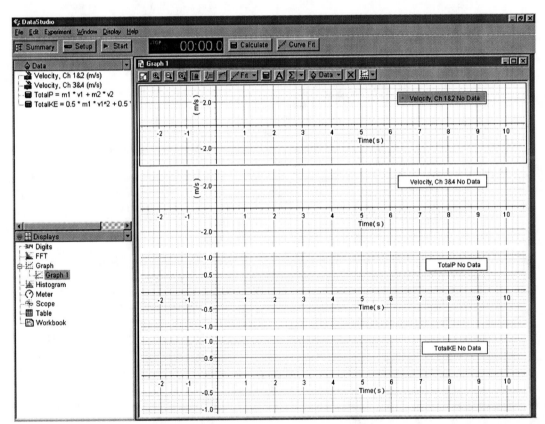

CI Figure 4   Data Studio setup.   Data for velocity of each car, total momentum, and total kinetic energy will appear simultaneously on four plots, with matching time axes. The graph window may be maximized to occupy the whole screen in order to display the experimental results better. (Data displayed using DataStudio Software. Reprinted courtesy of PASCO scientific.)

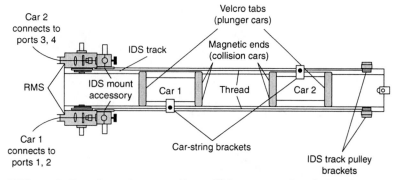

CI Figure 5   Experimental setup.   Two collision carts are installed on the same track. Each cart is connected to its own rotary motion sensor on one side and to its own IDS-RMS adapter (track pulley bracket) on the other side. An elastic collision can be performed by having the magnetic ends of the cars face each other. An inelastic collision can be performed by having the nonmagnetic sides face each other and putting clay or Velcro on the ends of the cars.

4. A string will make a loop starting from the cart-string bracket on top of the car, to the large pulley of the RMS, to the small pulley of the RMS/IDS adapter and back to the cart, as shown in ● CI Fig. 7.   Do this for both cars, as shown in the complete set up of CI Fig. 5. Adjust the height of the pulleys so that the strings are tense, not sagging, but the cars are able to move freely.

CASE 1: ELASTIC COLLISION BETWEEN TWO CARS OF (NEARLY) EQUAL MASS, WITH ONE INITIALLY AT REST

5. Set Car 2 somewhere on the middle of the track, at rest.
6. Set Car 1 all the way to the end of the track.
7. Press the START button and then give Car 1 a good push toward Car 2.
8. Press the STOP button after the collision, before Car 2 bounces at the end of the track. (Several practice runs, and the help of a partner, may be needed.)

9. Click anywhere on the Velocity Ch1&2 graph, and then press the Scale-to-Fit button on the graph toolbar (The Scale-to-Fit button is the leftmost button of the graph toolbar.) This will make the data scale to the length of the graph on the screen. Repeat for all the other three graphs.
10. If any of the graphs of velocity is reading negative values, switch the yellow and black cables of the corresponding rotary motion sensor in the interface so that the yellow cord connects to where the black cord was, and vice versa. Repeat the data collection process and use the new data in the rest of the analysis.
11. Print the graph. If no printer is available, make a careful drawing of the graph, paying special attention to dips and peaks in the graphs. Attach the graph to the laboratory report.
12. Click anywhere on the Velocity Ch1&2 graph, and then press the Smart-Tool button on the graph toolbar.

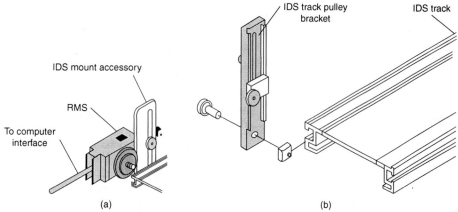

CI Figure 6   Mounting the RMS and the IDS track pulley to the track.   (a) This figure shows how to mount the rotary motion sensor to one end of the track, using the mount accesory. (b) This figure shows how the IDS/RMS adapter (the track pulley bracket) should be mounted to the track.

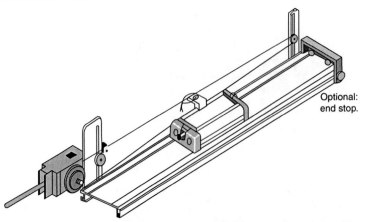

Optional: end stop.

CI Figure 7 Example of one side of the experimental setup This diagram illustrates one of the carts completely set up. Notice the string connecting the pulleys to the cart-bracket is to have tension, but not be so tight that the cart cannot move freely.

(The Smart-Tool is a button of the graph toolbar labeled XY.) A set of crosshairs will appear on the graph. Repeat for each of the other graphs to get a set of crosshairs on each graph. The crosshairs can be dragged around to determine the exact (x,y) value of any point in the graphs.

13. Use the smart-tools to find the time t_0 that corresponds to the moment just before the collision. Report the value of t_0 in the laboratory report. (*Hint:* Use the velocity graphs and think of what the cars were doing just before the collision.)

14. In the graph printout, mark the time t_0 in all graphs by drawing a single, vertical line from top to bottom of the page crossing time t_0.

15. Use the smart-tools to find the time t_f that corresponds to the moment just after the collision ended. Report the value of t_f in the laboratory report (*Hint:* The collision does not end at the same time as when it started, look carefully! Again, think of what the cars were doing right after the collision.)

16. In the graph printout, mark the time t_f in all graphs by drawing a single, vertical line from top to bottom of the page crossing time t_f. The two vertical lines now separate the before-collision from the after-collision moments.

17. Determine how long (in time) the collision lasted.

18. Use the smart-tools to determine, at time t_0:
    • the velocity of Car 1
    • the velocity of Car 2
    • the total momentum of the system
    • the total kinetic energy of the system
    Enter the results in CI Data Table 1.

19. Use the smart-tools to determine, at time t_f:
    • the velocity of Car1
    • the velocity of Car 2

• the total momentum of the system
• the total kinetic energy of the system
Enter the results in CI Data Table 1.

20. Calculate the change in velocity of each car, the change in momentum of each car, the change in the total momentum of the system, and the change in the total kinetic energy of the system. Enter the results in CI Data Table 1.

CASE 2: INELASTIC COLLISION BETWEEN TWO CARS OF (NEARLY) EQUAL MASS, WITH ONE INITIALLY AT REST

21. Switch the cars on the track so that their magnetic ends are facing away from each other. The easiest way to do this without altering the strings is to unscrew the cart-string brackets from the carts but not from the strings. The cars can then be switched under the brackets and the brackets installed back.*

22. Place a small piece of clay on the colliding end of both cars. (*Note:* Velcro strips and sticky masking tape also work well for this. Some PASCO carts already come with Velcro strips attached.)

23. Set Car 2 somewhere on the middle of the track, at rest.

24. Set Car1 all the way to the end of the track.

25. Press the START button, and then give Car 1 a good push toward Car 2.

26. Press the STOP button after the collision, before the cars reach the end of the track and bounce. (The cars must stick together after the collision. Several practice runs, and the help of a partner, may be needed to get the hang of it.)

27. Repeat Steps 10 to 20, for this set of data, but enter the results in CI Data Table 2.

---

* Some PASCO carts have magnets on both ends. These won't work. A new set of carts with no magnets (plunger carts) will be needed, which means new masses must be measured and entered in the Calculator, if this is the case.

C I   EXPERIMENT 5

Conservation of Linear Momentum

CI *Laboratory Report*

CASE 1: ELASTIC COLLISION BETWEEN TWO CARS OF (NEARLY) EQUAL MASS, WITH ONE INITIALLY AT REST

Car 1: $m_1 = $ _____

Car 2: $m_2 = $ _____

Collision started at $t_o = $ _____   Total collision time $\Delta t = t_f - t_o = $ _____

Collision ended at $t_f = $ _____

CI   DATA TABLE 1

Purpose: To analyze an elastic collision between two objects of nearly identical mass.

	Just before the collision	Just after the collision	Changes	
Velocity of Car 1, v_1			Δv_1	
Velocity of Car 2, v_2			Δv_2	
Total momentum, P_{Total}			ΔP	
Total kinetic energy, K_{Total}			ΔK	

Don't forget units

(continued)

CASE 2: INELASTIC COLLISION BETWEEN TWO CARS OF (NEARLY) EQUAL MASS, WITH ONE INITIALLY AT REST

Car 1: $m_1 =$ _____

Car 2: $m_2 =$ _____

Collision started at $t_o =$ _____   Total collision time $\Delta t = t_f - t_o =$ _____

Collision ended at $t_f =$ _____

CI DATA TABLE 2

Purpose: To analyze an inelastic collision between two objects of nearly identical mass.

	Just before the collision	Just after the collision	Changes	
Velocity of Car 1, v_1			Δv_1	
Velocity of Car 2, v_2			Δv_2	
Total momentum, P_{Total}			ΔP	
Total kinetic energy, K_{Total}			ΔK	

EXPERIMENT 5   *Laboratory Report*

CI QUESTIONS

1. How well do the results support the law of conservation of momentum, considering the possible sources of uncertainty?

2. Which collision took a longer time, the elastic or the inelastic collision? Discuss the possible reasons.

3. Was the kinetic energy of the system conserved? Discuss by comparing the results for the elastic collision and the inelastic collision.

4. During the inelastic collision, the kinetic energy was obviously not conserved. What do you think happened to the "lost" energy?

(continued)

5. During the collision, both cars changed their momentum. How does the change in momentum of each car compare to that of the other? Does one car change more than the other? What do you think would happen if the cars had different mass? (If time is available, try it.)

6. For an object to undergo a change in its momentum, a net force needs to be applied. The amount of change in momentum produced by the force depends on the length of the time during which the force acts and is called the *impulse*. That is,

$$\text{Impulse} = \Delta p = F\Delta t$$

where the force F is assumed to be constant, or to be an "average force." For each of the collisions, calculate the average force acting on the cars during the collision, and compare them.

7. Suppose a ball falls on your head. What is better for you (less damage), for the ball to bounce straight back off your head, or for it to stop and stick to you? Justify your answer.

Name _____ Section _____ Date _____

Lab Partner(s) _____

Projectile Motion:
The Ballistic Pendulum

/TI/ *Advance Study Assignment*

Read the experiment and answer the following questions.

A. *The Ballistic Pendulum*

1. In determining the magnitude of the initial velocity of the ballistic pendulum projectile, what conservation laws are involved and in which parts of the procedure?

2. Why is it justified to say that the momentum in the horizontal direction is conserved over the collision interval? Is momentum conserved before and after the collision? Explain.

3. Is the center of mass of the pendulum-ball system at the center of the ball? If not, where and how is the center of mass located?

(continued)

B. *Determination of the Initial Velocity of a Projectile from Range-Fall Measurements*

4. After the horizontal projectile leaves the gun, what are the accelerations in the
    x and y directions?

5. How is the location where the ball strikes the floor determined?

6. Besides the range, what else is needed to determine the magnitude of the initial velocity of
    the ball?

C. *Projectile Range Dependence on the Angle of Projection*

7. For a given initial velocity, how does the range of a projectile vary with the angle of
    projection θ?

8. Theoretically, the angle of projection for maximum range is 45°. Does this set a limit on
    the range? Explain.

Projectile Motion:
The Ballistic Pendulum

INTRODUCTION AND OBJECTIVES

Projectile motion is the motion of an object in a plane (two dimensions) under the influence only of gravity. The kinematic equations of motion describe the components of such motion and may be used to analyze projectile motion. In most textbook cases, the initial velocity of a projectile (speed and angle of projection) is given and the motion is described through the equations of motion.

However, in this laboratory experiment, the unknown initial velocity will be determined from experimental measurements. This will be done (1) through the use of the ballistic pendulum and (2) from range-fall distance measurements. The dependence of the projectile range on the angle of projection will also be investigated so as to obtain an experimental indication of the angle of projection that gives the maximum range.

These procedures will greatly assist you in understanding some of the most basic physical principles. After performing the experiment and analyzing the data, you should be able to do the following:

1. Explain the use of conservation laws (linear momentum and energy) in determining the initial velocity of a projectile using the ballistic pendulum.
2. Describe the components of motion and how they are used in determining the velocity of a projectile with range-fall measurements.
3. Tell how the range of a projectile varies with the angle of projection.

EQUIPMENT NEEDED

- Ballistic pendulum
- Sheets of plain paper (and carbon paper)*
- Meterstick

* May or may not be needed.

- Protractor
- Laboratory balance
- Masking tape
- Wooden blocks
- 1 sheet of Cartesian graph paper
- Safety glasses

THEORY

A. The Ballistic Pendulum

Types of ballistic pendula apparatus are shown in ● Fig. 1. The ballistic pendulum is used to measure the initial velocity of a horizontally projected object (a metal ball) fired from a spring gun. The projectile is fired into a stationary, hollow pendulum bob suspended by a rod, and the pendulum and the embedded projectile swing upward.

A catch mechanism stops the pendulum at its highest position of swing. By measuring the vertical distance that the center of mass of the pendulum-ball system rises, one can compute the initial velocity of the projectile through the use of the conservation of linear momentum and the conservation of mechanical energy (neglecting rotational considerations).

Consider the schematic diagram of a ballistic pendulum shown in ● Fig. 2. A projectile of mass m with an initial horizontal velocity of v_{x_o} is fired into and becomes embedded in a stationary pendulum of mass M.

The horizontal momentum is conserved during collision (if the external forces are vertical). To a good approximation, this applies over the time interval of the collision. Therefore, the horizontal component of total momentum is taken to be the same immediately before and after collision. The velocity of the pendulum bob is initially zero, and the combined system $(m + M)$ has a velocity of magnitude V just after collision. Hence, we may write for the horizontal direction

$$\underset{\text{(before)}}{m v_{x_o}} = \underset{\text{(after)}}{(m + M)V} \tag{1}$$

After the collision, the pendulum with the embedded projectile swings upward (momentum of the system no longer conserved) and stops. The center of mass of the system is raised a maximum vertical distance $h = h_2 - h_1$. By the conservation of mechanical energy, the increase in potential energy is equal to the kinetic energy of the system

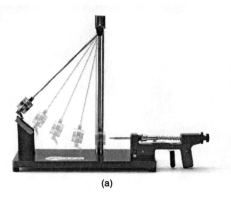

(a)

(c)

(b)

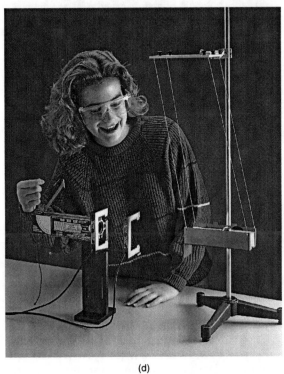

(d)

Figure 1 Ballistic pendula.   Types of ballistic pendula. [Courtesy of (a) and (b) Sargent-Welch, (c) Bernard O. Beck & Co., and (d) PASCO Scientific.]

just after collision (the friction of the support is considered negligible). Hence,

$$\tfrac{1}{2}(m + M)V^2 \;=\; (m + M)gh \qquad (2)$$

kinetic energy   change in
just after collision   potential energy

Solving Eq. 7.2 for V, we obtain

$$V = \sqrt{2gh} \qquad (3)$$

Substituting this expression into Eq. 7.1 and solving for v_{x_0} yields

$$\boxed{v_{x_0} = \frac{m + M}{m}\sqrt{2gh}} \qquad (4)$$

(initial speed)

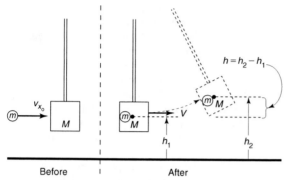

Figure 2 **Ballistic pendulum action.** Ideally, the horizontal linear momentum is conserved during collision. After collision, work is done against gravity, and kinetic energy is converted into potential energy. (Rotational considerations neglected.)

Hence, by measuring m, M, and h, we can compute the initial velocity of the projectile.

B. Determination of the Initial Speed of a Horizontal Projectile from Range-Fall Measurements

If a projectile is projected horizontally with an initial velocity of magnitude v_{x_o} from a height of y, it will describe an arc as illustrated in ● Fig. 3. The projectile will travel a horizontal distance x (called the *range* of the projectile) while falling a vertical distance y.

The initial vertical velocity is zero, $v_{y_o} = 0$, and the acceleration in the y direction has a magnitude of $a_y = g$ (acceleration due to gravity). There is no horizontal acceleration, $a_x = 0$; hence the components of the motion are described by

$$x = v_{x_o} t \qquad (5)$$

and

$$-y = -\tfrac{1}{2}gt^2 \qquad (6)$$

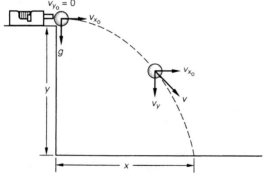

Figure 3 **Range-fall.** The configuration for range-fall measurements. See text for description.

Eliminating t from these equations and solving for v_{x_o}, we have (neglecting air resistance)

$$v_{x_o} = \sqrt{\frac{gx^2}{2y}} = \left(\frac{g}{2y}\right)^{\frac{1}{2}} x \qquad (7)$$

Hence, by measuring the range x and the distance of fall y, one can compute the initial speed of the projectile.

C. Projectile Range Dependence on the Angle of Projection

The projectile path for a general angle of projection θ is shown in ● Fig. 4. The components of the initial velocity have magnitudes of

$$v_{x_o} = v_o \cos \theta$$
$$v_{y_o} = v_o \sin \theta \qquad (8)$$

At the top of the arc path, $v_y = 0$, and since

$$v_y = v_{y_o} - gt$$
$$= v_o \sin \theta - gt$$

(downward taken as negative), we have

$$v_o \sin \theta - gt_m = 0$$

or

$$t_m = \frac{v_o \sin \theta}{g} \qquad (9)$$

where t_m is the time for the projectile to reach the maximum height of y_m.

If the projectile returns to the same elevation as that from which it was fired, then the total time of flight t is

$$t = 2t_m = \frac{2v_o \sin \theta}{g} \qquad (10)$$

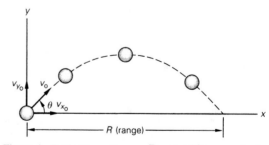

Figure 4 **Projectile motion.** For an arbitrary projection angle above the horizontal, the range R of a projectile depends on the initial velocity—that is, on the speed and angle of projection.

During the time t, the projectile travels a distance R (range) in the x direction:

$$R = v_{x_o} t = \frac{2v_o^2 \sin\theta \cos\theta}{g}$$

But using the trigonometric identity $2\sin\theta\cos\theta = \sin 2\theta$, we find that the **range** or maximum distance in the x direction is

$$\boxed{R = \frac{v_o^2 \sin 2\theta}{g}} \qquad (11)$$

(range)

From Eq. 11, we see that the range of the projectile depends on the angle of projection θ. The maximum range R_{max} occurs when $\sin 2\theta = 1$. Since $\sin 90° = 1$, by comparison

$$2\theta = 90° \qquad \text{or} \qquad \theta = 45°$$

Hence, a projectile has a maximum range for $\theta = 45°$, and

$$\boxed{R_{max} = \frac{v_o^2}{g}} \qquad (12)$$

(maximum range, $\theta = 45°$)

which provides another convenient method to determine experimentally the initial speed of a projectile.

(*Note:* This development neglects air resistance, but the equations give the range to a good approximation for relatively small initial speeds and short projectile paths. Why?)

EXPERIMENTAL PROCEDURE

Caution: With projectiles involved, it is recommended that safety glasses be worn during all procedures.

A. The Ballistic Pendulum

1. Obtain the projectile ball, which may be in the pendulum bob. (*Note:* When removing the ball from the pendulum bob of some types of ballistic pendula, be sure to push up on the spring catch that holds the ball in the pendulum so as not to damage it.)

   Place the projectile ball on the ball rod of the spring gun, and cock the gun by pushing on the ball. Both ball and rod may move backward, or the ball may slip over the rod, depending on the type of ballistic pendulum. *Caution: In either case, be careful not to bruise or hurt your hand when cocking the gun. Also, keep your fingers away from the projectile end of the gun.*

Fire the projectile into the pendulum to see how the apparatus operates. If the catch mechanism does not catch on the notched track, you should adjust the pendulum suspension to obtain the proper alignment.

2. A preset pointer or a dot on the side of the pendulum bob indicates the position of the center of mass of the pendulum-ball system. With the pendulum hanging freely, measure the height h_1 of the pointer above the base surface (Fig. 2) and record it in Data Table 1.

3. Fire the ball into the freely hanging stationary pendulum and note the notch at which the catch mechanism stops on the curved track. Counting upward on the curved track, record the notch number in Data Table 1. Repeat this procedure four times, and for each trial record the notch number in the data table. (Alternatively, the height may be measured each time. See procedure 4.)

4. Determine the average of these observations, which is the average highest position of the pendulum. Place the catch mechanism in the notch corresponding most closely to the average, and measure the height h_2 of the pointer or dot above the base surface used for the h_1 measurement (Fig. 2).

   Note: To minimize frictional losses, the catch mechanism may be disabled by tying it up with thread or using a rubber band. The mechanism then acts as a pointer to indicate the highest notch, which is observed by a lab partner. Holding some reference object, such as a pencil, by the notched track helps to determine the proper notch number.

5. Loosen the screw of the pendulum support and carefully remove the pendulum. Weigh and record the masses of the ball (m) and the pendulum (M). *Note:* The mass of the pendulum is that of the bob and the support rod. Do not attempt to remove the support rod from the bob. Consult your instructor for an explanation if a different model is used.

6. From the data, compute the magnitude of the initial velocity using Eq. 4 ($g = 9.80$ m/s^2 = 980 cm/s^2).

B. Determination of the Initial Velocity of a Projectile from Range-Fall Measurements

7. With the pendulum removed or in the upper catch mechanism notch so as not to interfere with the projectile, position the apparatus near one edge of the laboratory table as shown in Fig. 3.

   Fire the ball from the gun, and note where the ball strikes the floor. (The range of the ball is appreciable, so you will probably have to shoot the ball down an aisle. Be careful not to hit anyone with the ball, particularly the instructor.)

8. Place a sheet of paper where the ball hits the floor. Tape the paper to the floor (or weight it down) so that it will not move. When the ball strikes the paper, the indentation mark will enable you to determine the range of the projectile.* Also mark the position of the apparatus on the table (e.g., using a piece of tape as a reference). It is important that the gun be fired from the same position each time.

9. Fire the ball five times and measure the horizontal distance or range x the ball travels for each trial (see Fig. 3). [If the faint indentation marks cannot be found, cover the paper on the floor with a sheet of carbon paper (carbon side down). The ball will then make a carbon mark on the paper on impact.]

   Record the measurements in Data Table 2 and find the average range. Also measure the height y of the ball rod from the floor and record it in the data table. The height y is measured from the bottom of the ball (as it rests on the gun) to the floor.

10. Using Eq. 7, compute the magnitude of the initial velocity of the ball ($g = 9.80$ m/s^2 = 980 cm/s^2). Compare this to the velocity determined in part A, and compute the percent difference.

C. Dependence of Projectile Range on the Angle of Projection

11. With the ballistic pendulum apparatus on the floor (with pendulum removed), elevate the front end so that it can be fired at an angle θ relative to the horizontal. Your instructor will tell you how to do this. Aim the projectile down an aisle or hallway, *being careful not to aim at anything or anybody.*

12. Using a protractor to set the angles of projection, fire the projectile at angles of 20°, 30°, 40°, 45°, 50°, 60°, and 70° with two or three trials for each angle. The projectile should be aimed so that it lands as close as possible to the same spot for the trials of a particular angle.

    Station one or more lab partners near where the projectile strikes the floor. They are to judge the average range of the two or three trials. Measure the average range for each angle of projection, and record the data in Data Table 3.

    Suggestion: It is convenient to measure the distance from the gun to a position near where the ball lands and to mark this position. The range measurement then can be made relative to the measured mark, instead of from the starting point each time. Also, it is convenient to shoot toward a wall at the end of the hall or aisle or to lay a meterstick on the floor perpendicularly to the line of flight, in order to stop the ball from rolling.

13. Plot the range versus the angle of projection, and draw a smooth curve that fits the data best. As you might expect, the points may be scattered widely because of the rather crude experimental procedure. Even so, you should be able to obtain a good idea of the angle for the maximum range. Determine this angle from the graph, and record it in Data Table 3.

---

* The range will be measured from the position below the center of the ball just as it leaves the gun to the marks that are on the paper on the floor. This will be determined by putting the ball on the gun without loading the spring.

Name _____ Section _____ Date _____

Lab Partner(s) _____

Projectile Motion:
The Ballistic Pendulum

TI/ *Laboratory Report*

A. The Ballistic Pendulum

DATA TABLE 1
(Modify Data Table if it does not apply to your ballistic pendulum.)

Purpose: To determine the magnitude of initial projectile velocity.

Trials	Notch number of pendulum catch	Height h_2 of pointer with pendulum catch in closest-to-average notch number
		_____
1		Height h_1 of pointer with pendulum freely suspended
2		_____
3		$h = h_2 - h_1$ _____
4		Mass of ball m
5		_____
		Mass of pendulum M (bob and support)
Average		_____

Calculations
(show work)

Computed v_{x_0} _____
(units)

Don't forget units

(continued)

B. Determination of the Initial Velocity of a Projectile from Range-Fall Measurements

DATA TABLE 2

Purpose: To determine the magnitude of initial projectile velocity.

Trial	Range
1	
2	
3	
4	
5	
Average	

Vertical distance of fall, y _____

Computed v_{x_o} _____

(units)

Percent difference between
results of parts A and B _____

Calculations
(show work)

C. Dependence of Projectile Range on the Angle of Projection

DATA TABLE 3

Purpose: To investigate projection angle from maximum range.

Angle of projection	Average range
20°	
30°	
40°	
45°	
50°	
60°	
70°	

Angle of projection for
maximum range from graph _____

EXPERIMENT 6   *Laboratory Report*

/TI/ QUESTIONS

A. *The Ballistic Pendulum*

1. Is the collision between the ball and the pendulum elastic or inelastic? Justify your answer by calculating the kinetic energy of the system before collision using the value of v_{x_0} found in the experiment and the kinetic energy just after collision using the experimental value of h in Eq. 2.

2. Using the results of Question 1 that would apply if the collision were inelastic, find the fractional kinetic energy loss during the collision. Express the "loss" as a percent. What became of the "lost energy"?

3. Expressing the kinetic energy in terms of momentum ($K = \frac{1}{2}mv^2 = p^2/2m$), prove using symbols, not numbers, that the fractional loss during the collision is equal to $M/(m + M)$.

4. Compute the fractional energy loss from the experimental mass values using the equation developed in Question 3, and compare this to the result in Question 2. Explain the difference, if any.

(continued)

5. Is the friction of the pendulum (pointer, support axis, etc.) a random or systematic error? Will this source of error cause your calculated velocity to be less than or greater than the actual velocity?

B. Determination of the Initial Velocity of a Projectile from Range-Fall Measurements

6. What effect does the force of gravity have on the horizontal velocity of the projectile? Explain.

7. What affect does air resistance have on the range of the projectile?

C. Dependence of Projectile Range on the Angle of Projection

8. Using experimental data, compute the magnitude of the initial velocity v_0 of the projectile from Eq. 12, and compare this to the results of parts A and B of the procedure.

9. If, for a given initial velocity, the maximum range is at a projection angle of $45°$, then there must be equal ranges for angles above and below this. Show this explicitly.

Name _____ Section _____ Date _____

Lab Partner(s) _____

Centripetal Force

/TI/ *Advance Study Assignment*

Read the experiment and answer the following questions.

1. Define centripetal force.

2. What supplies the centripetal force for (a) a satellite in orbit around the Earth, (b) the mass in uniform circular motion in this experiment?

3. An object moving in *uniform* circular motion is accelerating. How can this be, since uniform motion implies constant motion?

4. For an object in uniform circular motion, on what parameters does the determination of the centripetal force depend when we use $F = ma$?

(continued)

5. If the centripetal force acting on an object in uniform motion suddenly ceased to act or went to zero, what would happen to the object? That is, what would be its subsequent motion?

6. Suppose that the centripetal force acting on an object in circular motion were increased to a new value, and the object remained in a circular path with the same radius. How would its motion be affected?

7. Explain how the centripetal force is directly determined for each apparatus described in the experiment.

Centripetal Force

INTRODUCTION AND OBJECTIVES

The Earth revolves about the Sun, and atomic electrons move around the nucleus. What keeps these objects in orbit? The answer is **centripetal (center-seeking) force.** The centripetal force is supplied by gravitational and electrical interactions, respectively, for each of these cases.

The study of centripetal force in the laboratory is simplified by considering objects in uniform circular motion. An object in uniform circular motion moves with a constant speed (a scalar) but has a changing velocity (a vector) because of the continuing change in direction. This change in velocity results from centripetal acceleration due to a centripetal force.

In the experimental situation(s) of this experiment, the centripetal force will be supplied by a spring and can be readily measured. However, the magnitude of the centripetal force can also be determined from other experimental parameters (e.g., the frequency of rotation of the object, its mass, and its radius of orbit). Centripetal force will be experimentally investigated by measuring these parameters and comparing the calculated results with the direct measurement of the spring force, which mechanically supplies the center-seeking centripetal force.

After performing the experiment and analyzing the data, you should be able to do the following:

1. Explain why a centripetal force is necessary for circular motion.
2. Describe how the magnitude of the centripetal force for uniform circular motion may be determined from motional parameters.
3. Summarize what determines the magnitude of the centripetal force necessary to keep an object in uniform circular motion.

EQUIPMENT NEEDED

A. *Centripetal Force Apparatus with Variable-Speed Rotor and Counter*

- Laboratory timer or stopwatch
- Weight hanger and slotted weights
- Vernier caliper
- Support rod and clamp
- String
- Safety glasses

B. *Manual Centripetal Force Apparatus*

- Laboratory timer or stopwatch
- Meterstick
- Weight hanger and slotted weights
- String
- Laboratory balance
- Safety glasses

THEORY

An object in uniform circular motion requires a centripetal, or center-seeking, force to "hold" it in orbit. For example, when one swings a ball on a rope in a horizontal circle around one's head (● Fig. 1), the **centripetal force** $F_c = ma_c$ is supplied by the person and transmitted to the ball through the rope. In the absence of the centripetal force (e.g., if the rope breaks or if the person releases the rope), the ball would no longer be held in orbit and would initially fly off in the direction of its tangential velocity v.

An object in uniform circular motion moves with a constant speed. Even though the object's speed is constant, its velocity is changing because the direction of the motion is continually changing. This change in velocity results from a centripetal acceleration a_c that is due to the applied centripetal force F_c. The direction of the acceleration (and force) is always toward the center of the object's circular path, and it can be shown (see your textbook) that the magnitude of the acceleration is given by

$$a_c = \frac{v^2}{r} \tag{1}$$

(centripetal acceleration)

where v is the tangential or orbital speed of the object and r is the radius of the circular orbit. By Newton's second law, $F = ma$, the magnitude of the centripetal force is

$$F_c = ma_c = \frac{mv^2}{r} \tag{2}$$

(centripetal force)

Figure 1 Centripetal acceleration. An object in uniform circular motion must have a centripetal acceleration with a magnitude of $a_c = v^2/r$ directed toward the center of the circular path. In the case of swinging a ball on a rope around one's head, the centripetal force $F_c = ma_c$ is supplied by the person and transmitted through the rope. (Courtesy of Tony Freeman/PhotoEdit.)

where m is the mass of the object. In terms of distance and time, the orbital speed v is given by $v = 2\pi r/T$, where $2\pi r$ is the circumference of the circular orbit and T is the period.

Notice that this general expression describes the centripetal force acting on an object in uniform circular motion in terms of the properties of the motion and orbit. It is equal to the expression of a physical force that actually supplies the centripetal action. For example, in the case of a satellite in uniform circular motion around the Earth, the centripetal force is supplied by gravity, which is generally expressed $F_g = Gm_1m_2/r^2$, and $F_c = F_g$. Similarly, for an object being held in uniform circular motion by the tension force of a string, the tension force (F_t) is equal to Eq. 2 (that is, $F_t = mv^2/r$).

The centripetal force given by Eq. 2 can also be expressed in terms of the angular speed ω or frequency f of rotation, using the expressions $v = r\omega$ and $\omega = 2\pi f$:

$$F_c = \frac{mv^2}{r} = \frac{m(r\omega)^2}{r} = mr\omega^2$$

and

$$F_c = mr(2\pi f)^2 = 4\pi^2 mrf^2 \qquad (3)$$

where ω is in radians per second and f is in hertz (cycles per second). In this experiment it is convenient to think of f as being in revolutions per second.

EXPERIMENTAL PROCEDURE

A. *Centripetal Force Apparatus with Variable-Speed Rotor*

1. The centripetal force apparatus mounted on a variable-speed rotor is shown in ● Fig. 2.* **Before turning on the rotor,**
   (a) By means of the threaded collar on the centripetal force apparatus, adjust the spring to a minimum tension (0–5 on the scale above the threaded collar).
   (b) By means of the milled screw head near the base of the rotor, move the rubber friction disk to near the center of the driving disk. (The driving disk can be pushed back so that the friction disk can be moved freely.) This will give a low angular starting speed when the rotor is turned on (but don't turn it on yet!).
       The speed of the rotor is increased or decreased by moving the friction disk in (up) or out (down), respectively, along the radius of the driving disk.
       Caution: Excessive speeds can be dangerous. Do not go beyond the speeds needed.
   (c) *Make certain* that the force apparatus is locked securely in the rotor mount by means of the locking screw. Have the instructor check your setup at this point.

2. Referring to ● Fig. 3 : When the motor is turned on and adjusted to the proper speed, the cylindrical mass m in the centripetal force apparatus in contact with pointer P will cause the pointer to rise and horizontally point toward the index screw I. In this condition, the mass will be in uniform circular motion around the axis of rotation through I.
   *Caution: When taking measurements, be careful not to come in contact with the rotating apparatus. The rotor should **not** be operated without a belt guard covering the belt and pulleys. See Fig. 2(c).*

3. Put on your safety glasses and turn on the rotor. Adjust the speed until the pointer rises and is opposite the head of the index screw I (Fig. 3). Observe this with your eyes on a level with the index screw. (*Caution: Why is it a good precaution to wear safety glasses while doing this?*) The pointer will be slightly erratic, and as a particular speed is reached, it will "jump" and may point slightly above the index screw I. If so, adjust the speed so that the pointer is horizontally toward the index screw I. Do not exceed this speed. The pointer

---

* The following procedures apply particularly to the belt-driven rotor model.

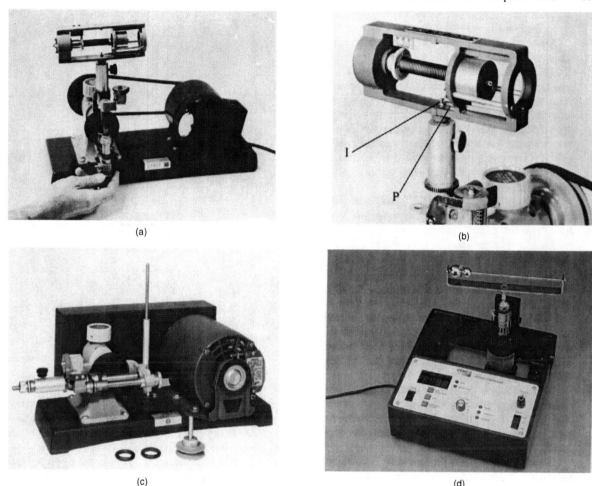

(a)

(b)

(c)

(d)

Figure 2   Centripetal force apparatus.   (a) A model for which the speed of the rotor is adjusted by moving a rubber friction disk by means of milled screw head, as illustrated in the photo. (b) For this apparatus, when the centripetal force is equal to the spring force, the pointer P will rise and point horizontally toward the tip of the index screw I. See also Fig. 9.3. (c) Motor with belt guard and rotating arm in horizontal storage position. *Note:* The belt guard has been removed in (a) and (b) for more complete illustration. ***Caution:** When in operation, the motor should always be equipped with a belt guard for safety.* (d) A self-contained centripetal force apparatus that eliminates any belt-guard problem. The apparatus has a digital readout. (Photos courtesy of Central Scientific Co., Inc.)

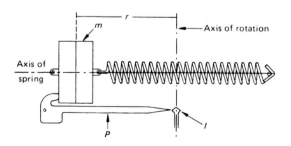

Figure 3   Pointer and screw index.   When the apparatus is rotating, the mass acting against the pointer P will cause it to rise and point toward the index screw I.

should be aimed at the head of the index screw when the rotor is spinning at higher speeds, too. (Why?)

Do not lock the friction disk. Rather, observe and adjust the speed of the rotor *continuously* during each timed interval in order to keep the pointer as steady as possible. Continuous adjustment is necessary because the rotor speed varies when the counter is engaged. Because the pointer will point horizontally at excessive speeds and induce experimental error, an alternative technique is to adjust the rotor speed continually so that the pointer is not quite horizontal—that is, so that it is aimed midway or just below the head of the index screw.

Experiment with your apparatus and see which technique is better, trying to maintain the pointer horizontally at the critical "jump" speed or aiming the pointer at a lower position on the screw at a slightly slower speed.

4. Practice engaging the counter and adjusting the rotor speed. (*Do not* engage the counter too forcefully or you will overly slow down the rotor, yet don't engage the counter so lightly that you accidentally cause the rotor to lose contact with the rotor gear.) When you are satisfied with your technique, record the (initial) counter reading in Data Table 1.

Then, using a laboratory timer or stopwatch, measure (count) the number of rotations for a 1-minute interval. One lab partner should engage the counter for the timing interval while the other adjusts the rotor speed.

Repeat this procedure for four more 1-minute intervals, but *do not* use the previous final counter reading for the next initial interval reading. Advance the counter to a new arbitrary initial reading for each trial.

Also, share the action. One lab partner should be the "speed controller" who constantly watches and adjusts the rotor speed as described in procedure 3. Another partner should be the "timer" who engages the counter and times the interval. If there are three lab partners, the third may handle the counter engagement and disengagement in response to the timer's instructions. Rotate team responsibilities periodically. (Why might such rotation produce better experimental results?)

5. Subtract the counter readings to find the number of rotations for each timed interval. (They should be similar.) Then compute the average number of rotations N of the five 1-minute intervals (average rotations per minute).

Divide the average value by 60 (1 min = 60 s) to obtain the average rotation frequency in rotations (cycles) per second, or hertz (Hz).

6. Without altering the spring tension setting, remove the centripetal force apparatus from the rotor and suspend

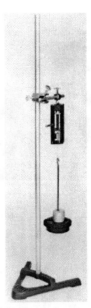

Figure 4 Spring tension. Arrangement for the application of gravitational force to measure the spring tensions. (Courtesy of Sargent-Welch.)

it from a support as shown in ● Fig. 4. Suspend enough mass on the hanger to produce the same extension of the spring as when on the rotor (pointer aimed at the index screw position).

Record this mass M' (includes mass of hanger) in the Laboratory Report below Data Table 1. Also record the mass of the cylinder m in the force apparatus (stamped on the end of the cylinder).

Add the masses to find the total suspended mass, $M = M' + m$, and compute the direct measure of $F_c =$ weight of total suspended mass $= Mg$.

With the spring at the same tension setting and the apparatus still hanging from the support with the same mass M' suspended, use a vernier caliper to measure the distance r, or the radius of the circular rotational path, and record. This is the distance between the axis of rotation (line through the index screw) and the center of mass of the cylinder (see Fig. 3).

The distance is conveniently measured between a line scribed on the upper part of the force apparatus frame above the index screw and a line scribed on the center of the cylinder.

7. Using Eq. 3, compute the magnitude of the centripetal force. Compare this with the directly measured value given by the weight force required to produce the same extension of the spring by computing the percent difference.

8. Change the spring tension to a maximum setting (about the 20 mark on the scale above the threaded collar) and repeat procedures 3 through 7, recording your results in Data Table 2.

B. Manual Centripetal Force Apparatus

9. A type of hand-operated centripetal force apparatus is shown in ● Fig. 5. By rolling the rotor between the thumb and fingers, the operator sets a suspended mass bob into circular motion, with the centripetal force being supplied by a spring. The horizontal support arm

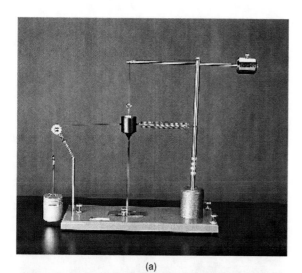

(a)

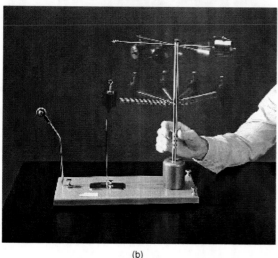

(b)

Figure 5   Hand-operated centripetal force apparatus. (a) The suspended weights, used to determine the centripetal force supplied by the spring, are not attached to the bob when the apparatus is operationally rotating. (b) Apparatus in action. See text for description. (Courtesy of Sargent-Welch.)

is counterbalanced for ease of operation; the position of the counterbalance is not critical.

A pulley mounted to the base of the apparatus is used to make direct measurement of the spring tension supplying the centripetal force for uniform circular motion of a particular radius indicated by the distance between the vertical pointer rod P and the axis of rotation.

10. Remove the bob and determine its mass on a laboratory balance. Record the mass value in Data Table 3. Adjust the position of the vertical pointer rod, if possible, to the smallest possible radius (distance between the pointer tip and the center of vertical rotor shaft). Measure this distance and record.

11. Attach the bob to the string on the horizontal support arm, and with the bob hanging freely (spring unattached), adjust the support arm so that the bob is suspended directly over the pointer. Attach the spring to the bob and practice rolling the rotor between your thumb and fingers so that the bob revolves in a circular path and passes over the pointer on each revolution in uniform circular motion. (Adjust the position of the counterbalance on the support arm if necessary for ease of operation.) Make sure the locking screws are tight, and be careful of the rotating counterweight. *Caution: Safety glasses should be worn. This is always a good practice in a laboratory with moving equipment.*

While one lab partner operates the rotor, another lab partner with a laboratory timer or stopwatch times the interval for the bob to make about 25 revolutions. The number of counted revolutions may have to be varied depending on the speed of the rotor. Count enough revolutions for an interval of at least 10 s. Record the data in Data Table 3. Practice the procedure before making an actual measurement.

12. Repeat the counting-timing procedure twice. Compute the time per revolution of the bob for each trial, and determine the average time per revolution of the three trials.

From the data, calculate the average speed of the bob. Recall $v = c/t = 2\pi r/T$, where c is the circumference of the circular orbit, r is the radius of the orbit, and T is the average time per revolution or period. Then, using Eq. 2, calculate the centripetal force.

13. Attach a string to the bob opposite the spring and suspend a weight hanger over the pulley. Add weights to the hanger until the bob is directly over the pointer. Record the weight Mg in the data table. (Do not forget to add the mass of the weight hanger.) This weight is a direct measure of the centripetal force supplied by the

spring during rotation. Compare this with the calculated value and compute the percent difference of the two values.

14. *Variation of mass.* Unscrew the nut on the top of the bob, insert a slotted mass of 100 g or more under it, and retighten the nut. Repeat procedures 11 through 13 for determining the period of rotation and comparing the computed value of the centripetal force with the direct measurement of the spring tension. (*Question:* Does the latter measurement need to be repeated?) Record your findings in Data Table 4.

15. *Variation of radius.* Remove the slotted masses from the bob, and if pointer *P* is adjustable, move it farther away from the axis of rotation to provide a larger path radius. Measure and record this distance in Data Table 5. Repeat procedures 11 through 13 for this experimental condition.

16. *Variation of spring tension (optional).* Replace the spring with another spring of different stiffness. Repeat procedures 11 through 13, recording your findings in Data Table 6.

EXPERIMENT 7

Centripetal Force

TI/ *Laboratory Report*

A. Centripetal Force Apparatus with Variable-Speed Rotor

DATA TABLE 1

Purpose: To determine rotational frequency for computation of centripetal force.

Minimum spring tension: scale reading _____

Trial	Counter readings		Difference in readings (rotations/min)
	Final	Initial	
1			
2			
3			
4			
5			
Average number of rotations N			

Computation of centripetal force (show work)

Average rotational frequency $(f = N/60)$ _____

Suspended mass M' _____

Cylinder mass m _____

Total suspended mass $(M = M' + m)$ _____

Direct measure of F_c $(F_c = Mg)$ _____

Radius of circular path r _____

Computed F_c _____

Percent difference _____

Don't forget units

(continued)

DATA TABLE 2

Purpose: To determine rotational frequency for computation of centripetal force.

Minimum spring tension:
scale reading _____

| Trial | Counter readings | | Difference in readings (rotations/min) |
	Final	Initial	
1			
2			
3			
4			
5			
		Average number of rotations N	

Computation of centripetal force (show work)

Average rotational frequency
$(f = N/60)$ _____

Suspended mass M' _____

Cylinder mass m _____

Total suspended mass
$(M = M' + m)$ _____

Direct measure of F_c
$(F_c = Mg)$ _____

Radius of circular path r _____

Computed F_c _____

Percent difference _____

EXPERIMENT 7   *Laboratory Report*

B. Manual Centripetal Force Apparatus

DATA TABLE 3

Purpose: To determine period of revolution for computation of centripetal force.

	Trial 1	Trial 2	Trial 3
Number of revolutions			
Total time ( )			
Time/revolution ( )			

Computation of centripetal force
(attach additional sheet)

Mass of bob _____

Radius of circular path _____

Average time per revolution _____

Average speed of bob (v) _____

Computed value of
centripetal force _____

Direct measurement of
centripetal force _____

Percent difference _____

DATA TABLE 4

Purpose: To observe the effect of varying mass.

	Trial 1	Trial 2	Trial 3
Number of revolutions			
Total time ( )			
Time/revolution ( )			

Computation of centripetal force
(attach additional sheet)

Mass of bob plus slotted mass _____

Radius of circular path _____

Average time per revolution _____

Average speed of bob (v) _____

Computed value of
centripetal force _____

Direct measurement of
centripetal force _____

Percent difference _____

(continued)

DATA TABLE 5
Purpose: To observe the effect of varying radius.

	Trial 1	Trial 2	Trial 3
Number of revolutions			
Total time ( )			
Time/revolution ( )			

Mass of bob _____

Radius of circular path _____

Average time per revolution _____

Average speed of bob (v) _____

Computed value of
centripetal force _____

Direct measurement of
centripetal force _____

Percent difference _____

Computation of centripetal force
(attach additional sheet)

DATA TABLE 6 (Optional)
Purpose: To observe the effect of varying spring
tension.

	Trial 1	Trial 2	Trial 3
Number of revolutions			
Total time ( )			
Time/revolution ( )			

Mass of bob _____

Radius of circular path _____

Average time per revolution _____

Average speed of bob (v) _____

Computed value of
centripetal force _____

Direct measurement of
centripetal force _____

Percent difference _____

Computation of centripetal force
(attach additional sheet)

EXPERIMENT 7   *Laboratory Report*

TI/ QUESTIONS

1. How does the centripetal force vary with the radius of the circular path? Consider
   (a) constant frequency, and (b) constant speed. Was this substantiated by experimental
   results?

2. If the centripetal force on an object in uniform circular motion is increased, what is the
   effect on (a) the frequency of rotation f (with r constant), and (b) f and r when both are free
   to vary?

3. Does the centripetal force acting on an object in uniform circular motion do work on the
   object? Explain.

(continued)

4. In part A of the experimental procedure, the counter was advanced to a new arbitrary initial reading for the time interval trials. This is because if consecutive counter readings had been used (that is, if the final reading for one interval had been taken as the initial reading for the next interval), then in averaging the differences, all readings except the first and last would have been "thrown out" by the averaging process. That is, the result is essentially one 5-minute interval. Prove this explicitly. (*Hint:* Let the counter readings be n_0, n_1, n_2, . . . , and then $\Delta n_1 = n_1 - n_0$, $\Delta n_2 = n_2 - n_1$, etc. Perform the averaging process on these values.)

5. Figure 1 shows a student swinging a ball in a circle about his head. (a) Is the rope between the hand and the ball horizontal? (b) Is the force applied by the student on the rope equal to the centripetal force? (Use a diagram to illustrate and justify your answers.)

EXPERIMENT 8

Friction

/TI/ *Advance Study Assignment*

Read the experiment and answer the following questions.

1. State the three general empirical rules used to describe friction.

2. What is the normal force, and why is it used instead of the load?

3. Why is it important to have the string parallel to the horizontal surface in the procedures where suspended weights are used?

4. What is the coefficient of friction, and in what units is it expressed? Distinguish between μ_s and μ_k. Which is generally greater?

(continued)

5. Explain how graphs of weight versus normal force in procedures A and B give the coefficients of friction.

CI *Advance Study Assignment*

Read the experiment and answer the following question.

1. Under what conditions is the tension in the string pulling horizontally on the cart equal in magnitude to the frictional force?

Friction

OVERVIEW

Experiment examines friction using complementary TI and CI approaches. The TI procedures are concerned with determination of the coefficients of friction, μ_s and μ_k, with an option of investigating the dependence of μ on various parameters, such as different materials, lubrication, and so on.

The CI procedures extend the investigation by examining the effect of speed on sliding friction.

INTRODUCTION AND OBJECTIVES

In general, the term **friction** refers to the force or resistance to motion between contacting material surfaces. (Internal friction occurs in liquids and gases.) The friction between unlubricated solids is a broad and complicated topic, because it depends on the contacting surfaces and the material properties of the solids. Three general empirical "rules" are often used to describe friction between solid surfaces. These are that the frictional force is

1. Independent of the surface area of contact.
2. Directly proportional to the **load,** or the contact force that presses the surfaces together.
3. Independent of the sliding speed.

    Let's take a look at these rules:

1. Intuitively, one would think that friction would depend on the roughness or irregularities of the surfaces or, more specifically, on the *actual* area of contact of the irregularities of the surfaces. This would seem to contradict rule 1.
2. The contact area of the surfaces, however, should depend on the force that presses the surfaces together, or the load. Increasing this force should increase the amount of contact surface and hence the friction. Rule 2 then seems logical.
3. Is it consistent that the friction between a sliding object and a surface be independent of the sliding speed? It would seem that the rate at which the surface irregularities met, which is dependent on the sliding speed, should have some effect.

    With such thoughts in mind, in this experiment we will investigate the validity of the foregoing empirical rules. Although we apply these rules in analyzing friction, you might find experimentally that they are very general and, at best, approximations when applied to all materials and all situations.

    After performing the experiment and analyzing the data, you should be able to do the following:

TI OBJECTIVES

1. Comment on the validity of the empirical rules of friction.
2. Describe how coefficients of friction are determined experimentally.
3. Tell why the normal reaction force of a surface on an object is used to determine the frictional force rather than the weight of the object.

CI OBJECTIVES

1. Verify that friction is proportional to the normal force.
2. Indicate whether or not friction is independent of sliding speed.

Friction

- Board with attached pulley
- Rectangular wooden block with hook (e.g., piece of 2 × 4 lumber or commercially available block)
- Weight hanger and set of weights
- String
- Protractor
- Laboratory balance
- Table clamp and support

- Meterstick
- Masking tape
- 2 sheets of Cartesian graph paper

(*Optional*)

- Plastic block
- Aluminum block
- Wheel cart
- Lubricating powder

TI THEORY (general, TI and CI)

It is sometimes assumed that the *load*, or the contact force that presses the surfaces together, is simply the weight of the object resting on a surface. Consider the case of a block resting on a horizontal surface as illustrated in ● TI Fig. 1a. The force that presses the surfaces together is the downward weight force of the block (magnitude mg), which is the load. However, on an inclined plane, only a component of the weight contributes to the load, the component perpendicular to the surface. (See TI Fig. 3, where the magnitude of the load is $mg \cos \theta$.)

In order to take such differences into account, the frictional force f is commonly taken to be directly proportional to the normal force N, which is the force of the surface *on* the block—that is, $f \propto N$ (see TI Fig. 1). In the absence of other perpendicular forces, the normal force is equal in magnitude to the load, $N = mg$ in TI Fig. 1 and $N = mg \cos \theta$ in TI Fig. 3, which avoids any confusion between weight and load.

With $f \propto N$, we may write in equation form

or

$$f = \mu N$$
$$\mu = f/N$$

(TI 1)

where the Greek letter mu (μ) is a dimensionless constant of proportionality called the **coefficient of friction.**

When a force F is applied to the block parallel to the surface and no motion occurs, we say that the applied force is balanced by an opposite force of static friction f_s that is exerted on the block by the table along the surface of contact (TI Fig. 1b, $\Sigma F_x = F - f_s = ma = 0$). As the

magnitude of the applied force is increased, f_s increases to a *maximum* value given by (TI Fig. 1c)

$$f_{s_{max}} = \mu_s N$$

(TI 2)

(static friction)

where μ_s is the coefficient of static friction.* The maximum force of static friction is experimentally approximated by the smallest force applied parallel to the surface that will just set the block into motion.

At the instant the applied force F becomes greater than $f_s = \mu_s N$, however slightly, the block is set into motion, the motion is opposed by the force of kinetic (sliding) friction f_k, and (TI Fig. 1d)

$$f_k = \mu_k N$$

(TI 3)

(kinetic friction)

where μ_k is the coefficient of kinetic (sliding) friction.

In general, $\mu_k N < \mu_s N$, and the unbalanced force causes the block to accelerate ($\Sigma F_x = F - f_s = ma$). However, if the applied force is reduced so that the block moves with a uniform velocity ($a = 0$), then $F = f_k = \mu_k N$.

Usually, for a given pair of surfaces, $\mu_k < \mu_s$. That is, it takes more force to overcome static friction (get an object moving) than to overcome kinetic friction (keep it moving).

* These conditions of f_s are sometimes written $f_s \leq \mu_s N$; that is, f_s is less than or equal to the maximum value of $\mu_s N$. (See TI Fig. 1.)

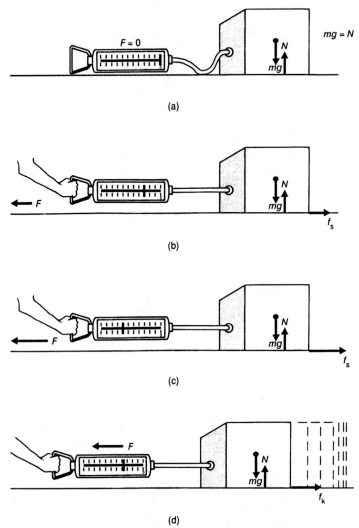

TI Figure 1   Friction.   The applied force is balanced by the force of static friction f_s (a–c, $\Sigma F_x = F - f_s = ma = 0$) until a maximum value is reached. A slightly greater force (d) sets the block into motion ($\Sigma F_x = F - f_k = ma$), with the applied force being opposed by the force of kinetic friction, f_k.

Both coefficients may be greater than 1, but they are usually less than 1. The actual values depend on the nature and roughness of the surfaces.

TI/ EXPERIMENTAL PROCEDURE

A. Determination of μ_s

1. Determine the mass of the wooden block on a laboratory balance, and record it in the Laboratory Report.

2. Clean the surfaces of the board and block so they are free from dust and other contaminants. Place the board

with the pulley near the edge of the table so that the pulley projects over the table's edge (● TI Fig. 2). Attach one end of a length of string to the wooden block and the other end to a weight hanger. Place the block flat on the board, and run the string over the pulley so that the weight hanger is suspended over the end of the table. Be sure that the string is parallel to the board.

3. With the rectangular block lying on one of its sides of larger area, add weights to the hanger until the block just begins to move. (*Note:* If the 50-g hanger causes the block to move, add some weights to the block and

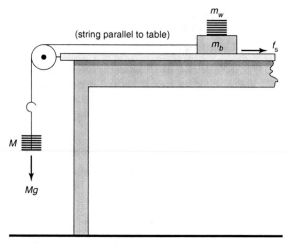

TI Figure 2   Coefficient of static friction.   Experimental setup to determine μ_s. See text for description.

add this mass to the mass of the block, m_b.) Determine the required suspended mass within 1 gram. Record the weight force (Mg) required to move the block in TI Data Table 1. This is equal in magnitude to the frictional force, $f_{s_{max}}$.

Suggested experimental technique:

(a) Keep the block in the middle of the plane.

(b) Lift the block, gently lower it onto the plane, *restrain* it from moving for a count of five (*do not* press it against the plane), and then release the block. If the block moves, the suspended mass M is too large; if it doesn't move, M is too small; if the block moves about half the time, M is about right.

4. Repeat procedure 3 with m_w = 100-, 200-, 300-, 400-, and 500-g masses, respectively, added to the block. Record the results in TI Data Table 1.

5. Plot the weight force just required to move the block (or the maximum force of static friction, $F = f_s$) versus the normal force N of the surface on the block [$N = (m_b + m_w)g$]. Draw a straight line that best fits the data. Include the point (0, 0). Why?

Since $f_s = \mu_s N$, the slope of the straight line is μ_s. Determine the slope and record it in TI Data Table 1.

B. Determination of μ_k

HORIZONTAL BOARD

6. When the block moves with a uniform speed, its acceleration is zero. The weight force F and the frictional force f_k are then equal and opposite ($\Sigma F_x = F - f_k = ma = 0$, and $F = f_k$).

7. Using the larger side (surface area) of the block and the series of added masses as in part A, add mass to the

weight hanger until a slight push on the block will cause it to move with a uniform speed. It may be helpful to tape the weights to the block. The required weight force for the motion in each case should be less than that for the corresponding case in part A. Why? Record the data in TI Data Table 2.

Suggested experimental technique:

(a) Begin with the block at one end of the plane, and give it a push so that it slides across the entire plane.

(b) Observe the behavior of the block in the same region as before, namely in the middle of the plane. This is where the block should be observed for constant speed.

8. Plot the weight force (or the force of kinetic friction, $F = f_k$) versus the normal force N for these data on the same graph as for part A. Draw a straight line that best fits the data.

Since $f_k = \mu_k N$, the slope of the straight line is μ_k. Determine the slope and record it in TI Data Table 2. Calculate the percent decrease of μ_k from the μ_s value.

ELEVATED BOARD (INCLINED PLANE)

9. Elevate the pulley end of the board on a support to form an inclined plane (● TI Fig. 3, see Fig. 10.3 for a similar setup). Note in the figure that the magnitude of the normal force is equal to a *component* of the weight force.

With the block lying on a *side* of its larger surface area, determine the angle θ of incline that will allow the block to slide down the plane with a constant speed after being given a slight tap. (No suspended weight is used in this case.) *Note:* The maximum angle before slipping *without* tapping gives μ_s, whereas the angle of constant velocity *with* tapping gives μ_k.

10. Using a protractor, measure the angle θ and record it in TI Data Table 3. Also, with a meterstick, measure the length L of the base (along the table) and the height h of the inclined plane. Record the ratio h/L in TI Data Table 3.

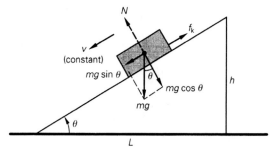

TI Figure 3   Coefficient of kinetic friction.   Experimental setup to determine μ_k. See text for description.

11. Repeat this procedure for the block with the series of added masses as in the previous procedure for the horizontal board and record in TI Data Table 3. It may be helpful to tape the masses to the block.

12. Using your calculator, find the tangents of the θs, and record. Compute the average of these values and the average of the ratios h/L. These averages should be similar. Why?

13. Compare the average value of tan θ with the value of μ_k found in the procedure for the horizontal board. It can be shown theoretically (Question 3) that

tan $\theta = \mu_k$ in this case. Compute the percent difference of the experimental values.

C. Dependences of μ (optional)*

14. Use the *inclined plane method* to investigate the dependence of μ on area, material, velocity, rolling, and lubrication. The experimental setups are described in TI Data Table 4. Answer the questions listed after the data table.

* This experimental procedure and modifications were suggested by Professor I. L. Fischer, Bergen Community College, New Jersey.

Name _____ Section _____ Date _____
Lab Partner(s) _____

T I EXPERIMENT 8

Friction

/TI/ *Laboratory Report*

Note: Attach graphs to Laboratory Report

Mass of block m_b _____

A. Determination of μ_s

/TI/ **DATA TABLE 1**

Purpose: To investigate $f_s = \mu_s N$, where N depends on $m_b + m_w$, by measuring μ_s on a level plane (see TI Fig. 2).

m_w	0					
$N = (m_b + m_w)g*$						
$f_s = F = Mg$						

* It is convenient to express the force in terms of mg, where g is left in symbol form [e.g., $(0.250)g\ N$], even when graphing.

Calculations
(show work)

μ_s _____
(from graph)

Don't forget units

(continued)

B. Determination of μ_k

/TI/ **DATA TABLE 2**

Purpose: To investigate $f_k = \mu_k N$, where N depends on $m_b + m_w$, by measuring μ_k on a level plane.

m_w	0					
$N = (m_b + m_w)g$						
$f_k = F = Mg$						

Calculations
(show work)

μ_k _____
(from graph)

Percent decrease of
μ_k relative to μ_s _____

/TI/ **DATA TABLE 3**

Purpose: To investigate $\mu_k = \tan \theta$, where θ is independent of $m_b + m_w$, by measuring μ_k by the inclined plane method (see TI Fig. 3).

m_w	0					
θ						Average
h/L						
$\tan \theta$						

Calculations
(show work)

Percent difference between
$\tan \theta = \mu_k$ and μ_k from TI Data Table 2 _____

EXPERIMENT 8   *Laboratory Report*

C. *Dependences of μ (optional)*

$\boxed{TI}$ **DATA TABLE 4**

Purpose: To investigate dependences of μ by various measurements using the inclined plane method and other materials (if available).

No.	Conditions	θ	$\mu = \tan\theta$
1	Wooden block on larger area, static (μ_s)		
2	Wooden block on smaller area, static (μ_s)		
3	Wooden block on smaller area, kinetic (μ_k)		
Other materials			
4	Plastic block		
5	Aluminum block, moving slowly		
6	Aluminum block, moving faster		
7	Wheeled cart		
8	Aluminum block with dry lubricating powder		
9	Plastic block with dry lubricating powder		

Answer the following questions on a separate sheet of paper and attach it to the TI Laboratory Report.

(a) Compare No. 1 with TI Data Table 1: Is the inclined plane method valid for μ_s?
(b) Compare No. 2 with No. 1 and No. 3 with TI Data Table 3: Does μ depend on area?
(c) Compare Nos. 3, 4, and 5: Does μ_k depend on material?
(d) Compare No. 5 with No. 6: Does μ_k depend on velocity?
(e) Compare No. 7 with anything: How does rolling friction compare with other types of friction?
(f) Compare Nos. 8 and 9 with Nos. 5 and 4: What is the effect of adding the lubricant?

$\boxed{TI}$ QUESTIONS

1. Explain why $f_s \leq \mu_s N$; that is, why is f_s less than or equal to $\mu_s N$?

(continued)

2. Speculate, in terms of the microscopic surface irregularities, about why $\mu_k < \mu_s$ and what effect a lubricant has on the coefficient of friction.

3. (a) Prove that $\tan \theta$ is equal to μ_k when the block slides down the incline with a constant speed. (Use symbols, not numbers.)

    (b) If θ is the maximum angle of incline just before the block moves, what is μ_s in terms of θ?

4. Suppose that the block were made to move up the inclined plane with a uniform speed by suspending masses on a string over the pulley. Derive an equation for the coefficient of kinetic friction for this case in terms of the suspended masses, the mass of the block, and the angle of decline. (Neglect any friction and mass effects of the pulley.)

5. On the basis of your experimental results, draw and justify conclusions about the validity of the empirical rules for friction. What does this tell you about applying general rules to all materials and about the nature of friction?

CI EXPERIMENT 8

Friction

 EQUIPMENT NEEDED

- 1 wooden block (The block used in the TI procedure can be used here also. Another option is the "Friction Block" included in the PASCO Classic Dynamics System.)
- Additional blocks as needed to make the string horizontal when connected to the force sensor (Two PASCO cars (ME–9430 or 9454), stacked upside down on top of each other and on top of the friction block will make a tower of the correct height.)

- 1 straight, smooth track (PASCO dynamics track)
- 1 force sensor (PASCO CI-6537)
- 1 constant-speed motorized car (PASCO ME-9781)
- Extra weights to load the sliding object (200-g or 500-g pieces will work fine) (The PASCO Classic Dynamics System includes mass bars that can be used in this part.)
- Graph paper

THEORY

In this experiment, we will study two of the general empirical rules used to describe the friction between solid surfaces. In the first part, we will examine the relationship between friction and the normal force to verify that they are proportional to each other. In the second part, we will examine the effect of the speed of the object on the amount of frictional force. In both cases, a force sensor will be used to measure the frictional force between a sliding wooden block and a track.

● CI Figure 1 illustrates the experimental situation. The sliding object is a wooden block. Other blocks are shown added as needed so that the string is *horizontal* when connected to a force sensor riding on a motorized car. As an alternative, the figure also shows the setup using the suggested PASCO equipment, where a stack of cars is used to make the object the correct height. Other alternatives include

using a single 2 × 4 board with a nail that makes it possible to attach the string at the proper height (not pictured).

● CI Figure 2 shows a free-body diagram of a block as it slides with constant speed along a level track. The horizontal forces are F, the tension of the string, and f, the frictional force provided by the track. With the speed constant, there is no acceleration. From Newton's second law, we have

$$\Sigma F_x = F - f = ma = 0$$

or

$$F = f$$

In this experiment, the force sensor will directly measure F, the tension in the string. Notice that as long as the car moves at a constant speed, the magnitude of F is equal to the magnitude of the frictional force acting on the sliding block.

On the other hand, the vertical forces balance each other out, so the magnitude of the normal force N can be determined as the magnitude of the weight of the object: $N = mg$.

SETTING UP DATA STUDIO

Note: The force sensor needs to be calibrated before use. Refer to the user's manual for instructions on how to calibrate the sensor. The procedures described below assume that the force sensor has been properly calibrated.

1. Open Data Studio and choose "Create Experiment."
2. From the list of sensors, choose a Force Sensor. Connect the force sensor to channel A of the interface, as shown on the screen.
3. Double-click on the Force Sensor icon. The Sensor Properties window will open. Under "General," set the

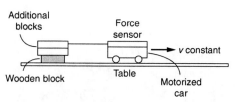

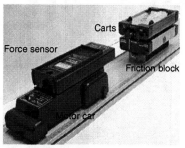

CI Figure 1 The experimental setup. A wooden block slides on a flat surface while being pulled by a motorized car that moves at a constant speed. Additional blocks can be added as necessary on top of the wooden block so that the string is horizontal when connected to the force sensor. The force sensor rides on the motorized car. As an alternative, PASCO dynamic cars can be stacked on top of a friction block to achieve the same effect.

129

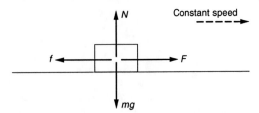

CI Figure 2   Free-body diagram of the sliding block.   The horizontal forces are F, the tension on the string, and f, the friction from the surface. The force sensor measures F. At constant speed, the horizontal force vectors are equal and opposite, and $F = f$. The force sensor readings can be taken to be the friction as long as the block slides at constant speed.

sample rate to 200 Hz, fast. Click OK to get back to the main window.

4. Create a digits display by double-clicking on "Digits" in the displays list (lower left of the screen). The display window that opens will show the force readings from the sensor when data are collected.

5. On the digits display's toolbar there is a drop menu with the sigma symbol (Σ). Press it and choose "Mean." This will show the average of a series of measurements on the display.

6. The size of the display window can be adjusted for easier viewing, if needed. The bigger the screen, the more digits you will be able to see once data are collected. For the purpose of this experiment, keep the size such that only two decimal places are shown. (Wait until data are collected to adjust this. There have to be data on the display before any change can be noticed.)

7. ● CI Figure 3 shows what the screen will look like after the setup is complete and data are taken.

CI EXPERIMENTAL PROCEDURE

A. The Effect of the Load

1. Measure the mass of the wooden block and of any other block or car that will be placed on top of it to add height, as illustrated in CI Fig. 1.   Record the total mass in Trial 1 of CI Data Table 1.

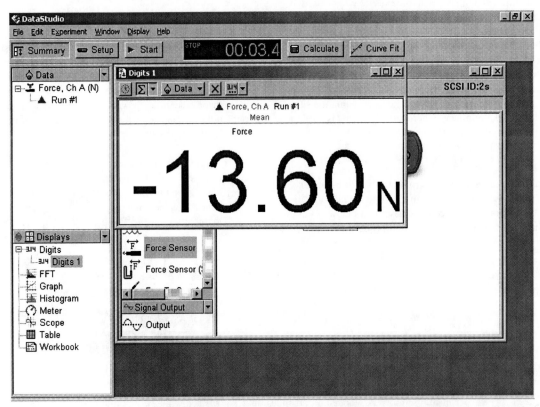

CI Figure 3   Data Studio setup.   A digits display will show the force reading of the sensor. Once data are collected, the size of the display window is adjusted to show two decimal places. (Data displayed using DataStudio Software. Reprinted courtesy of PASCO scientific.)

2. Set up the equipment as shown in CI Fig. 1. It is important that the string connecting the force sensor to the pile of objects be *horizontal*. If using additional blocks instead of the PASCO cars, tape the blocks together so that they will not fall off.

3. Set the motorized car for a medium speed, and do not change it during the experiment.

4. ***Trial 1: The object with no extra load.***

   a. With the string slack, press the TARE button on the side of the force sensor to zero the sensor.

   b. Turn the motorized car on.

   c. Wait until the string tenses before pressing the START button to begin collecting data. Let the car move, pulling along the pile of blocks (the "object"), for about 20 cm, and then press the STOP button.

   d. Stop the car.

   e. Report the average fictional force reading in CI Data Table 1. Do not worry if the sensor reading is negative. That is a convention for direction (pull or push). In this experiment, we need only the magnitude.

5. ***Trials 2, 3, 4 and 5: The object with a load.***

   a. Place a load on top of the sliding object and record the new mass of the sliding object in CI Data Table 1.

   b. Repeat the data collection process as described in steps (a) to (e) for Trial 1.

   c. Repeat by continuing to add mass on top of the object until the table is complete.

6. Calculate the normal force for each trial by determining the weight of the object plus load in each case. Record the results in CI Data Table 1.

7. Use a full page of graph paper to make a plot of friction versus normal force. Determine the slope of the best-fitting line for the plot, and enter the result in the table. Attach the graph to the Laboratory Report.

B. *The Effect of the Speed*

1. Set up the equipment as shown in CI Fig. 1. It is important that the string connecting the force sensor to the pile of objects be *horizontal*. If using additional blocks instead of the PASCO cars, tape the blocks together so that they will not fall off.

2. Set the motorized car for a slow speed.

3. Turn on the motorized car. Wait until the string tenses before pressing the START button to begin collecting data. Let the car move, pulling along the block, for about 20 cm, and then press the STOP button.

4. Stop the car.

5. Report the average frictional force reading in CI Data Table 2.

6. Increase the speed of the motorized car, and measure the average frictional force again. Repeat by increasing the speed for each trial until the table is complete.

C I   **EXPERIMENT 8**

Friction

 *Laboratory Report*

A. The Effect of the Load

 **DATA TABLE 1**

Purpose: To investigate the effect of changing the load on an object (and thus changing the normal force) on the magnitude of the frictional force.

	Trial	Total mass of sliding object	Frictional force (sensor reading)	Normal force $N = mg$
The object with no load	1			
The object with increasing load	2			
	3			
	4			
	5			

Slope of graph = _____

Don't forget units

(continued)

B. The Effect of Speed on Friction

C1 DATA TABLE 2

Purpose: To investigate the effect of speed on the frictional force.

Different speed trials (from low speed to high)	Average frictional force
1	
2	
3	
4	
5	
6	
7	
8	
9	
10	

EXPERIMENT 8

CI QUESTIONS

1. Is it true that the frictional force is directly proportional to the normal force? Discuss the experimental evidence.

2. What is the physical significance of the slope of the graph of friction versus normal force?

3. Is there a clear pattern for the frictional force as the speed of the object increases? (Compare to the pattern observed when increasing the load.) What can be concluded about the effect of the speed? Discuss.

4. Why was it so important that the string connecting the sensor and the object remain horizontal during the experiment? Discuss what would happen if it did not.

5. Refer to Step 3 of the Experimental Procedure for part A, which says, "Set the motorized car for a medium speed, and do not change it during the experiment." Given the results of part B of the experiment, discuss whether changing the speed would have made a difference in the results of experiment A.

EXPERIMENT 8

EXPERIMENT 9

Work and Energy
/TI/ *Advance Study Assignment*

Read the experiment and answer the following questions.

1. Distinguish between the conservation of mechanical energy and the conservation of total energy.

2. Is mechanical energy conserved in real situations? Is the total energy conserved? Explain.

3. Discuss the relationship between work and energy for a car moving with a constant speed (a) up an incline and (b) down an incline.

4. Under what conditions would the frictional forces be expected to be equal in magnitude for a car moving up an incline and a car moving down an incline?

(continued)

5. Is the force of friction the same for different angles of incline if all other parameters are equal? Explain by specifically considering the angles used in the experiment.

6. What are possible sources of error in this experiment?

Work and Energy

Introduction and Objectives

Work and **energy** are intimately related, as emphasized in a common definition of energy as the ability to do work. That is, an object or system possessing energy has the *capability* of doing work. When work is done by a system, energy is expended—the system loses energy. Conversely, when there is work input to a system, the system gains energy.

In an ideal conservative system, energy is transferred back and forth between kinetic energy and potential energy. In such a system, the sum of the kinetic and potential energies is constant, as expressed by the *law of conservation of mechanical energy*. However, in actual systems, friction is always present, and these systems are nonconservative. That is, some energy is lost as a result of the work done against frictional forces. Even so, the *total* energy is conserved (*conservation of total energy*).

In this experiment you will make use of the conservation of energy to study the relationship between work and

energy in the cases of a car rolling up and down an inclined plane. The ever-present frictional forces and the work done against friction will be investigated and taken into account so as to provide a better understanding of the concept of work-energy. To simplify matters, experimental conditions with constant speeds will be used so that only the relationship between work and changes in gravitational potential energy will have to be considered.

After performing this experiment and analyzing the data, you should be able to:

1. Explain how work and energy are related.
2. Describe how frictional work can be determined experimentally using either force-distance or energy considerations.
3. Better appreciate the nonconservative aspects of real situations and the difference between the conservation of mechanical energy and the conservation of total energy.

Equipment Needed

- Inclined plane with pulley and Hall's carriage (car)
- Weight hanger and slotted weights
- String

- Meterstick
- Protractor (if plane not so equipped)
- Laboratory balance

Theory

A. *Work of Friction: Force-Distance Method*

Car Moving Up the Plane

The situation for a car moving up an inclined plane with a constant velocity is illustrated in ● Fig. 1. Since the car is not accelerating, the force up the plane (F) must be equal in magnitude to the sum of the forces down (parallel to) the plane:

$$F = F_\parallel + f$$

where f is the force of friction and $F_\parallel = m_c g \sin \theta$ is the component of the car's weight parallel to the plane. (See Fig. 1.)

Since the magnitude of F is equal to the weight w_1 of the suspended mass (m_1), we may write

$$w_1 = F_\parallel + f$$

Then, solving for f and expressing the other forces in terms of the experimental parameters,

$$f = w_1 - F_1$$

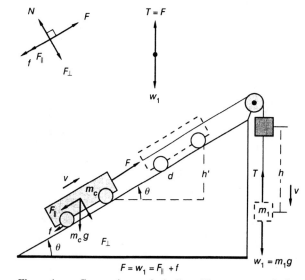

Figure 1   **Car moving up the incline with a constant velocity.** With no acceleration, the net force on the car is zero, and $F = F_\parallel + f = w_1$ (see free-body diagrams).

or

$$f = m_1g - m_cg \sin \theta \qquad (1)$$

(car moving up)

CAR MOVING DOWN THE PLANE

The situation for a car moving down an inclined plane with the same constant speed is illustrated in ● Fig. 2. Again, since the car is not accelerating, the sum of the forces up the plane must be equal in magnitude to the force down the plane, and

$$F = F_\parallel - f$$

where, in this case, the direction of f is up the plane. Since $F = w_2$,

$$F_\parallel = w_2 + f$$

and, expressing f as before,

$$f = m_cg \sin \theta - m_2g \qquad (2)$$

(car moving down)

Then, in either case, the frictional work is given by

$$W_f = fd \qquad (3)$$

where d is the distance the car moves.

If the car moves approximately at the same constant speed in each case, it might be assumed that the magnitude of the frictional force f would be the same in each case

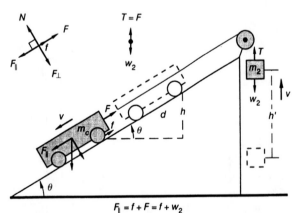

$$F_\parallel = f + F = f + w_2$$

Figure 2   Car m̄ vinḡ dw̄n the incline with the same constant speed as in Fig. 1. With no acceleration, the force on the car is zero, and $F_\parallel = F + f = w_2 + f$ (see free-body diagrams).

(same angle of incline and load). This will be investigated experimentally.

B. Work of Friction: Energy Method

Another way of looking at the frictional work is in terms of energy.

CAR MOVING UP THE PLANE

For the case of the car moving up the plane, by the conservation of energy, the *decrease* in the potential energy of the descending weight on the weight hanger $\Delta U_w = m_1gh$ is equal to the *increase* in the potential energy of the car $\Delta U_c = m_cgh'$ *plus* the energy lost to friction, which is equal to the work done against the force of friction W_f. That is,

$$\Delta U_w = \Delta U_c + W_f$$

or

$$W_f = \Delta U_w - \Delta U_c$$

and

$$W_f = m_1gh - m_cgh' \qquad (4)$$

(car moving up)

CAR MOVING DOWN THE PLANE

Similarly, for the case of the car moving down the plane, by the conservation of energy, the *decrease* in the potential energy of the descending car is equal to the *increase* in the potential energy of the ascending weight *plus* the work done against the force of friction:

$$\Delta U_c = \Delta U_w + W_f$$

or

$$W_f = \Delta U_c - \Delta U_w$$

and

$$W_f = m_cgh' - m_2gh \qquad (5)$$

(car moving down)

In terms of the experimental parameters, the methods for determining W_f are equivalent.

EXPERIMENTAL PROCEDURE

1. Using a laboratory balance, determine the mass of the car m_c and record it in the Laboratory Report.

2. Arrange the inclined plane and the car as shown in ● Fig. 3 with an angle of incline of $\theta = 30°$. Make certain that the pulley is adjusted so that the string at-

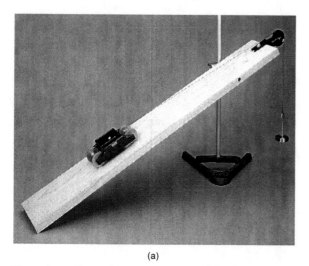

(a)

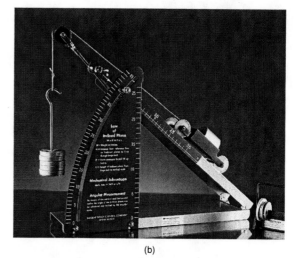

(b)

Figure 3   Types of inclined planes.   (a) Inclined plane with board and stand. (b) Calibrated incline plane. (Courtesy of Sargent-Welch.)

tached to the car is parallel to the plane. (Should the car accelerate up the plane by the weight of the weight hanger alone, place some weights in the car so that the car is initially stationary. Add the additional mass to that of the car in Data Table 1.)

3. Add enough weights to the weight hanger so that the car moves up the incline with a slow uniform speed when the car is given a slight tap. Record the total suspended mass in Data Table 1.

4. With the car positioned near the bottom of the incline, mark the position of the car's front wheels and give the car a slight tap to set it into motion. Stop the car near the top of the plane after it moves up the plane (with a constant speed), and measure the distance of d it moved up the plane as determined by the stopped position of the car's front wheels. Or measure the height h the weight hanger descends. This corresponds to the situation in Fig. 1. The lengths d and h are the same. Record this length in Data Table 1 as d.

5. With the car near the top of the plane, remove enough weights from the weight hanger so that the car rolls down the inclined plane with a slow uniform speed on being given a slight tap. Use as close to the same speed as for the upward case as is possible. This corresponds to the situation in Fig. 2 . Record the total suspended mass in Data Table 1. For convenience, use the same d (or h) as in procedure 4.

6. Compute the frictional force f (Eqs. 1 and 2) and work done against friction W_f (Eq. 3) for each case. Show your calculations and record the results in Data Table 1.

7. Compare the frictional work for the two cases by computing the percent difference.

8. Adjust the angle of the inclined plane to $\theta = 45°$ and repeat procedures 3 through 7, recording your measurements in Data Table 2.

Name _____ Section _____ Date _____

Lab Partner(s) _____

Work and Energy

$\boxed{\text{TI}}$ *Laboratory Report*

Angle of incline _____

DATA TABLE 1

Purpose: To determine work done against friction.

Mass of car m_c _____

		Suspended mass ( )	d ( )	f ( )	W_f ( )
Car moving up incline	m_1				
Car moving down incline	m_2				

Calculations
(show work)

Percent difference in W_f _____

Don't forget units

(continued)

Angle of incline _____

DATA TABLE 2

Mass of car m_c _____

Purpose: To determine work done against friction.

		Suspended mass (　)	d (　)	f (　)	W_f (　)
Car moving up incline	m_1				
Car moving down incline	m_2				

Calculations
(show work)

Percent difference in W_f _____

EXPERIMENT 9   *Laboratory Report*

/TI/ QUESTIONS

1. What was the work done by the suspended weight when the car (a) moved up the incline
    and (b) moved down the incline? (*Show your calculations.*)

                                                        $\theta = 30°$ $\theta = 45°$

                            Car moving up incline _____   _____

                            Car moving down incline _____   _____

2. What was the work done by gravity acting on the car when it (a) moved up the incline and
    (b) moved down the incline? (*Show your calculations.*)

                                                        $\theta = 30°$ $\theta = 45°$

                            Car moving up incline _____   _____

                            Car moving down incline _____   _____

3. (a) For the car going up the incline, what percentage of the work done by the suspended
    weight was lost to friction? (b) For the car moving down the incline, what percentage of
    the work done by gravity was lost to friction? (*Show your calculations.*)

                                                        $\theta = 30°$ $\theta = 45°$

                            Car moving up incline _____   _____

                            Car moving down incline _____   _____

(continued)

4. Suppose the car accelerated up and down the incline. How would this affect the experimental determinations?

5. Is the assumption justified that f would be the same for both up and down cases for the same constant speed? If not, speculate as to why there is a difference.

6. Assuming that $f = \mu N$ (see Experiment 9), show that the coefficient of (rolling) friction for the car moving down the inclined plane with a constant speed is given by $\mu = \tan \theta - \dfrac{m_2}{m_c \cos \theta}$. (Use symbols, not numbers.)

EXPERIMENT 10

Torques, Equilibrium, and Center of Gravity

/TI/ *Advance Study Assignment*

Read the experiment and answer the following questions.

1. What conditions must be present for (a) translational equilibrium and (b) rotational equilibrium of a rigid body?

2. If these conditions for equilibrium are satisfied, is the rigid body necessarily in static equilibrium? Explain.

3. Write a definition and a mathematical expression for torque.

Don't forget units

(continued)

4. If torque is a vector, with specific direction in space, what is meant by clockwise and counterclockwise torques? If the sums of these torques on a rigid body are equal, what does this imply physically?

5. What defines the center of gravity of a rigid body, and how is it related to the center of mass?

6. Define the term *linear mass density*. Also, what is implied if it is assumed that the linear mass density of an object is constant?

Torques, Equilibrium, and Center of Gravity

INTRODUCTION AND OBJECTIVES

In introductory physics, forces act on "objects." That is, we consider an object as a particle, which generally responds linearly to a force. In reality, an object is an extended collection of particles, and where a force is applied makes a difference. Rotational motion becomes relevant when we analyze the motion of a solid extended object or a rigid body. A **rigid body** is an object or system of particles in which the distances between particles are fixed and remain constant. A quantity of liquid water is *not* a rigid body, but the ice that would form if the water were frozen is.

Actually, the concept of a rigid body is an idealization. In reality, the particles (atoms and molecules) of a solid vibrate constantly. Also, solids can undergo deformations. Even so, most solids can be considered to be rigid bodies for the purposes of analyzing rotational motion.

An important condition of rigid bodies in many practical applications is **static equilibrium.** Examples include beams in bridges and beam balances. When a rigid body such as a beam or a rod is "in balance," it is at rest, or in static equilibrium. In particular, the beam is in rotational static equilibrium; it does not rotate about some point or axis of rotation.

The criterion for rotational static equilibrium is that the sum of the torques, or moments of force acting on a rigid body, be equal to zero. To study torques and rotational equilibrium, we will use a "beam" balance in the form of a meterstick and suspended weights. The mass of an object will be determined experimentally by the "moment-of-force" method, and the experimental value will be compared to the mass of the object as measured on a laboratory balance. Also, the concepts of center of gravity and center of mass will be investigated.

After performing this experiment and analyzing the data, you should be able to:

1. Explain mechanical equilibrium and how it is applied to rigid bodies.
2. Distinguish between center of mass and center of gravity.
3. Describe how a laboratory beam balance measures mass.

EQUIPMENT NEEDED

- Meterstick
- Support stand
- Laboratory balance
- String and one knife-edge clamp *or* four knife-edge clamps (three with wire hangers)
- Four hooked weights (50 g, two 100 g, and 200 g)
- Unknown mass with hook

THEORY

A. Equilibrium

The conditions for the mechanical equilibrium of a rigid body are

$$\Sigma \mathbf{F} = 0 \tag{1a}$$

$$\Sigma \boldsymbol{\tau} = 0 \tag{1b}$$

That is, the (vector) sums of the forces $\mathbf{F}$ and torques $\boldsymbol{\tau}$ acting on the body are zero.

The first condition, $\Sigma \mathbf{F} = 0$, is concerned with **translational equilibrium** and ensures that the object is at a particular location (not moving linearly) or that it is moving with a uniform linear velocity (Newton's first law of motion). In this experiment, the rigid body (the meterstick) is restricted from linear motion, so this is not a consideration.

To be in static equilibrium, a rigid body must also be in rotational static equilibrium. Although the sum of the forces on the object may be zero and it is not moving linearly, it is possible that it may be rotating about some fixed axis of rotation. However, if the sum of the torques is zero, $\Sigma \boldsymbol{\tau} = 0$, the object is in **rotational equilibrium,** and either it does not rotate (static case) or it rotates with a uniform angular velocity. (Forces produce linear motion, and torques produce rotational motion.)

A **torque** (or moment of force) results from the application of a force acting at a distance from an axis of rotation (● Fig. 1). The magnitude of the torque is equal to the product of the force's magnitude F and the perpendicular

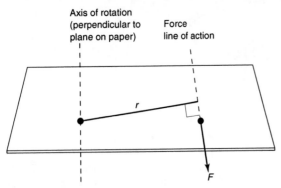

Figure 1   Torque. The magnitude of a torque is equal to the product of the magnitude of the force F and the perpendicular distance (lever arm) r from the axis of rotation to the force's line of action; that is, $\tau = rF$.

distance r from the axis of rotation to the force's line of action (a straight line through the force vector arrow). That is,

$$\tau = rF \qquad (2)$$

The perpendicular distance r is called the **lever arm** or **moment arm**. The unit of torque can be seen to be the meter-newton (m-N). Notice that these units are the same as those of work, newton-meter (N-m) = joule (J). The unit of torque is usually written meter-newton (m-N) to emphasize the distinction.

Torque is a vector quantity that points along the axis of rotation that is perpendicular to the plane of the **r** and **F** vectors. By convention, if the torque tends to rotate the body in a counterclockwise direction (as viewed from above), then the torque is positive and the torque vector points toward you along the axis of rotation. If the torque tends to rotate the body in a clockwise direction, then the torque is negative and the vector points away from you.

For example, in ● Fig. 2, taking the axis of rotation at the 50-cm position, F_1 and F_2 produce counterclockwise torques and F_3 and F_4 produce clockwise torques, but no

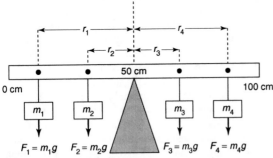

Figure 2   Torques in different directions. The forces F_1 and F_2 give rise to counterclockwise torques, and F_3 and F_4 give rise to clockwise torques, as applied to the meterstick.

rotation takes place if the torques are balanced and the system is in rotational static equilibrium.

It is convenient to sum the torques using magnitudes and directional signs, as determined by the counterclockwise (cc) and clockwise (cw) convention. In this case, the condition for rotational equilibrium (Eq. 1b) becomes

$$\Sigma \tau_{cw} - \Sigma \tau_{cc} = 0$$

or

$$\Sigma \tau_{cc} = \Sigma \tau_{cw} \qquad (3)$$

(sum of counterclockwise torques =
                                    sum of clockwise torques.)

Hence, we may simply equate the magnitudes of the cc and cw torques. For example, for the meterstick in Fig. 2, we have

	Counterclockwise	Clockwise
	$\tau_1 + \tau_2$	= $\tau_3 + \tau_4$

or

$$F_1 r_1 + F_2 r_2 = F_3 r_3 + F_4 r_4$$

The forces are due to weights suspended from the rod, and with $F = mg$,

$$m_1 g r_1 + m_2 g r_2 = m_3 g r_3 + m_4 g r_4 \qquad (4)$$

and, canceling g,

$$m_1 r_1 + m_2 r_2 = m_3 r_3 + m_4 r_4$$

Example 1   Let $m_1 = m_3 = 50$ g, $m_2 = m_4 = 100$ g in Fig. 2, where m_1, m_2, and m_3 are at the 10-, 40-, and 60-cm marks or positions, respectively, on the meterstick*. Where would m_4 have to be suspended for the stick to be in static equilibrium?

Solution   In static equilibrium, the sum of the torques is zero, or the sum of the counterclockwise torques is equal to the sum of the clockwise torques (Eq. 3),

$$\Sigma \tau_{cc} = \Sigma \tau_{cw}$$

In terms of forces and lever arms, we have (writing the forces first)

$$F_1 r_1 + F_2 r_2 = F_3 r_3 + F_4 r_4$$

where the forces are $F_i = m_i g$. The lever arms are measured from the 50-cm position of the meterstick, which is the pivot point, or the location of the axis of rotation. In gen-

---

* The official abbreviation for gram is g, and the commonly used symbol for acceleration due to gravity is g. The gravity g is written in italics, and the gram g is not. Look closely to avoid confusion.

eral, $r_i = (50 \text{ cm} - x_i)$, where x_i is the centimeter location of a mass. Hence,

$$m_1 g(50 \text{ cm} - 10 \text{ cm}) + m_2 g(50 \text{ cm} - 40 \text{ cm})$$
$$= m_3 g(60 \text{ cm} - 50 \text{ cm}) + m_4 g r_4$$

and, canceling the gs,

$$m_1(40 \text{ cm}) + m_2(10 \text{ cm}) = m_3(10 \text{ cm}) + m_4 r_4$$

Then, putting in the mass values,

$$(50 \text{ g})(40 \text{ cm}) + (100 \text{ g})(10 \text{ cm})$$
$$= (50 \text{ g})(10 \text{ cm}) + (100 \text{ g})r_4$$

and solving for r_4,

$$r_4 = \frac{2500 \text{ g-cm}}{100 \text{ g}} = 25 \text{ cm}$$

Hence, for rotational equilibrium m_4 is 25 cm from the support position (axis of rotation), or at the 75-cm position on the meterstick (measured from the zero end).

Here it is assumed that the meterstick is uniform (uniform mass distribution) so that the torques caused by the masses of the portions of the meterstick are the same on both sides of the support and therefore cancel.

B. Center of Gravity and Center of Mass

The gravitational torques due to "individual" mass particles of a rigid body define what is known as the body's center of gravity. The **center of gravity** is the "balance" point, the point of the body about which the sum of the gravitational torques about an axis through this point is zero. For example, consider the meterstick shown in ● Fig. 3. If the uniform meterstick is visualized as being made up of individual mass particles and the point of support is selected such that $\Sigma\tau = 0$, then

$$\Sigma\tau_{cc} = \Sigma\tau_{cw}$$

or

$$\sum_{cc}(m_i g)r_i = \sum_{cw}(m_i g)r_i$$

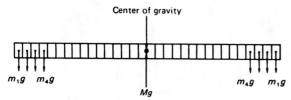

Center of gravity

Figure 3   Center of gravity. A rod may be considered to be made up of individual masses in rotational equilibrium when the vertical support is directly through the center of gravity.

and

$$(m_1 r_1 + m_2 r_2 + m_3 r_3 + \cdots)_{cc}$$
$$= (m_1 r_1 + m_2 r_2 + m_3 r_3 + \cdots)_{cw}$$

where g cancels. When the meterstick is in equilibrium, it is supported by a force equal to its weight, and the support force is directed through the center of gravity. Hence, it is as though all of the object's weight (Mg) is concentrated at the center of gravity. That is, if you were blindfolded and supported an object at its center of gravity on your finger, weightwise you would not be able to tell, from its weight alone, whether it was a rod or a block or an irregularly shaped object of equal mass. For a uniform meterstick, the center of gravity would be at the 50-cm position. (Why?)

If an object's weight is concentrated at its center of gravity, so should its mass be concentrated there, and we often refer to an object's **center of mass** instead of its center of gravity. *These points are the same as long as the acceleration due to gravity g is constant* (uniform gravitational field). Notice how g can be factored and divided out of the previous *weight* equations, leaving *mass* equations.

Also, it should be evident that for a symmetric object with a uniform mass distribution, the center of gravity and center of mass are located at the center of symmetry. For example, if a rod has a uniform mass distribution, its center of gravity is located at the center of the rod's length. For a uniform sphere, the centers are at the center of the sphere.

LINEAR MASS DENSITY

In part of the experiment, the masses of certain lengths of the meterstick will need to be known. These may be obtained from the **linear mass density** μ of the stick—that is, the mass per unit length

$$\boxed{\mu = \frac{m}{L}} \tag{5}$$

with units of grams/centimeter or kilograms/meter. For example, suppose a meterstick is measured to have a mass of 50 g on a balance. Then, since the stick is 100 cm long ($L = 100$ cm), the linear mass density of the stick is $\mu = m/L = 50 \text{ g}/100 \text{ cm} = 0.50 \text{ g/cm}$. If the mass distribution of the stick were uniform, then every centimeter would have a mass of 0.50 g. However, metersticks are not uniform, so this is an average value.

Example 2   If a meterstick has a linear mass density of 0.50 g/cm, what is the mass of a 16-cm length of the stick?

Solution Since $\mu = m/L$, we have $m = \mu L$, and for $\mu = 0.50$ g/cm and $L = 16$ cm,

$$m = \mu L = (0.50 \text{ g/cm})(16 \text{ cm}) = 8.0 \text{ g}$$

EXPERIMENTAL PROCEDURE

A. *Apparatus with Support Point at Center of Gravity*

1. A general experimental setup is illustrated in ● Fig. 4, where the masses or weights are suspended by clamp weight hangers. The hooked masses may also be suspended from small loops of string, which can be slid easily along the meterstick. The string allows the position of a mass to be read easily and may be held in place by a small piece of masking tape.
   (a) Determine the mass of the meterstick (without any clamps) and record it in the Laboratory Report.
   (b) Weights may be suspended by loops of string or clamps with weight hangers. The string method is simpler; however, if you choose or are instructed to use weight hangers, weigh the three clamps together on a laboratory balance and compute the average mass of a clamp. Record it in the Laboratory Report.

2. With a knife-edge clamp on the meterstick near its center, place the meterstick (without any suspended weights) on the support stand. Make certain that the knife edges are on the support stand. (The tightening screw head on the clamp will be down.)

   Adjust the meterstick through the clamp until the stick is balanced on the stand. Tighten the clamp screw, and record in Data Table 1 the meterstick reading or the distance of the balancing point x_0 from the zero end of the meterstick.

3. *Case 1: Two known masses.*
   (a) With the meterstick on the support stand at x_0, suspend a mass $m_1 = 100$ g at the 15-cm position on the meterstick—that is, 15 cm from the zero end of the meterstick.
   (b) Set up the conditions for static equilibrium by adjusting the moment arm of a mass $m_2 = 200$ g suspended on the side of the meterstick opposite m_1.

Record the masses and moment arms in Data Table 1. If clamps are used instead of string, do not forget to add the masses of the clamps. Remember the moment arms are the distances from the pivot point to the masses (i.e., $r_i = |x_i - x_0|$).
   (c) Compute the torques and find the percent difference in the computed values (i.e., compare the clockwise torque with the counterclockwise torque).

4. *Case 2: Three known masses.*
   Case (a)
   (i) With the meterstick on the support stand at x_0, suspend $m_1 = 100$ g at the 30-cm position and $m_2 = 200$ g at the 70-cm position. Suspend $m_3 = 50$ g and adjust the moment arm of this mass so that the meterstick is in static equilibrium. Record the data in Data Table 1.
   (ii) Compute the torques and compare as in procedure 3.
   Case (b)
   (i) Calculate theoretically the lever arm (r_3) for the mass $m_3 = 50$ g for the system to be in equilibrium if $m_1 = 100$ g is at the 20-cm position and $m_2 = 200$ g is at the 60-cm position. (Remember to add the masses of the hanger clamps if used.) Record this value in the data table.
   (ii) Check your results experimentally, and compute the percent error of the experimental value of r_3, taking the previously calculated value as the accepted value.

5. *Case 3: Unknown mass—The balance principle.* A balance (scale) essentially uses the method of moments to compare an unknown mass with a known mass. Some balances have constant and equal lever arms, and others do not (see Experiment 2, Fig. 2.1). This procedure will illustrate the balance principle.
   (a) With the meterstick on the support stand at x_0, suspend the unknown mass (m_1) near one end of the meterstick (for example, at the 10-cm position).

Figure 4   Torque apparatus.   Example of experimental setup and equilibrium conditions. (Courtesy of Sargent-Welch.)

Suspend from the other side of the meterstick an appropriate known countermass m_2 (for example, 200 g) and adjust its position until the meterstick is "in balance" or equilibrium. Record the value of the known mass and the moment arms in Data Table 1.

(b) Remove the unknown mass and determine its mass on a laboratory balance.

(c) Compute the value of the unknown mass by the method of moments and compare it with the measured value by calculating the percent error.

6. *Case 4: Instructor's choice (optional).* Your instructor may have a particular case he or she would like you to investigate. If so, the conditions will be given. Space has been provided in the data table for reporting your findings.

B. Apparatus Supported at Different Pivot Points

In the previous cases, the mass of the meterstick was not explicitly taken into account since the fulcrum or the position of the support was at the meterstick's center of gravity or center of mass. In effect, the torques due to the mass of the meterstick on either side of the support position canceled each other. The centers of gravity of the lengths of the stick on either side of the support are equidistant from the support (for example, at the 25-cm and 75-cm positions for a uniform stick) and have equal masses and moment arms.

For the following cases, the meterstick will not be supported at its center-of-gravity position (x_o) but at some other pivot points (designated in general by x'_o; for example, see ● Fig. 5). In these cases, the mass of the meterstick needs to be taken into account. To illustrate this very vividly, let's start off with a case with only one suspended mass.

7. *Case 5: Meterstick with one mass.* Suspend a mass $m_1 = 100$ g at or near the zero end of the meterstick (Fig. 5). Record the mass position x_1 in Data Table 2. If a string loop is used, a piece of tape to hold

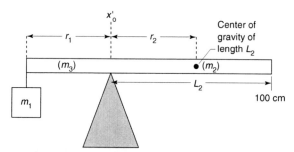

Figure 5 Equilibrium. A meterstick in equilibrium with one suspended mass. See text for description.

the string in position helps. Move the meterstick in the support clamp until the system is in equilibrium. (This case is analogous to the solitary seesaw—sitting on one side of a balanced seesaw with no one on the other side.) Record the support position x'_o in Data Table 2.

Since the meterstick is in balance (static equilibrium), the point of support must be at the center of gravity of the system; that is, the torques (clockwise and counterclockwise) on either side of the meterstick must be equal. But where is the mass or force on the side of the meterstick opposite the suspended mass? The balancing torque must be due to the mass of length L_2 of the meterstick (Fig. 5). To investigate this:

(a) Using the total mass m of the meterstick (measured previously) as m_2, with a moment arm r_2 (see the diagram in Data Table 2), compute the counterclockwise and clockwise torques, and compare them by computing the percent difference. Record it in Data Table 2.

(b) Now the masses of the lengths of meterstick will be taken into account. Compute the average linear mass density of the meterstick (see Theory, Section B) and record it in the data table.

If we assume that the mass of the meterstick is uniformly distributed, the center of mass (or center of gravity) of the length of meterstick L_2 on the *opposite* side of the support from m_1 is at its center position (see Fig. 5). Compute the mass m_2 of this length of stick (see Example 2) and record. Also, record the center position of L_2, where this mass is considered concentrated (x_2), and find the length of the lever arm r_2. It should be evident that $r_2 = L_2/2$.

Compute the torque due to m_2 and record it as τ_{cw}. From the linear mass density compute the m_3 of the portion of the meterstick remaining to the left of the pivot. Calculate the torque due to this portion of the meterstick, add it to the torque due to mass m_1 to find the total counterclockwise torque, and record it as τ_{cc}. Compare the torque *differences* with those found in Case 5(a).

8. *Case 6: Center of gravity.*

(a) With a mass $m_1 = 100$ g positioned at or near one end of the meterstick as in Case 5, suspend a mass $m_2 = 100$ g on the opposite side of the support stand at the 60-cm position. Adjust the meterstick in the support-stand clamp until the stick is in balance. This locates the center of gravity x'_o of the system. Record in Data Table 2 and find r_1 and r_2.

(b) Repeat the procedure with m_2 positioned at 70 cm.

(c) Repeat the procedure with m_2 positioned at 80 cm. Notice how the position of the center of gravity moves as the mass distribution is varied.

(d) Based on the experimental data, what would you predict the position of the center of gravity x_o' of the system would be if m_2 were moved to the 90-cm position? Record your prediction in the data table.

Using your prediction, compute the counterclockwise and clockwise torques, taking into account the mass of the meterstick as in procedure 7(c). Compare the torques by computing the percent difference.

Experimentally determine the position of the center of gravity of the system, and compute the percent difference between the experimental and predicted values.

Name _____ Section _____ Date _____

Lab Partner(s) _____

Torques, Equilibrium, and Center of Gravity

/TI/ *Laboratory Report*

A. Apparatus with Point of Support at Center of Gravity

Mass of meterstick _____   Total mass of clamps _____

Average mass of one clamp m_c _____

Balancing position (center of gravity) of meterstick x_o _____

DATA TABLE 1

Diagram*	Values (add m_c to masses if clamps used)		Moment (lever) arms	Results[†]
Case 1	m_1 _____ $x_1 = 15$ cm		r_1 _____	τ_{cc} _____
	m_2 _____ x_2 _____		r_2 _____	τ_{cw} _____
				Percent diff. _____
Case 2(a)	m_1 _____ $x_1 = 30$ cm		r_1 _____	τ_{cc} _____
	m_2 _____ $x_2 = 70$ cm		r_2 _____	τ_{cw} _____
	m_3 _____ x_3 _____		r_3 _____	Percent diff. _____
Case 2(b)	m_1 _____ $x_1 = 20$ cm		r_1 _____	r_3 _____ (calculated)
	m_2 _____ $x_2 = 60$ cm		r_2 _____	
	m_3 _____ x_3 _____			r_3 _____ (measured)
				Percent error _____

* Draw a diagram to illustrate each case, using the Case 1 diagram as an example.
[†] Attach a sheet to the Laboratory Report showing calculations for each use.

Don't forget units

(continued)

Diagram*	Values (add m_c to masses if clamps are used)	Moment (lever) arms	Results[†]
Case 3	x_1 —————— (known) m_2 —————— (known) x_2 —————— (from expt.)	r_1 —————— r_2 ——————	m_1 —————— (measured) m_1 —————— (calculated) Percent error ——————
Case 4 (instructor's option)			

* Draw a diagram to illustrate each case, using the Case 1 diagram as an example.
† Attach a sheet to the Laboratory Report showing calculations for each use.

B. Apparatus Supported at Different Pivot Points

DATA TABLE 2　　　　　　　　　　Linear mass density of meterstick, $\mu = m/L$ ——————————

Diagram*	Values (add m_c if applicable)	Moment (lever) arms	Results[†]
Case 5(a)	m_1 —————— m_2 —————— x_1 —————— x_2 —————— x'_0 ——————	r_1 —————— r_2 ——————	τ_{cc} —————— τ_{cw} —————— Torque differences (show below table)
Case 5(b)	m_1 —————— m_2 —————— m_3 —————— x_1 —————— x_2 —————— x_3 —————— x'_0 ——————	r_1 —————— r_2 —————— r_3 ——————	τ_{cc} —————— τ_{cw} —————— Torque differences (show below table)

* Draw a diagram to illustrate each case, using the Case 5(a) diagram as an example. Put the mass of a length of stick in parentheses as in that diagram.
† Attach a sheet to the Laboratory Report showing calculations for each use.

EXPERIMENT 10   *Laboratory Report*

Diagram*	Values (add m_c if applicable)	Moment (lever) arms	Results†
Case 6(a)	m_1 _____ $x_1 = 0$ cm m_2 _____ $x_2 = 60$ cm x'_o _____	r_1 _____ r_2 _____	
Case 6(b)	same except $x_2 = 70$ cm x'_o _____	r_1 _____ r_2 _____	
Case 6(c)	same except $x_2 = 80$ cm x'_o _____	r_1 _____ r_2 _____	
Case 6(d)	same except $x_2 = 90$ cm x'_o _____ (predicted)	τ_{cc} _____ τ_{cw} _____ Percent diff. _____	x'_o _____ (measured) Percent diff. _____

* Draw a diagram to illustrate each case, using the Case 5(a) diagram as an example. Put the mass of a length of stick in parentheses as in that diagram.
† Attach a sheet to the Laboratory Report showing calculations for each use.

/TI/ QUESTIONS (Answer those that are applicable.)

1. Explain how the condition $\Sigma \mathbf{F} = 0$ is satisfied for the meterstick in part A of the experiment.

(continued)

2. Why are clockwise and counterclockwise referred to as "senses," rather than directions?

3. Suppose in a situation like Case 2(a) in the experiment, m_1 = 200 g were at the 20-cm position and m_2 = 100 g at the 65-cm position. Would there be a problem in experimentally balancing the system with m_3 = 50 g? Explain. If so, how might the problem be resolved?

4. Describe the effects of taking the mass of the meterstick into account when the balancing position is not near the 50-cm position.

5. *(Optional)* A uniform meterstick is in static rotational equilibrium when a mass of 220 g is suspended from the 5.0-cm mark, a mass of 120 g is suspended from the 90-cm mark, and the support stand is placed at the 40-cm mark. What is the mass of the meterstick?

Name _____ Section _____ Date _____

Lab Partner(s) _____

EXPERIMENT 11

Simple Harmonic Motion

/TI/ *Advance Study Assignment*

Read the experiment and answer the following questions.

1. What are Hooke's law and simple harmonic motion, and how are they related?

2. What is the physical significance of the spring constant? What does it tell you?

3. How is the spring constant of a spring determined in this experiment?

4. In the equation of motion for simple harmonic motion (Eq. TI 2), what physically determines A and T?

(continued)

5. How is the period of a mass oscillating on a spring related to the spring constant? (Express your answer mathematically and verbally.)

◧ *Advance Study Assignment*

Read the experiment and answer the following questions.

1. What are the requirements for an object to move with simple harmonic motion?

2. Why is simple harmonic motion an idealization?

3. What is a simple pendulum?

4. Under what conditions can a pendulum be considered a simple harmonic oscillator?

5. Why is it important to start taking data when the pendulum is still at rest in its equilibrium position?

Simple Harmonic Motion

OVERVIEW

Experiment   considers simple harmonic motion (SHM) with TI and/or CI procedures. The TI procedure examines Hooke's law, using rubber-band and spring elongations. SHM is investigated through the period of oscillation of a mass on a spring.

The CI procedure investigates the SHM of a simple pendulum and the resulting conversion of energy (kinetic and potential) that occurs during the motion. An electronic sensor measures the angular speed, $\omega = \Delta\theta/\Delta t$, of the pendulum, from which the tangential speed is computed and the energies calculated.

INTRODUCTION AND OBJECTIVES

An elastic material tends to return to its original form or shape after being deformed. Hence, elasticity implies a restoring force that can give rise to vibrations or oscillations. For many elastic materials, the restoring force is proportional to the amount of deformation, if the deformation is not too great.

This is best seen for a coil spring. The restoring force F exerted by a stretched (or compressed) spring is proportional to the stretching (compressing) distance x, or $F \propto x$. In equation form, we have what is known as **Hooke's law,**

$$F = -kx$$

where x is the displacement of one end of the spring from its unstretched ($x = 0$) position, k is a positive constant of proportionality, and the minus sign indicates that the displacement and force are in opposite directions. The constant k is called the **spring constant** and is a relative indication of the "stiffness" of the spring.

A particle or object in motion under the influence of a linear restoring force such as that described by Hooke's law undergoes what is known as **simple harmonic motion (SHM).** This periodic oscillatory motion is one of the common types found in nature. The period of oscillation of an object in simple harmonic motion is related to the constant of proportionality in Hooke's law.

In this experiment, Hooke's law will be investigated, along with the parameters and description of simple harmonic motion.

After performing this experiment and analyzing the data, you should be able to:

TI OBJECTIVES

1. Tell how Hooke's law is represented graphically, and cite an example of an elastic object that does *not* follow Hooke's law.
2. Explain why simple harmonic motion (SHM) is simple and harmonic.
3. Better understand how the period of a weight oscillating on a spring varies with the weight's mass and the spring constant.

CI OBJECTIVES

1. Explain the energy conversion that happens during the simple harmonic motion of a pendulum.
2. Experimentally verify the law of conservation of mechanical energy.

Simple Harmonic Motion

- Coil spring
- Wide rubber band
- Slotted weights and weight hanger

- Laboratory timer or stopwatch
- Meterstick
- Laboratory balance
- 2 sheets of Cartesian graph paper

TI THEORY

A. Hooke's Law

The fact that for many elastic substances the restoring force that resists the deformation is directly proportional to the deformation was first demonstrated by Robert Hooke (1635–1703), an English physicist and contemporary of Isaac Newton. For one dimension, this relationship—known as Hooke's law—is expressed mathematically as

or

$$F = -k\Delta x = -k(x - x_0) \qquad \textbf{(TI 1)}$$

$$F = -kx \qquad (\text{with } x_0 = 0)$$

where Δx is the linear deformation or displacement of the spring and x_0 is its initial position (using sign convention for vectors). The minus sign indicates that the force and displacement are in opposite directions. (For experimental convenience, the minus sign may be neglected.)

For coil springs, the constant k, called the **spring constant,** depends on the shear modulus of the wire, the radius of the wire, the radius of the coils, and the number of coils.

The spring constant is sometimes called the "stiffness constant," since it gives an indication of the relative stiffness of a spring—the greater the k, the greater the stiffness. As can be seen from TI Eq. 1, k may have units of N/m or lb/in.

According to Hooke's law, the elongation of a spring as a whole is directly proportional to the magnitude of the stretching force.* For example, as illustrated in ● TI Fig. 1, if a spring has an initial length y_0, and a suspended weight of mass m stretches the spring so that its length is y_1, then in equilibrium the weight force is balanced by the spring force and

$$F_1 = mg = k(y_1 - y_0)$$

---

* The restoring spring force and the stretching force are equal in magnitude and opposite in direction (Newton's third law).

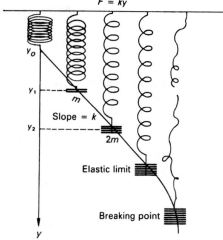

TI Figure 1   Hooke's law. An illustration in graphical form of spring elongation versus force. The greater the force, the greater the elongation, $F = -ky$. This Hooke's law relationship holds up to the elastic limit.

Here we use y to indicate the vertical direction instead of x as in TI Eq. 1 which is usually used to mean the horizontal direction. Similarly, if another mass m is added and the spring is stretched to a length y_2, then

$$F_2 = 2mg = k(y_2 - y_0)$$

and so on for more added weights. The linear relationship of Hooke's law holds, provided that the deformation or elongation is not too great. Beyond the elastic limit, a spring is permanently deformed and eventually breaks with increasing force.

Notice that Hooke's law has the form of an equation for a straight line:

$$F = k(y - y_0)$$

or

$$F = ky - ky_0$$

which is of the general form $y = x + b$

B. Simple Harmonic Motion

When the motion of an object is repeated in regular time intervals or periods, it is called **periodic motion.** Examples include the oscillations of a pendulum with a path back and forth along a circular arc and a mass oscillating linearly up and down on a spring. The latter is under the influence of the type of force described by Hooke's law, and its motion is called **simple harmonic motion (SHM)**—simple because the restoring force has the simplest form and harmonic because the motion can be described by *harmonic functions* (sines and cosines).

As illustrated in ● TI Fig. 2, a mass oscillating on a spring would trace out a wavy, time-varying curve on a moving roll of paper. The equation for this curve, which describes the oscillatory motion of the mass, can be written

$$y = A \cos \frac{2\pi t}{T} \qquad \textbf{(TI 2)}$$

where T is the period of oscillation and A is the amplitude or maximum displacement of the mass.

The amplitude A depends on the initial conditions of the system (i.e., how far the mass was initially displaced from its equilibrium position). If the mass were initially ($t = 0$) pulled below its equilibrium position (to $y = -A$) and released, the equation of motion would be $y = -A \cos 2\pi t/T$, which satisfies the initial condition at $t = 0$, that is, $\cos 0 = 1$ and $y = -A$. The argument of the cosine (i.e., $2\pi t/T$) is in radians rather than degrees.

In actual practice, the amplitude decreases slowly as energy is lost to friction, and the oscillatory motion is slowly "damped." In some applications, the simple harmonic motion of an object is intentionally damped (e.g., the spring-loaded needle indicator of an electrical measurement instrument or the dial on a common bathroom scale). Otherwise, the needle or dial would oscillate about the equilibrium position for some time, making it difficult to obtain a quick and accurate reading.

The period of oscillation depends on the parameters of the system and, for a mass on a spring, is given by

$$T = 2\pi \sqrt{\frac{m}{k}} \qquad \textbf{(TI 3)}$$

(period of mass oscillating on a spring)

/TI/ EXPERIMENTAL PROCEDURE

A. Rubber-Band Elongation

1. Hang a rubber band on a support and suspend a weight hanger from the rubber band. Add an appropriate weight to the weight hanger (e.g., 100–300 g) and record the total suspended weight ($m_1 g$) in TI Data Table 1. [It is convenient to leave g the acceleration due to gravity in symbolic form; that is, if $m_1 = 100$ g or 0.100 kg, then weight $= m_1 g = (0.100$ kg$)g$ N.]* Fix a meterstick vertically alongside the weight hanger

---

* Be careful not to confuse the symbol for the acceleration due to gravity, g (italic), and the abbreviation for gram, g.

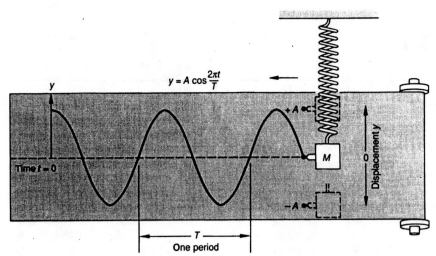

TI Figure 2   Simple harmonic motion.   A marker on a mass oscillating on a spring traces out a curve, as illustrated, on the moving paper. The curve may be represented as a function of displacement (magnitude y) versus time—that is, as $y = A \cos 2\pi t/T$, where $y = A$ at $t = 0$.

and note the position of the bottom of the weight hanger on the meterstick. Record this as y_1 in the data table.

2. Add appropriate weights (e.g., 100 g) to the weight hanger one at a time, and record the total suspended weight and the position of the bottom of the weight hanger on the meterstick after each elongation (y_2, y_3, etc.). The weights should be small enough so that seven or eight weights can be added without overstretching the rubber band.

3. Plot the total suspended weight force versus elongation position (*mg* versus *y*), and draw a smooth curve that best fits the data points.

B. Spring Elongation

4. Repeat procedures 1 and 2 for a coil spring and record the results in TI Data Table 2. A commercially available Hooke's law apparatus is shown in ● TI Fig. 3. Choose appropriate mass increments for the spring stiffness.

5. Plot *mg* versus *y* on the same sheet of graph paper used in procedure 3 (double-label axes if necessary) and draw a straight line that best fits the data. Determine the slope of the line (the spring constant *k*) and record it in the data table. Answer TI Questions 1 through 3 following the data tables.

C. Period of Oscillation

6. **(a)** On the weight hanger suspended from the spring, place a mass just great enough to prevent the spring from oscillating too fast and to prevent the hanger from moving relative to the end of the spring during oscillations when it is pulled down (e.g., 5 to 10 cm) and released. Record the total mass in TI Data Table 3.

   (b) Using a laboratory timer or stopwatch, release the spring weight hanger from the predetermined initial displacement and determine the time it takes for the mass to make a number (5 to 10) of complete oscillations or cycles. The number of cycles timed will depend on how quickly the system loses energy or is damped. Make an effort to time enough cycles to get a good average period of oscillation. Record in the data table the total time and the number of oscillations.

TI Figure 3 Hooke's law apparatus. The variables of Hooke's law ($F = mg$ and x) are measured using spring elongation. (Courtesy of Sargent-Welch.)

Divide the total time by the number of oscillations to determine the average period.

7. Repeat procedure 6 for four more mass values, each of which is several times larger than the smallest mass, and record the results in TI Data Table 3. The initial displacement may be varied if necessary. (This should have no effect on the period. Why?)

8. Plot a graph of the average period squared (T^2) versus the mass (*m*) and draw a straight line that best fits the data points. Determine the slope of the line and compute the spring constant *k*. [Note from TI Eq. 3 that *k* is not simply equal to the slope; rather, $k = (2\pi)^2/\text{slope}$.] Compare this value of *k* with that determined from the slope of the spring elongation graph in part B by computing the percent difference, and finish answering the TI Questions.

T I   EXPERIMENT 11

Simple Harmonic Motion

TI/ *Laboratory Report*

A. Rubber-Band Elongation

TI/ **DATA TABLE 1**

	Total suspended weight* ( )		Scale reading ( )
$m_1 g$		y_1	
$m_2 g$		y_2	
$m_3 g$		y_3	
$m_4 g$		y_4	
$m_5 g$		y_5	
$m_6 g$		y_6	
$m_7 g$		y_7	
$m_8 g$		y_8	

B. Spring Elongation

TI/ **DATA TABLE 2**

	Total suspended weight* ( )		Scale reading ( )
$m_1 g$		y_1	
$m_2 g$		y_2	
$m_3 g$		y_3	
$m_4 g$		y_4	
$m_5 g$		y_5	
$m_6 g$		y_6	
$m_7 g$		y_7	
$m_8 g$		y_8	

* It is convenient to leave g in symbol form, even when graphing.

Calculations
(show work)

k (slope of graph) _____

(units)

Don't forget units

(continued)

C. Period of Oscillation

/TI/ **DATA TABLE 3**

	Total suspended mass ( )	Total time ( )	Number of oscillations	Average period T ( )	T^2 ( )
m_1					
m_2					
m_3					
m_4					
m_5					

Calculations
(show work)

Slope of graph _____

Computed spring constant k _____

Percent difference (of k's in B and C) _____

/TI/ QUESTIONS

1. Interpret the intercepts of the straight line for the spring elongation in the mg-versus-y graph of part B.

2. Is the elastic property of the rubber band an example of Hooke's law? Explain.

3. Draw a horizontal line through the *y*-intercept of the straight-line graph of part B, and form
   a triangle by drawing a vertical line through the last data point.
   (a) Prove that the area of the triangle is the work done in stretching the spring.
      (*Hint:* $W = \frac{1}{2}kx^2$, and area of triangle $A = \frac{1}{2}ab$, or $\frac{1}{2}$ altitude times base.)

   (b) From the graph, compute the work done in stretching the spring.

4. Interpret the *x*-intercept of the straight line of the T^2-versus-*m* graph of part C.

(continued)

5. For a mass oscillating on a spring, at what positions do the (a) velocity and (b) acceleration of the mass have maximum values?

6. What is the form of the equation of motion for the SHM of a mass suspended on a spring when the mass is initially (a) released 10 cm above the equilibrium position; (b) given an upward push from the equilibrium position, so that it undergoes a maximum displacement of 8 cm; (c) given a downward push from the equilibrium position, so that it undergoes a maximum displacement of 12 cm? (*Hint:* Sketch the curve for the motion as in TI Fig. 2 and fit the appropriate trigonometric function to the curve.)

7. For case (a) in Question 6 only, what is the displacement y of the mass at times (a) $t = T/2$; (b) $t = 3T/2$; (c) $t = 3T$?

CI EXPERIMENT 11

CI EQUIPMENT NEEDED

- Rotary Motion Sensor (PASCO CI-6538)
- Mini-rotational accessory (PASCO CI-6691. This set

includes a brass mass and a light rod to make the pendulum.)
- Support rods and clamps.

CI THEORY

In this experiment, the simple harmonic motion of a pendulum will be investigated by examining the energy conversions that occur during the motion.

Simple harmonic motion is the motion executed by an object of mass m subject to two conditions:

- The object is subject to a force that is proportional to the displacement of the object that attempts to restore the object to its equilibrium position.
- No dissipative forces act during the motion, so there is no energy loss.

Notice that as it is described in theory, simple harmonic motion is an idealization because of the assumption of no frictional forces acting on the particle.

In this experiment, the simple harmonic motion of a pendulum will be investigated. A simple pendulum consists of a mass (called a bob) suspended by a "massless" string from a point of support. The pendulum swings in a plane.

The restoring force on a simple pendulum is the component of its weight that tends to move the pendulum back to its equilibrium position. As shown in ● CI Figure 1, that force is

$$F = mg \sin \theta \qquad \text{(CI 1)}$$

Note, however, that this force is not proportional to the angular displacement θ of the pendulum, as required for SHM, but is proportional to the $\sin \theta$ instead. A pendulum can be approximated to be in SHM motion only if the

angle θ small, in which case $\sin \theta \approx \theta$ (where θ is in radians). Thus

$$F = mg \sin \theta \approx mg \, \theta \qquad \text{(CI 2)}$$

Notice that in this approximation, the force is directly proportional to the displacement θ.

As the pendulum swings, kinetic energy is converted into potential energy as the pendulum rises. This potential energy is converted back to kinetic energy as the pendulum swings downward. The kinetic and potential energies of the pendulum at any moment during its motion can easily be determined. The kinetic energy of a pendulum of mass m moving with a linear speed v is given by

$$K = \tfrac{1}{2} mv^2 \qquad \text{(CI 3)}$$

The potential energy, measured with respect to the equilibrium position, depends on the height above the equilibrium at a particular time. That is,

$$U = mgh = mg(L - L \cos \theta) \qquad \text{(CI 4)}$$

(See ● CI Figure 2.)

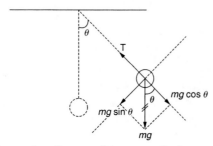

CI Figure 1   Forces acting on a swinging pendulum. The restoring force acting on a pendulum is the component $mg \sin \theta$ of gravity, which attempts to bring the pendulum back to the equilibrium position.

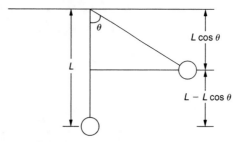

CI Figure 2   The elevation of a pendulum with respect to the equilibrium position. The elevation of a pendulum with respect to the equilibrium (lowest) position can be expressed in terms of L, the length of the pendulum, and of θ, as $L - L \cos \theta$. (The angular displacement has been exaggerated in the illustration. For simple harmonic motion, θ must be small.)

In this experiment, a sensor will keep track of the angular position, θ, of the pendulum as it swings. The sensor will also keep track of the angular speed, $\omega = \Delta\theta/\Delta t$, of the pendulum. The linear speed (v) can then be determined as $v = \omega L$, where L is the length of the pendulum and also the radius of the circular arc described by its motion. The kinetic and potential energies of the pendulum at any time can then be calculated.

BEFORE YOU BEGIN

1. Measure the mass of the pendulum bob (M) and record it in the laboratory report, in kilograms.
2. Measure the length of the pendulum (L), in meters, from the center of rotation to the center of the bob. Record it in the report.

This information will be needed during the setup of Data Studio.

SETTING UP DATA STUDIO

1. Open Data Studio and choose "Create Experiment."
2. From the sensor list, choose a Rotary Motion Sensor (RMS). Connect the sensor to the interface, as shown in the computer screen, to channels 1 and 2.
3. Double-click on the Rotary Motion Sensor icon. The Sensor Properties window will open. Select the following:
    Under General: set the sample rate to 20 Hz, Fast.
    Under Measurement: choose Angular Position (deg), Angular Position (rad), and Angular Velocity (rad/s); deselect any others.
    Under Rotary Motion Sensor: set the Divisions/Rotations to 1440, and set the Linear Calibration to Large Pulley (Groove).
Click OK to accept the choices and to close the Sensor Properties window.
4. Open the program's calculator by clicking on the Calculate button, on the top main menu. Usually a small version of the calculator opens, as shown in ● CI Fig. 3. Expand the calculator window by clicking on the button marked Experiment Constants.

5. The expanded window (shown in ● CI Fig. 4) is used to establish values of parameters that will remain constant throughout the experiment. In this case, these are the length of the pendulum (L) and the mass of the pendulum (M), which have already been measured. This is how to do it:
    a. Click on the lower New button (within the "Experiment Constants" section of the calculator window) and enter the name of the constant as L, the value as the length of the pendulum measured before, and the units as meters (m).
    b. Click the lower Accept button.
    c. Click on the New button again and enter the name of the constant as M, the value as the mass of the pendulum measured before, and the units as kilograms (kg).
    d. Click the lower Accept button.
    e. Close the experiment constants portion of the calculator window by pressing the button marked Experiment Constants again.

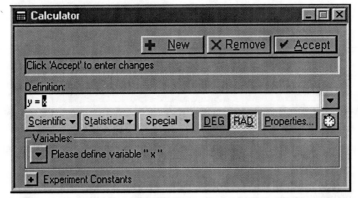

CI Figure 3   The Calculator Window.   This small version of the calculator window opens when the Calculate button is pressed. The calculator will be used to enter equations that handle the values measured by the sensor. The computer will perform the calculations automatically as the sensor takes data. (Data displayed using DataStudio Software. Reprinted courtesy of PASCO scientific.)

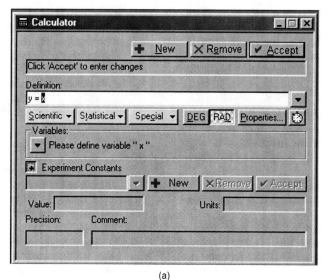

(a)

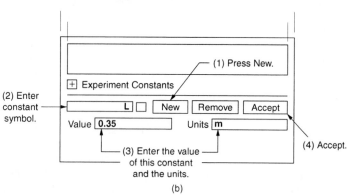

(b)

CI Figure 4   The expanded calculator window.   (a) After the button marked the Experiment Constants is pressed, the calculator window expands to full size. (b) The "Experiment Constants" section is the lower part of the expanded calculator window. This section is used to define parameters that are to remain constant during the experiment. The diagram shows the steps needed to enter experimental constants into the calculator. (Data displayed using DataStudio Software. Reprinted courtesy of PASCO scientific.)

7. **Calculation of the linear speed:**
   a. In the same calculator window, clear the definition box and enter the following equation:

   $$V = L * \text{smooth}(6, w)$$

   This is the calculation of the linear speed $v = \omega L$, which will be called V. Note that we are multiplying the length L of the pendulum by the angular speed, which we are calling w here. The smooth function is to produce a sharper graph.
   b. Press the Accept button after entering the formula. The variables L and w will appear in a list. L will

have the value defined before, but w will be waiting to be defined.
   c. To define the variable w, click on the drop menu button on the left side of the variable. A list of options will show, asking what type of variable this is.
   • Define w as a Data Measurement and, when prompted, choose Angular Velocity (rad/s).

8. **Calculation of the kinetic energy:**
   a. Still in the same calculator window, press the New button again to enter a new equation.
   b. Clear the definition box and enter the following equation: KE = 0.5 * M * v^2.   This   is   the

calculation of the kinetic energy $K = \frac{1}{2}mv^2$, that will be called KE.

   c. Press the <u>Accept</u> button after entering the formula. The variables M and v will appear in a list; M is the value entered before for the mass, and v is waiting to be defined.

   d. To define the variable v, click on the drop menu button on the left side of the variable. The list of options will show, asking what type of variable this is.

     • Define v as a Data Measurement and, when prompted, choose V, the equation defined previously.

9. **Calculation of the potential energy:**

   a. Press the <u>New</u> button once again to enter a new equation.

   b. Clear the definition box and enter the following equation: PE = M * 9.81 * (L − L* cos (smooth (6, x))). This is the calculation of the potential energy $U = mgh = mg(L - L \cos \theta)$, which will be called PE. Note that M is the mass, 9.81 is the value of g, and the variable x in this formula will stand for the angular position θ of the pendulum, in radians.

   c. Press the <u>Accept</u> button after entering the equation. The variables M, L, and x will appear in a list, with x waiting to be defined.

   d. Define x as a Data Measurement and, when prompted, choose Angular Position (rad). *Note:* Make sure that the choice is the one in radians, not the one in degrees!

   e. Press the <u>Accept</u> button.

10. Close the calculator window.

11. The data list on the upper left of the screen should now include icons for the three quantities that are calculated: V, KE and PE. A small calculator icon will show on the left of the calculated data quantities.

12. Create a graph by dragging the Angular Position (deg) data icon and dropping it on top of the "Graph" icon on the displays list. A graph of angular position (deg) versus time will open. The window will be called Graph 1.

13. Drag the KE equation icon and drop it somewhere on top of the graph created in Step 12. The graph will then split in two, with the graph of angular position versus time on top and the graph of KE versus time on the bottom. The graphs will have matching time axes.

14. Repeat Step 12 to create a second graph window. Graph 2 will also be a graph of angular position (deg) versus time.

15. Drag the PE equation icon and drop it on Graph 2. Graph 2 will then split in two, showing both the position and the PE of the pendulum at any time *t*, with matching time axes.

16. It is not necessary to be able to see both graph windows at the same time, but they can be moved around the screen so that both are visible. Their sizes may also be adjusted so that when they are active, they occupy the full screen individually. It is easy to change from viewing one to viewing the other by clicking on the particular graph to bring it to the front. ● CI Fig. 5 shows what the screen will look like after all the setup is finished.

CI EXPERIMENTAL PROCEDURE

1. Put the rotary motion sensor on a support rod. Install the mass on the light rod, and then install the pendulum on the front screw of the rotary motion sensor. A diagram of the equipment setup is shown in ● CI Fig. 6.

2. The rotary motion sensor will set its "zero" at the location of the pendulum when the START button is pressed. If we want the position $\theta = 0$ to correspond with the equilibrium position of the pendulum, it is *very important* that the START button be pressed while the pendulum is at rest in the equilibrium position.

3. *After* pressing the start button, displace the pendulum a small angle (~10°) to the side and let it go.

4. Collect data for about 5 or 6 seconds, and then press the STOP button.

5. Print the graphs and paste them to the Laboratory Report.

6. Read from any of the position graphs what was the maximum amplitude of the pendulum, and record it in CI Data Table 1.

7. Determine from the graph the period of oscillation of the pendulum, and record it in the table.

8. From the kinetic energy graph, look at the first clear complete cycle of the motion, and find the maximum kinetic energy during that cycle. Record it in the table. Record also the position of the pendulum when the maximum kinetic energy was reached.

9. From the potential energy graph, look at the first clear complete cycle of the motion, and find the maximum potential energy during that cycle. Record it in the table. Record also the position of the pendulum when the maximum potential energy was reached.

10. Repeat for the minimum values of kinetic and potential energies.

11. To further reinforce the idea of conversions between kinetic and potential energy, create a new graph ("Graph 3") by dragging the kinetic energy data icon and dropping it on top of the "Graph" icon on the displays list. Then drag the potential energy icon and drop it in the graph. This graph will show both KE and PE as functions of time.

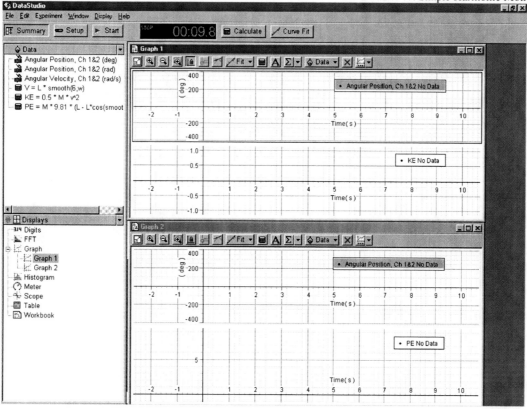

CI Figure 5   Data Studio setup.   Graph displays are generated for angular position, kinetic energy, and potential energy. The individual graph windows can be viewed together (as in this picture) or independently, if resized to fit the full screen. (Data displayed using DataStudio Software. Reprinted courtesy of PASCO scientific.)

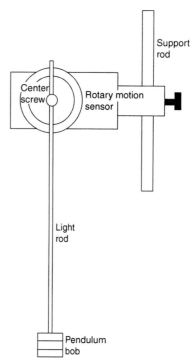

CI Figure 6   The experimental setup.   The light rod with the bob at the end is attached to the front screw of the rotary motion sensor.

C I   **EXPERIMENT 11**

Simple Harmonic Motion

CI *Laboratory Report*

CI   **DATA TABLE 1**

Purpose: To examine the variations of kinetic and potential energy as a pendulum swings.

Mass of pendulum, M _____ kg   Max. amplitude _____ °

Length, L _____ m   Period _____ s

	Value	Position of pendulum (deg)
KE max		
PE max		

KE min		
PE min		

Don't forget to attach the graphs to the Laboratory Report.

Don't forget units

(continued)

CI QUESTIONS

1. Compare the values of the maximum kinetic energy and the maximum potential energy. Discuss them in terms of the conservation of energy.

2. The following diagram illustrates three different positions of the pendulum as it moves in simple harmonic motion. (The angular displacement has been exaggerated for illustration purposes.) Label in the diagram which position corresponds to maximum KE, which to maximum PE, which to minimum KE, and which to minimum PE.

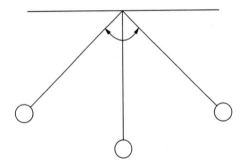

3. Was the amplitude of the pendulum constant? Explain.

4. The period of a simple pendulum in SHM is given by $T = 2\pi\sqrt{L/g}$. Use the measured length of the pendulum to calculate its period using this formula. Then compare to the period you determined from the graph. Discuss what causes the percent error.

5. *Optional Exercise:* Create a new calculation (in the Calculator window) that will determine the total energy of the pendulum. That is, calculate KE + PE. Then plot the total energy as a function of time. Was the total energy constant? Explain.

Name _____ Section _____ Date _____

Lab Partner(s) _____

Standing Waves in a String

/TI/ *Advance Study Assignment*

Read the experiment and answer the following questions.

1. How is wave speed related to frequency and wavelength? How is the period of oscillation related to wave speed?

2. What is a standing wave, and what are nodes and antinodes?

3. What are normal modes?

(continued)

4. How does the wavelength of a standing wave in a vibrating string vary as the tension force in the string and/or the linear mass density of the string varies?

5. Standing waves in a string can be produced by oscillating the string at the various natural frequencies. However, in this experiment the string vibrator has only one frequency. How, then, are standing waves with different wavelengths produced?

Standing Waves in a String

INTRODUCTION AND OBJECTIVES

A **wave** is the propagation of a disturbance or energy. When a stretched cord or string is disturbed, the wave travels along the string with a speed that depends on the tension in the string and its linear mass density. Upon reaching a fixed end of the string, the wave is reflected back along the string.

For a continuous disturbance, the propagating waves interfere with the oppositely moving reflected waves, and a standing- (or stationary-) wave pattern is formed under certain conditions. These standing-wave patterns can be visually observed, and the number of loops in a pattern depends on the length of the string and the wave speed, which is dependent on the tension in the string.

The visual observation and measurement of standing waves serve to provide a better understanding of wave properties and characteristics. In this experiment, we will study the relationship between the tension force and the wavelength in a vibrating string, as applied to the natural frequencies or normal modes of oscillation of the string.

After performing this experiment and analyzing the data, you should be able to:

1. Explain how standing waves are formed.
2. Distinguish between nodes and antinodes.
3. Tell what determines the natural frequencies of a vibrating string system.

EQUIPMENT NEEDED

- Electric string vibrator
- Clamps and support rod
- Pulley with rod support
- String

- Weight hanger and slotted weights
- Meterstick
- Laboratory balance
- 1 sheet of Cartesian graph paper

THEORY

A wave is characterized by its **wavelength** λ, **frequency** (of oscillation) f, in Hz or $1/s = s^{-1}$, and **wave speed** v. (See ● Fig. 1.) These quantities are related by the expression

$$\lambda f = v \tag{1}$$

(Check to see whether the equation is dimensionally correct.)

Waves in a stretched string are transverse waves; that is, the "particle" displacement is perpendicular to the

direction of propagation. In longitudinal waves, the particle displacement is in the direction of wave propagation (e.g., sound waves). The maximum displacements of the particle oscillation are $+A$ and $-A$. The magnitude of the maximum displacement, called the **amplitude** (A), is related to the energy of the wave. The **period** (of oscillation) T is related to the frequency of oscillation, $T = 1/f$.

When two waves meet, they interfere and the combined wave form is a superposition of the two interfering waves. The superposition of two waves of equal amplitude and frequency traveling in opposite directions gives rise to what is known as a **standing** or **stationary wave.**

The periodic constructive and destructive interference causes the formation of a standing-wave pattern as illustrated in ● Fig. 2. Notice that some of the "particles" on the axis are stationary. These positions are called *nodal points* or **nodes,** and the points of maximum displacement are called **antinodes.** It might be said that the energy is standing in the wave envelope, alternating between the kinetic and potential energies of the particles.

In a stretched string being oscillated or shaken at one end, waves traveling outward from the oscillator interfere with waves that have been reflected at the other fixed end. However, standing waves in a given length of string occur only for *certain* wave frequencies. That is, for a given stretching tension or force, the string must be driven or

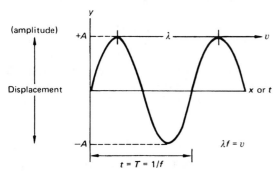

Figure 1   Wave description.   The parameters involved in describing a wave. See text for description.

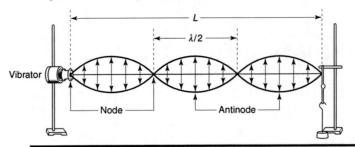

Figure 2 Standing wave. Periodic constructive and destructive interferences give rise to a standing wave form, as illustrated here. The length of one loop of the standing wave is equal to one-half the standing wave's wavelength. Note the positions of the nodes and antinodes.

oscillated with certain vibrational frequencies to produce standing waves.

The frequencies at which large-amplitude standing waves are produced are called **natural frequencies** or **resonant frequencies.** The resulting standing-wave patterns are called *normal,* or *resonant, modes of vibration.* In general, all systems that oscillate have one or more natural frequencies, which depend on such factors as mass, elasticity or restoring force, and geometry (boundary conditions).

Since the string is fixed at each end, a standing wave must have a node at each end. As a result, only an integral number of half wave-lengths may "fit" into the length L of the string, $L = \lambda/2, \lambda, 3\lambda/2, 2\lambda$, and so on, such that in general

$$L = n\left(\frac{\lambda_n}{2}\right) \quad \text{or} \quad \lambda_n = \frac{2L}{n}$$
$$n = 1, 2, 3, 4, \ldots \qquad (2)$$

Figure 2 illustrates the case for $L = 3\lambda/2$.

The wave speed in a stretched string is given by

$$v = \sqrt{\frac{F}{\mu}} \qquad (3)$$

(wave speed in a stretched string)

where F is the magnitude of the tension force in the string and μ is the linear mass density (mass per unit length, $\mu = m/L$) of the string. Using Eqs. 2 and 3 in $\lambda f = v$ (Eq. 1) yields

$$f_n = \frac{v}{\lambda_n} = \frac{n}{2L}\sqrt{\frac{F}{\mu}} \qquad n = 1, 2, 3, \ldots \qquad (4)$$

(resonant frequencies)

where f_n and λ_n are the frequency and wavelength, respectively, for a given integer n.

Setting $n = 1$ in Eq. 5 gives the lowest possible frequency, which is known as the **fundamental frequency**

$$f_1 = \frac{1}{2L}\sqrt{\frac{F}{\mu}} \qquad (5)$$

(fundamental frequency)

so Eq. 4 may be written in-terms-of-the fundamental frequency as

$$f_n = \frac{v}{\lambda_n} = \frac{n}{2L}\sqrt{\frac{F}{\mu}} = nf_1 \qquad n = 1, 2, 3, \ldots \qquad (6)$$

Moreover, only certain frequencies produce standing waves for a given string tension, density, and length.

As noted above, the lowest natural frequency f_1 (Eq. 5) is called the *fundamental frequency.* All other natural frequencies are integral multiples of the fundamental frequency: $f_n = nf_1$ (for $n = 1, 2, 3, \ldots$). The set of frequencies $f_1, f_2 = 2f_1, f_3 = 3f_1, \ldots$ is called a **harmonic series:** f_1 (the fundamental frequency) is the *first harmonic,* f_2 the *second harmonic,* and so on.

In this experiment, the electrically driven string vibrator has a fixed frequency, so the driving frequency cannot be varied to produce different normal-mode standing-wave patterns. However, varying the string tension can vary the wave speed to produce different standing-wave patterns. Since $v = \sqrt{F/\mu}$ (Eq. 3),

$$\lambda = \frac{v}{f} = \frac{1}{f}\sqrt{\frac{F}{\mu}} \qquad (7)$$

where f and μ are constant. Hence, by varying F, one can select the appropriate wavelengths that will "fit" into a given string length L to produce standing waves.

EXPERIMENTAL PROCEDURE

1. If one has not been provided, cut a piece of string long enough to be used in the experimental setup—long enough to be looped at each end so as to be attached to the vibrator and a weight hanger suspended from the end running over the pulley (● Fig. 3). The vibrator and pulley should be clamped to support posts at the opposite ends of the laboratory table to give an active string length of about 150 cm. (This length may vary for a given setup.)

   Measure the total length of the string, and determine its mass on a laboratory balance. Record these values in the data table, and compute the linear mass density $\mu = m/L_0$. (*Note:* L_0 is the total length of the string.)

2. Attach the string to the vibrator and suspend a weight hanger from the other end as shown in Fig. 3. Make certain that the string is aligned properly and that it is parallel to the table surface. Measure the distance between the vibrator arm and the point of contact of the string on the pulley. Record this length L in the data table.

   Turn on the vibrator. Try to produce different standing-wave patterns in the string by alternately lifting and carefully pulling down on the weight hanger. It is helpful to fold a thin strip of paper in half and hang it on the string to observe vibrating action. The number of loops should increase with less tension. (Why?) Also, try grasping the string at a node and antinode of a given pattern to see what happens.

3. When you are familiar with the operation of the apparatus, add enough weights to the weight hanger so that a standing-wave pattern of two loops is formed in the string (nodal point at the center). Adjust the tension by adding or removing some small weights until the loops are of maximum amplitude.

   If sufficiently small weights are not available, a fine adjustment can be made by loosening the clamp holding the vibrator rod and sliding it slightly back and forth so as to find the optimum string length between the ends that gives the maximum loop width or amplitude for a given tension.

   When this is accomplished, measure with a meterstick the distance from the point where the string contacts the pulley to the center nodal point. The meterstick can be held alongside the vibrating string, or you may find it more convenient to grasp the string at the nodal point with your fingers, shut off the vibrator, and measure the distance from the pulley contact to the nodal point along the nonvibrating string. Make certain not to pull the string toward the vibrator, for that would increase the length by raising the weight hanger.

Apply a slight tension in the string *away* from the vibrator if necessary.

Record this length L_1 and the total suspended mass in the data table. Since the length of one loop is one-half of a wavelength, $L_1 = \lambda/2$.

4. Remove enough weights from the weight hanger and adjust so that a standing-wave pattern of maximum amplitude with three loops (two nodal points in the string) is formed. Measure the distance from the pulley contact to the nodal point nearest the vibrator. (The fixed-end nodal point *at* the vibrator is not used because in vibrating up and down, it is not a "true" nodal point.)

   Record this length L_2 and the total suspended mass in the data table. Since the length of two loops is equal to one wavelength, $L_2 = \lambda$.

5. Repeat procedure 4 for consecutive standing-wave patterns up to eight measured loops if possible. [The weight hanger by itself may supply too much tension for higher-order patterns, so it may have to be removed and smaller weight(s) suspended.] Compute the wavelength for each case.

   It should become evident that in general, $\lambda = 2L_N/N$, or $L_N = N\lambda/2$, where N is the number of loops in a given L_N. Notice the similarity of the latter form of this equation to Eq. 2, wherein the length L is the total vibrating length of the string.

6. Notice that Eq. 7 can be rewritten as

$$\lambda = \frac{1}{f}\sqrt{\frac{F}{\mu}} = \left(\frac{1}{f\sqrt{\mu}}\right)\sqrt{F} \qquad \text{(7A)}$$

where f and μ are constants. It has the form of an equation of a straight line, $y = mx + b$, with $x = \sqrt{F}$, and $b = 0$.

   Plot the experimental data on a graph of λ versus $\sqrt{F}$. Draw the straight line that best fits the data, and determine the slope of the line. From this value and the previously determined value of μ, compute the average frequency f of the oscillations.*

   The string vibrator operates on 60-cycle ac current. The vibrating action is accomplished by means of an electromagnet operated by the input current. The vibrator arm is attracted toward an electromagnet during each half-cycle, or twice each cycle, so the vibrating frequency is $2 \times 60 = 120$ Hz (cycles per second). Using this as the accepted value of the vibrational frequency, compute the percent error of the experimentally determined value.

---

* If you have some scattered data points far from the straight line, see Question 2.

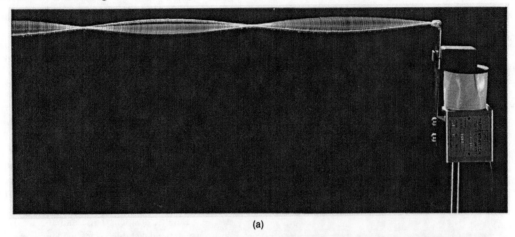

(a)

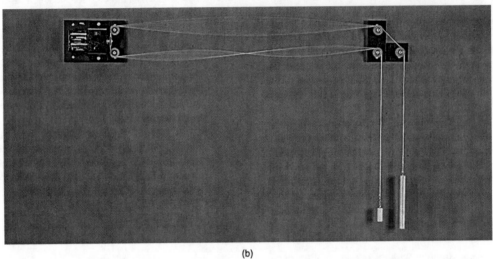

(b)

Figure 3   Standing wave apparatus.   (a) A string vibrator oscillates the string. Different standing waves are produced by varying the tension in the string. (b) A dual string vibrator. Different tensions produce different normal modes. (Courtesy of Sargent-Welch.)

Name _____ Section _____ Date _____

Lab Partner(s) _____

Standing Waves in a String
/TI/ *Laboratory Report*

Mass of string _____

Total length of string L_o _____

Linear mass density μ _____

Length of string between
vibrator and pulley L _____

DATA TABLE
Purpose: To determine the frequency of oscillation from normal modes.

Number of loops measured N	Suspended mass ( )	Tension force F^* ( )	Measured length L_N for N loops ( )	Wavelength λ ( )	$\sqrt{F}$ ( )
1		F_1	L_1		
2		F_2	L_2		
3		F_3	L_3		
4		F_4	L_4		
5		F_5	L_5		
6		F_6	L_6		
7		F_7	L_7		
8		F_8	L_8		

* For convenience, express the tension weight force in terms of g (e.g., if $m = 0.10$ kg, then $F = mg = 0.10 g$ N).

Calculations
(show work)

Slope of graph _____

Computed frequency f _____

Accepted frequency _____

Percent error _____

Don't forget units

(continued)

Calculations
(show work)

/TI/ QUESTIONS

1. The wavelength associated with F_1 and L_1 in the experiment does not correspond to the wavelength of the fundamental frequency of the string.
   (a) To which natural frequency does the wavelength associated with F_1 and L_1 correspond?

   (b) What tension in the string would be required to produce a standing wave with a wavelength corresponding to the fundamental frequency of the string? (*Hint:* Use Eq. 7.)

2. Theoretically, the vibrator frequency is 120 Hz. However, sometimes the vibrator resonates with the string at a "subharmonic" of 60 Hz.
   (a) If this were the case in all instances, how would it affect the slope of the graph?

(b) If you have some scattered data points far from the straight line on your graph, analyze the data for these points using Eq. 7 to determine the frequency.

3. How many normal modes of oscillation or natural frequencies does each of the following have: (a) a simple pendulum; and (b) a mass oscillating on a spring?

4. Stringed musical instruments, such as violins and guitars, use stretched strings. Explain (a) how tightening and loosening the strings tunes them to their designated tone pitch or frequency; (b) why the strings of lower tones are thicker or heavier; (c) why notes of higher pitch or frequency are produced when the fingers are placed on the strings.

5. (*Optional*) Consider a long whip antenna of the type used on automobiles for CB radios. Show that the natural frequencies of oscillation for the antenna are $f_m = mv/4L$, where $m = 1, 3, 5, \ldots$, v is the wave speed, and L is the length of the antenna. (*Hint:* The boundary conditions are a node and an antinode.)

EXPERIMENT 12

Name _____ Section _____ Date _____

Lab Partner(s) _____

The Thermal Coefficient of Linear Expansion

/TI/ *Advance Study Assignment*

Read the experiment and answer the following questions.

1. Distinguish between linear expansion and isotropic expansion.

2. What is the difference between the notation °C and the notation C°?

3. How is the thermal coefficient of linear expansion determined experimentally?

4. What are the units of the thermal coefficient of linear expansion?

Don't forget units

(continued)

5. What is meant by the fractional change in length?

The Thermal Coefficient of Linear Expansion

INTRODUCTION AND OBJECTIVES

With few exceptions, solids increase in size or dimensions as the temperature increases. Although this effect is relatively small, it is very important in applications involving materials that undergo heating and cooling. Unless these changes are taken into account, material and structural damage can result; for example, a piston may become too tight in its cylinder, a rivet could loosen, or a bridge girder could produce damaging stress.

The expansion properties of a material depend on its internal makeup and structure. Macroscopically, we express the thermal expansion in terms of temperature coefficients of expansion, which are experimental quantities that represent the change in the dimensions of a material per degree of temperature change. In this experiment, we will investigate the thermal expansion of some metals and determine their temperature coefficients of linear expansion.

After performing this experiment and analyzing the data, you should be able to:

1. Tell how the thermal coefficient of linear expansion describes such expansion.
2. Explain how the thermal coefficient of linear expansion is measured, and give an order of magnitude of its values for metals.
3. Describe and give examples of how thermal expansion considerations are important in applications of materials.

EQUIPMENT NEEDED

- Linear expansion apparatus and accessories
- Steam generator and stand
- Bunsen burner and striker or electric hot plate
- Rubber tubing
- Beaker
- Meterstick
- Thermometer (0°C to 110°C)
- Two or three kinds of metal rods (e.g., iron and aluminum)

THEORY

Changes in the dimensions and volumes of materials are common effects. The thermal expansion of gases is very obvious and is generally described by gas laws. But the thermal expansion of solids and liquids is no less important. In fact, we use such expansions to measure temperature in liquid and bimetallic thermometers.

In general for solids, a temperature increase leads to the thermal expansion of an object as a whole. This expansion results from a change in the average distance separating the atoms (or molecules) of a substance. The atoms are held together by bonding forces, which can be represented simplistically as springs in a simple model of a solid (● Fig. 1). The atoms vibrate back and forth; and with increased temperature (more internal energy), they become increasingly active and vibrate over greater distances. With wider vibrations in all dimensions, the solid expands as a whole. This may be different in different directions; however, if the expansion is the same in all directions, it is referred to as *isotopic expansion*.

The change in one dimension (length, width, or thickness) of a solid is called **linear expansion.** For small temperature changes, linear expansion is approximately proportional to ΔT, or the change in temperature $T - T_0$ (● Fig. 2). The *fractional* change in length is $(L - L_0)/L_0$, or $\Delta L/L_0$, where L_0 is the original length of the solid at the initial temperature. This ratio is related to the change in temperature by

$$\frac{\Delta L}{L_0} = \alpha \Delta T \qquad \text{or} \qquad \Delta L = \alpha L_0 \Delta T \qquad (1)$$

where $\Delta L = L - L_0$ and $\Delta T = T - T_0$ and α is the **thermal coefficient of linear expansion,** with units of inverse temperature—that is, $1/\text{C}°$. [It is convenient to write the unit of a ΔT as $\text{C}°$ (Celsius degree) so as to distinguish from a particular temperature T (°C, degree Celsius).] Note that with a temperature decrease and a contraction, ΔL would be negative, or a negative expansion.

As Eq. 1 shows, α is the fractional change in length per degree temperature change, $\Delta L/L_0$.* This thermal coefficient of expansion may vary slightly for different

---

* To help understand what is meant by *fractional change,* consider a money analogy. If you have $1.00 in the bank and get 5¢ interest, then the fractional change (increase) in your money is

$$\Delta \$/\$_0 = 5 \text{ cents}/100 \text{ cents} = 1/20 = 0.050 \text{ (or 5.0\%)}.$$

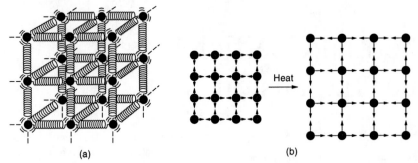

Figure 1 A springy solid. (a) The elastic nature of interatomic forces is indicated by simplistically representing them as springs, which, like the forces, resist deformation. (b) Heat causes the molecules to vibrate with greater amplitudes in the lattice, thereby increasing the volume of the solid (right). The arrows represent the molecular bonds, and the drawing is obviously not to scale. (James Shipman, Jerry Wilson, and Aaron Todd, *An Introduction to Physical Science,* Ninth Edition. Copyright © 2000 by Houghton Mifflin Company. Reprinted with permission.) From Wilson/Buffa, *College Physics,* Fifth Edition. Copyright © 2003. Reprinted by permission of Pearson Education.

temperature ranges, but this variation is usually negligible for common applications, and α is considered to be constant.

By Eq. 1, α is defined in terms of experimentally measurable quantities:

$$\alpha = \frac{\Delta L}{L_0 \Delta T} \qquad (2)$$

Hence, by measuring the initial length L_0 of an object (e.g., a metal rod) at an initial temperature T_0 and the change in its length ΔL for a corresponding temperature change ΔT, we can compute α.

This development may be extended to two dimensions. The linear expansion expression (Eq. 1) may be written

$$L = L_0(1 + \alpha \Delta T) \qquad (3)$$

and for an isotropic material, its area is $A = L \times L$, or

$$
\begin{aligned}
A &= L^2 \\
&= L_0^2(1 + \alpha \Delta T)^2 \\
&= A_0(1 + 2\alpha \Delta T + \alpha^2 \Delta T^2)
\end{aligned}
$$

where $A_0 = L_0^2$. Since typical α's are of the order of $10^{-6}/\text{C}°$, the α^2 term may be dropped with negligible error, and to a good approximation,

$$A = A_0(1 + 2\alpha \Delta T) \qquad (4)$$

Comparing this expression with Eq. 3, the thermal coefficient of area expansion is seen to be approximately twice the coefficient of linear expansion (i.e., 2α).

A similar development can be carried out for the coefficient of volume expansion, which is approximately equal to 3α.

EXPERIMENTAL PROCEDURE

1. A typical arrangement for determining thermal coefficients of linear expansion is shown in ● Fig. 3. The apparatus consists of a steam jacket with a micrometer attachment for measuring ΔL of a metal rod. A thermometer in the steam jacket measures the temperature of the rod. Steam is supplied to the jacket by a steam generator, and a beaker is used to catch the condensate.

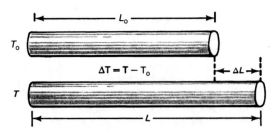

Figure 2 Linear thermal expansion. At the initial temperature T_0, the length of the rod is L_0. At some higher temperature T, the rod has expanded to a length L, and the change in length is $\Delta L = L - L_0$ for the temperature change ΔT.

2. Before assembling the apparatus, measure the lengths L_0 of the metal rods with a meterstick to the nearest 0.1 mm and record these lengths in the data table. Avoid handling the rods with your bare hands in order not to raise their temperature. Use a paper towel or cloth.

3. Assemble the apparatus, placing one of the rods in the steam jacket. Initially, have one end of the rod placed *firmly* against the fixed end screw and the other end not touching the micrometer screw.

   Carefully turn the micrometer screw until it just makes contact with the rod. Avoid mechanical backlash (and electrical spark-gap ionization, see below) by always turning the screw *toward* the rod just before reading. Do not force the screw. Record the micrometer setting. Do this three times and take the average as the initial setting. As soon as the initial micrometer is taken, read and record the initial temperature T_o.

   (The linear expansion apparatus may be equipped with an electrical circuit that uses a bell, light, or voltmeter to indicate when contact is made. The averaging process is unnecessary in this case.)

4. Turn the micrometer screw back from the end of the rod several millimeters to allow for the thermal expansion of the rod with increasing temperature. With the steam generator about one-half full, turn on the hot plate (or light the Bunsen burner) and boil the water so that steam passes through the jacket. The thermometer in the steam jacket should just touch the metal rod.

   Allow steam to pass through the jacket until the thermometer reading stabilizes (several minutes). When equilibrium has been reached, record the thermometer reading. Then carefully advance the micrometer screw until it touches the end of the rod, and record the micrometer setting. Do this three times, and take the average of the micrometer readings unless contact is indicated by electrical circuit. Turn off the heat source.

5. Repeat procedures 3 and 4 for the other metal rods.

   Caution: *Be careful not to burn yourself with the condensed hot water in the steam jacket or the hot rod when you remove it. Take proper precautions.*

6. Compute ΔL and ΔT, and find the coefficient of linear expansion for each metal. Compare these α's with the accepted values given in Appendix A, Table A3, by computing the percent errors.

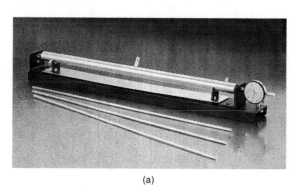

(a)

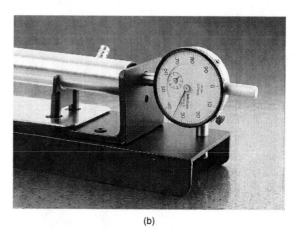

(b)

Figure 3   Linear thermal expansion apparatus.   (a) The heat of steam admitted to the steam jacket causes a metal rod to expand. Rods of different metals may be used. (b) The expansion is measured with a dial indicator. (Courtesy of Sargent-Welch.)

Name _____ Section _____ Date _____

Lab Partner(s) _____

The Thermal Coefficient of Linear Expansion

/TI/ *Laboratory Report*

DATA TABLE

Purpose: To determine the thermal coefficients of expansion of metal samples.

	Initial length L_0 ( )	Initial micrometer setting	Final micrometer setting	ΔL ( )	Initial temp. T_0 ( )	Final temp. T ( )	ΔT ( )	α meas. ( )	α accepted ( )
1. Type of rod _____									
2. Type of rod _____									
3. Type of rod _____									

Calculations
(show work)

Metal   Percent error

_____   _____

_____   _____

_____   _____

Don't forget units

(continued)

/TI/ QUESTIONS

1. What are the probable sources of error in this experiment? Which will cause the biggest error?

2. Would the numerical values of the thermal coefficients of linear expansion have been the same if the temperatures had been measured in degrees Fahrenheit? Explain, and give an example.

3. For a contraction with a negative fractional change, would the coefficient of thermal expansion be negative? Explain.

4. When a mercury-in-glass thermometer is placed in hot water, the thermometer reading first drops slightly and then rises. Explain why.

5. If flat strips of iron and brass were bonded together and this bimetallic strip were heated, what would be observed? Justify your answer, and draw a sketch of the situation. (*Hint:* See Appendix A, Table A3, for α's.)

6. A Pyrex graduated cylinder has a volume of exactly 200 mL at 0°C. If its temperature is increased to 100°C, will its volume increase or decrease? Compute the change in volume.

7. Assume a metal rod with an initial length L_0 is heated through a temperature increase of ΔT to a length L_1 and then cooled to its initial temperature—that is, through a temperature decrease of $-\Delta T$ (same ΔT increase and decrease). Call the final length of the rod L_2 after this thermal cycle.
   (a) Show that Eq. 3 implies that $L_2 = L_0 [1 - (\alpha\Delta T)^2]$, i.e., $L_2 \neq L_0$.

   (b) What is the implication if the rod were taken through a number of such thermal cycles?

(*continued*)

(c) Obviously something is wrong. Can you explain what it is? (*Hint:* Think of *basis,* or reference. For example, if you had an investment that appreciated 100% in value one day, and you lost 100% of your investment the next, would you still have any money left?)

Name _____ Section _____ Date _____

Lab Partner(s) _____

Specific Heats of Metals

/TI/ *Advance Study Assignment*

Read the experiment and answer the following questions.

1. Distinguish between heat capacity and specific heat.

2. Why is the specific heat of water equal to unity, i.e., 1.0 cal/g-C° or 1.0 kcal/kg-C°?

3. Given that the specific heat of one material is twice that of another, compare the relative amounts of heat required to raise the temperature of equal masses of each material by 1 C°.

4. Say the same amount of heat was added to samples of the materials in Question 3, and each sample had the same increase in temperature. Compare the relative masses of the samples.

(continued)

5. What is the method of mixtures, and how is it used to determine specific heat?

6. On what does the accuracy of the method of mixtures depend? That is, what are possible sources of error?

Specific Heats of Metals

INTRODUCTION AND OBJECTIVES

Different substances require different amounts of heat to produce a given temperature change. For example, about three and one-half times as much heat is needed to raise the temperature of 1 kg of iron through a given temperature interval ΔT as is needed to raise the temperature of 1 kg of lead by the same amount.

This material behavior is characterized quantitatively by **specific heat,** which is the amount of heat necessary to raise the temperature of a unit mass of a substance by one unit temperature interval, e.g., to raise 1 gram or 1 kilogram of a substance 1 Celsius degree. Thus, in the previous example, iron has a greater specific heat than that of lead.

The specific heat of a material is *specific,* or characteristic, for that material. As can be seen from the definition, the specific heat of a given material can be determined by adding a known amount of heat to a known mass of material and noting the corresponding temperature change. It is the purpose of this experiment to determine the specific heats of some common metals by calorimetry methods.

After performing this experiment and analyzing the data, you should be able to:

1. Tell what is meant by the specific heat of a substance, and compare the effects of different specific heats.
2. Calculate the heat necessary to raise the temperature of a given mass of a substance a particular number of degrees.
3. Describe and explain calorimetry and the method of mixtures.

EQUIPMENT NEEDED

- Calorimeter
- Boiler and stand
- Hot plate or Bunsen burner and striker
- Two thermometers (0 to 110°C)
- Two kinds of metal (shot form or slugs with attached strings)
- Laboratory balance
- Ice
- Safety glasses
- Strainer

THEORY

The change in temperature ΔT of a substance is proportional to the amount of heat ΔQ added (or removed) from it:

$$\Delta Q \propto \Delta T$$

In equation form, we may write

$$\Delta Q = C \, \Delta T \qquad (1)$$

where the constant of proportionality C is called the **heat capacity** of the substance.

However, the amount of heat required to change the temperature of an object is also proportional to the mass of the object. Hence, it is convenient to define a *specific heat capacity* (or simply **specific heat**) c:

$$c = \frac{C}{m} \qquad (2)$$

which is the heat capacity per unit mass of a substance. Thus, Eq. 1 becomes $\Delta Q = mc \, \Delta T$, or

$$\boxed{c = \frac{\Delta Q}{m \Delta T}} \qquad (3)$$

(specific heat)

The specific heat is then the amount of heat required to change the temperature of 1 g of a substance 1C°.*

The calorie (cal) unit of heat is defined as the amount of heat required to raise the temperature of 1 g of water

---

* It is convenient to write the unit of a ΔT as C° (Celsius degree) so as to distinguish it from a particular temperature T (°C, degrees Celsius).

201

1C°. By definition, then, water has a specific heat of 1 cal/g-C°.

$$c = \frac{\Delta Q}{m \Delta T} = \frac{1 \text{ cal}}{(1g)(1°C)} = 1 \text{ cal/g-C}°$$

[A kilocalorie (kcal) is the unit of heat defined as the amount of heat required to raise the temperature of 1 kg of water by 1C°. In these units, water has a specific heat of 1 kcal/kg-C°, or, in SI units, 4.18×10^3 J/kg-C°. Your instructor may recommend that you use one of these units.]

The specific heat of a material can be determined experimentally by measuring the temperature change of a given mass of material produced by a quantity of heat. This is done indirectly by a calorimetry procedure known as the **method of mixtures.** If several substances at various temperatures are brought together, the hotter substances lose heat and the colder substances gain heat until all the substances reach a common equilibrium temperature. If the system is insulated so that no heat is lost to or gained from the surroundings, then, by the conservation of energy, the heat lost is equal to the heat gained.

In this experiment, hot metal is added to water in a calorimeter cup, and the mixture is stirred until the system is in thermal equilibrium. The calorimeter insulates the system from losing heat (● Fig. 1). Heat is lost by the metal and is gained by the water and cup and stirrer. In equation form, we may write

$$\text{heat lost} = \text{heat gained}$$

or

$$\Delta Q_{metal} = \Delta Q_{water} + \Delta Q_{cup \text{ and stirrer}}$$

and

$$\begin{aligned} m_m c_m (T_m - T_f) \\ = m_w c_w (T_f - T_w) + m_{cs} c_{cs} (T_f - T_w) \\ = (m_w c_w + m_{cs} c_{cs})(T_f - T_w) \end{aligned} \quad (4)$$

where T_f is the final intermediate equilibrium temperature of the system. The other subscripts indicate the masses, specific heats, and initial temperatures of the respective components. Hence, Eq. 4 may be used to determine the specific heat c_m of the metal if all the other quantities are known.

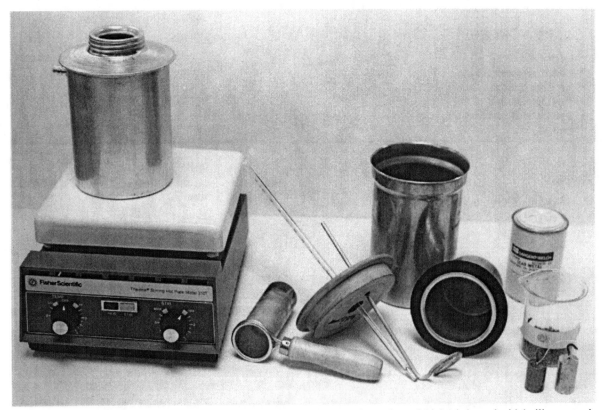

Figure 1   Apparatus for measurement of specific heats.   Metal shot or a piece of metal (right) is heated with boiling water in the container on the hot plate. The metal is then placed in a known amount of water in the calorimeter, which insulates the system from losing heat. The inner calorimeter cup is shown with its dark, insulating ring laying in front of the outer cup. A thermometer and stirrer extend through the calorimeter cover.

EXPERIMENTAL PROCEDURE

1. Weigh out 400 to 500 g (0.4 to 0.5 kg) of one kind of dry metal shot. [Do this by first determining the mass of the empty boiler cup (in which the metal shot is heated) and then adding an appropriate amount of metal shot to the cup and reweighing.]

   Record the mass of the metal m_m and the room temperature T_r in the data table. Your instructor may prefer to use a solid piece of metal with a string attached instead of metal shot. In this case it is necessary to weigh only the piece of metal.

2. Insert a thermometer well into the metal shot (or into the cup with a piece of metal, if used), place the cup and shot into the boiler, and start heating the boiler water.

   *Caution: If a mercury thermometer is used, special care must be taken. If the thermometer should break and mercury spill into the hot metal, immediately notify your instructor. The cup should be removed from the room (to an exhaust hood or outdoors). Mercury fumes are **highly** toxic.*

   The boiler should be about half full of water. Keep steam or water from dampening the dry metal by shielding the cup with a cardboard lid (with a hole for the thermometer).

3. While the boiler is heating, determine and record the mass of the inner calorimeter cup and the stirrer (without the ring). Record the total mass m_{cs}. Also, note and record the type of metal and specific heat of the cup and stirrer, which is usually stamped on the cup.* (The specific heat may be found in Appendix A, Table A4, if it is not stamped on the cup.)

4. Fill the calorimeter cup about one-half to two-thirds full of cold tap water, and weigh the cup, stirrer, and water to determine the mass of the water m_w.

   (If a solid piece of metal is used, which usually has less mass than the recommended amount of shot, less water should be used so as to obtain an appreciable

   ΔT temperature change. This may also be the case at high elevations, where the temperature of boiling water is substantially less than 100°C).

   Place the calorimeter cup with the water and stirrer in the calorimeter jacket, and put on the lid, with a thermometer extending into the water.

5. After the water in the boiler boils and the thermometer in the metal has stabilized (allow several minutes), read and record the temperature of the metal T_m.

   Start with the water and stirrer in the cup at a temperature T_w several degrees below room temperature T_r. Adjust the temperature of the inner calorimeter cup and its contents by placing it in a beaker of ice water. Measure and record the temperature T_w.

6. Remove the thermometer from the metal. Then remove the lid from the calorimeter and quickly, but carefully, lift the cup with the hot metal from the boiler and pour the metal shot into the calorimeter cup with as little splashing as possible so as not to splash out and lose any water. (If a solid piece of metal is used, carefully lower the metal piece into the calorimeter cup by means of the attached string.)

   Replace the lid with the thermometer, and stir the mixture gently. The thermometer should not touch the metal. While stirring, watch the thermometer and record the temperature when a maximum equilibrium temperature is reached (T_f).

   For best results, the final temperature T_f should be above room temperature T_r by about as many degrees as T_w was below it. If this is not approximately the case, repeat procedures 4 through 6, adjusting T_w until the relationship $T_f - T_r \approx T_r - T_w$ is satisfied.

7. Repeat procedures 1 through 6 for another kind of metal sample. Make certain that you use fresh water in the calorimeter cup. (Dump the previous metal shot and water into a strainer in a sink so that it may be dried and used by others doing the experiment later.)

8. Compute the specific heat of each metal, using Eq. 4. Look up the accepted values in Appendix A, Table A4, and compute the percent errors.

---

* If the cup and stirrer are not of the same material, they must be treated separately, and the last term in Eq. 4 becomes $(m_w c_w + m_c c_c + m_s c_s)(T_f - T_w)$.

Name _____ Section _____ Date _____

Lab Partner(s) _____

Specific Heats of Metals

/TI/ *Laboratory Report*

DATA TABLE

Purpose: To determine the specific heats of metal samples.   Room temperature T_r _____

Type of metal	Mass of metal m_m ( )	Mass of calorimeter and stirrer m_{cs} ( )	Specific heat of calorimeter and stirrer c_{cs} ( )	Mass of water m_w ( )	T_m ( )	T_w ( )	T_f ( )

Calculations (show work) | Type of metal | c_m (experimental) | c_m (accepted) | Percent error |

_____   _____   _____   _____

_____   _____   _____   _____

Don't forget units

(continued)

$\boxed{TI}$ QUESTIONS

1. (a) The percent errors of your experimental values of the specific heats may be quite large. Identify several sources of experimental error.

   (b) Why does it improve the accuracy of the experiment if $T_f - T_r \approx T_r - T_w$?

2. The specific heat of aluminum is 0.22 cal/g-C°. What is the value of the specific heat in (a) kcal/kg-C°, (b) J/kg-C°? (Show your calculations.)

3. (a) If wet shot had been poured into the calorimeter cup, how would the experimental value of the specific heat have been affected?

EXPERIMENT 14   *Laboratory Report*

(b) If some water had splashed out as you were pouring dry shot into the cup, how would the experimental value of the specific heat have been affected?

4. In solar heating applications, heat energy is stored in some medium until it is needed (e.g., to heat a home at night). Should this medium have a high or a low specific heat? Suggest a substance that would be appropriate for use as a heat-storage medium, and explain its advantages.

5. Explain why specific heat is *specific* and how it gives a relative indication of molecular configuration and bonding.

EXPERIMENT 14

EXPERIMENT 15

Archimedes' Principle: Buoyancy and Density

$\boxed{TI}$ *Advance Study Assignment*

Read the experiment and answer the following questions.

1. Describe the physical reason for the buoyant force in terms of pressure.

2. Show that the buoyant force is given by $F_b = \rho_f\, g V_f$ using the development in the Theory section.

3. Give the conditions on densities that determine whether an object will sink or float in a fluid.

4. Distinguish between density and specific gravity, and explain why is it convenient to express these quantities in cgs units?

(continued)

5. Describe how the density of an object less dense than water can be determined using Archimedes' principle. How about the density of a liquid?

6. Why is it important to make certain that no air bubbles adhere to objects during the submerged weighing procedures? How would the experimental results be affected if bubbles were present?

Archimedes' Principle: Buoyancy and Density

INTRODUCTION AND OBJECTIVES

Some objects float and others sink in a given fluid—a liquid or a gas. The fact that an object floats means it is "buoyed up" by a force greater than or equal to its weight. Archimedes (287–212 B.C.), a Greek scientist, deduced that the upward buoyant force acting on a floating object is equal to the weight of the fluid it displaces. Thus, an object sinks if its weight exceeds that of the fluid it displaces.

In this experiment, Archimedes' principle will be studied in an application: determining the densities and specific gravities of solid and liquid samples.

After performing this experiment and analyzing the data, you should be able to:

1. Tell whether an object will sink or float in a fluid, knowing the density of each.
2. Distinguish between density and specific gravity.
3. Describe how the densities of objects that sink or float may be determined experimentally.

EQUIPMENT NEEDED

- Triple-beam pan balance with swing platform (or single-beam double-pan balance with swing platform and set of weights); see Fig. 2.
- Overflow can (or graduated cylinder and eye dropper)
- Two beakers

- Metal cylinder, irregularly shaped metal object, or metal sinker
- Waxed block of wood
- Saltwater solution or alcohol
- String
- Hydrometer and cylinder

THEORY

When placed in a fluid, an object either floats or sinks. This is most commonly observed in liquids, particularly water, in which "light" objects float and "heavy" objects sink. But the same effect occurs for gases. A falling object is sinking in the atmosphere, whereas other objects float (● Fig. 1).

Things float because they are buoyant, or are buoyed up. That is, there must be an upward force that is greater

Figure 1 Gas buoyancy. Archimedes' principle applies to fluids—a liquid *or* a gas. Here, a helium-filled blimp floats in air. (Courtesy of Bill Aron/PhotoEdit.)

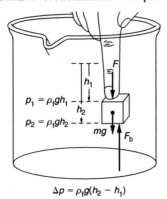

$$\Delta p = \rho_f g(h_2 - h_1)$$

Figure 2   Buoyancy. A buoyant force arises from the difference in pressure at different depths. The pressure at the bottom of the submerged block (p_2) is greater than that at the top (p_1), so there is a (buoyant) force directed upward (the arrow is shifted for clarity).

than (or equal to) the downward force of the object's weight. The upward force resulting from an object being wholly or partially immersed in a fluid is called the **buoyant force.** How the buoyant force arises can be understood by considering a buoyant object being held under the surface of a liquid (● Fig. 2). The pressures on the upper and lower surfaces of the block are given by the pressure-depth equations $p_1 = \rho_f g h_1$ and $p_2 = \rho_f g h_2$, respectively, where ρ_f is the density of the fluid. Thus there is a pressure difference $\Delta p = p_2 - p_1 = \rho_f g(h_2 - h_1)$, which gives an upward force (the buoyant force). In this case, the buoyant force is balanced by the applied force and the weight of the block.

It is not difficult to derive an expression for the magnitude of the buoyant force. If both the top and bottom areas of the block are A, the buoyant force (F_b) is given by $F_b = \Delta p A = \rho_f g V_f$, where V_f is the volume of the fluid displaced. But $\rho_f V_f$ is simply the mass of the fluid displaced by the block, m_f (recall that $\rho = m/V$). Hence the magnitude of the buoyant force is equal to the weight of the fluid displaced by the block. This general result is known as **Archimedes' principle:**

> *An object immersed wholly or partially in a fluid experiences a buoyant force equal in magnitude to the weight of the volume of fluid that it displaces.*

Thus the magnitude of the buoyant force depends only on the weight of the fluid displaced by the object, *not* on the weight of the object.

Archimedes' principle shows that an object

1. **will float** in a fluid if the density of the object ρ_o is less than the density of the fluid ρ_f that is, $(\rho_o < \rho_f)$;
2. **will sink** if the object's density is greater than that of the fluid, $(\rho_o > \rho_f)$ and

3. **will float in equilibrium** at any submerged depth where it is placed if its density is equal to that of the fluid, $(\rho_o = \rho_f)$.

This can be shown mathematically as follows. The weight of an object is $w_o = m_o g = \rho_o g V_o$, where V_o is the volume of the object and $\rho_o = m_o/V_o$. Similarly, the weight of the fluid displaced by the object, or the buoyant force, is $F_b = w_f = m_f g = \rho_f g V_f$. If the object is completely submersed in the fluid, then $V_o = V_f$, and dividing one equation by the other yields

$$\frac{F_b}{w_o} = \frac{\rho_f}{\rho_o} \qquad \text{or} \qquad F_b = \left(\frac{\rho_f}{\rho_o}\right) w_o \tag{1}$$

Hence

1. If $\rho_o < \rho_f$, then $F_b > w_o$, and the object will be buoyed up to the surface and float.
2. If $\rho_o > \rho_f$, then $F_b < w_o$, and the object will sink.
3. If $\rho_o = \rho_f$, then $F_b = w_o$, and the object is in equilibrium.

SPECIFIC GRAVITY AND DENSITY

Specific gravity will be used in the study and determination of density. The **specific gravity** of a solid or liquid is defined as the ratio of the weight of a given volume of the substance to an equal volume of water:

$$\text{specific gravity (sp. gr.)} = \frac{w_s}{w_w}$$

$$= \frac{\text{weight of a substance (of given volume)}}{\text{weight of an equal volume of water}} \tag{2}$$

where the subscripts s and w refer to the substance and water, respectively.

Specific gravity is a density-type designation that uses water as a comparison standard. Since it is a weight ratio, specific gravity has no units. Conveniently, the numerical value of a substance's specific gravity is the same as the magnitude of its density *in cgs units*. This can be seen as follows:

$$\text{sp. gr.} = \frac{w_s}{w_w} = \frac{w_s/V_s}{w_w/V_w} = \frac{m_s g/V_s}{m_w g/V_w}$$

$$= \frac{m_s/V_s}{m_w/V_w} = \frac{\rho_s}{\rho_w}$$

$$\boxed{\text{sp. gr.} = \frac{\rho_s}{\rho_w}} \tag{3}$$

Since, for practical purposes, the density of water is 1 g/cm^3 over the temperature range in which water is liquid,

$$\text{sp. gr.} = \frac{\rho_s}{\rho_w} = \frac{\rho_s(\text{g/cm}^3)}{1\ (\text{g/cm}^3)} = \rho_s \qquad (4)$$

where ρ_s is the numerical value of the density of a substance in g/cm^3.

For example, the density of mercury is 13.6 g/cm^3, and mercury has a specific gravity of 13.6. A specific gravity of 13.6 indicates that mercury is 13.6 times more dense than water, $\rho_s = (\text{sp. gr.})\rho_w$, or that a sample of mercury will weigh 13.6 times as much as an equal volume of water.

Archimedes' principle can be used to determine the specific gravity and density of a *submerged* object:

$$\text{sp. gr.} = \frac{w_o}{w_w} = \frac{w_o}{F_b} \qquad (5)$$

where w_o is the weight of the object, w_w is the weight of the water it displaces, and, by Archimedes' principle, $w_w = F_b$.

For a heavy object that sinks, the net force as it does so is equal to $w_o - F_b$. (Why?) If attached to a scale while submerged, it would have a measured *apparent* weight w_o' and $w_o' = w_o - F_b$. Thus $F_b = w_o - w_o'$, and Eq. 5 may be written

$$\text{sp. gr.} = \frac{w_o}{w_w} = \frac{w_o}{w_o - w_o'}$$

or, in terms of mass measured on a balance ($w = mg$),

$$\text{sp. gr.} = \frac{m_o}{m_o - m_o'} = \rho_o \qquad (6)$$
(of a heavy object that sinks)

where ρ_o is the magnitude of the density of the object in g/cm^3. This provides us with an experimental method to determine the specific gravity and density of an object that sinks.

To measure the specific gravity and density of an object that floats, or is less dense than water, using Archimedes' principle, it is necessary to use another object of sufficient weight and density to submerge the light object completely.

$w_1 = w_o + w_s'$ is the measured weight (mass) of the object and the sinker, with only the sinker submerged, and $w_2 = w_o' + w_s'$ is the measured weight when both are submerged. Then

$$w_1 - w_2 = (w_o + w_s') - (w_o' + w_s') = w_o - w_o'$$

or, in terms of mass,

$$m_1 - m_2 = m_o - m_o'$$

and the specific gravity and density can be found from Eq. 6. That is,

$$\text{sp. gr.} = \frac{m_o}{m_1 - m_2} = \rho_o \qquad (7)$$
(of a light object that floats)

The specific gravity and density of a liquid can also be found using Archimedes' principle. A heavy object is weighed first in air (w_o) and then when submerged in liquid (w_o'). Then $(w_o - w_o')_l$ is the weight of the volume of liquid the object displaces, by Archimedes' principle. When we carry out a similar procedure for the object in water, $(w_o - w_o')_w$ is the volume of water the object displaces.

Then, by the definition of specific gravity (Eq. 2),

$$\text{sp. gr.} = \frac{(w_o - w_o')_l}{(w_o - w_o')_w} = \rho_\ell \qquad (8)$$
(of a liquid)

where ρ_ℓ is the density of the liquid in g/cm^3.

You may have been thinking that there are easier ways to determine the density or specific gravity of a solid or liquid. This is true, but the purpose of the experiment is to familiarize you with Archimedes' principle. You may wish to check your experimental results by determining the densities and specific gravities of the solid samples by some other method. The specific gravity of the liquid sample will also be determined using a hydrometer.

EXPERIMENTAL PROCEDURE

A. *Direct Proof of Archimedes' Principle*

1. Weigh the metal sample and record its mass m_o and the type of metal in the Laboratory Report. Also, determine the mass of an empty beaker m_b and record. Fill the overflow can with water, and place it on the balance platform. Attach a string to the sample and suspend it from the balance arm as illustrated in ● Fig. 3.*

2. The overflow from the can when the sample is immersed is caught in the beaker. Take a mass reading m_o' of the submerged object. Make certain that no bubbles adhere to the object. (It is instructive to place the overflow can on a second balance, if available, and note that the "weight" of the overflow can does not change as the sample is submerged.)

---

* You may use an alternative method if no overflow can is available. Attach a string to the sample and place it in a graduated cylinder. Fill the cylinder with water until the sample is completely submerged. Add water (with an eyedropper) until the water level is at a specific reference mark on the cylinder (e.g., 35 mL). Remove the sample, shaking any drops of water back into the cylinder, and weigh the cylinder and water (m_b). Refill the cylinder to the reference mark and weigh it again ($m_w + m_b$). The mass of the "overflow" water is then the difference between these measurements.

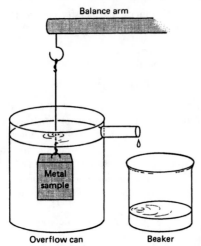

Figure 3 Archimedes' principle. The arrangement for proving Archimedes' principle. The weight of the displaced liquid that overflows into the beaker is equal to the reduction in weight of the metal sample when it is submerged, which is equal to the buoyant force.

Next weigh the beaker and water so as to determine the mass of the displaced water m_w (If the can does not fit on the balance platform, first suspend and immerse the object in the full overflow can, and catch the overflow in the beaker and find m_w. Then attach the sample to the balance arm and suspend it in a beaker of water that will fit on the balance platform to find m_o'.)

3. The buoyant force is then the difference between the object's true weight and its submerged weight, $F_b = m_o g - m_o' g$. According to Archimedes' principle, the magnitude of the buoyant force F_b should equal the weight of the displaced water:

$$F_b + w_w + m_w g$$

or

$$F_b = (m_o - m_o')g = m_w g$$

Compute the buoyant force, and compare it with the weight of the displaced water by finding the percent difference.

B. *Density of a Heavy Solid Object* $(\rho_o > \rho_w)$

4. Determine the specific gravity and density of the metal sample. This can be computed using the data from part A.

C. *Density of a Light Solid Object* $(\rho_o < \rho_w)$

5. Determine the specific gravity and density of the wooden block by the procedure described in the Theory section. First, measure the mass of the wooden block alone (in air). Then set up as in Fig. 3.

   Tie the sinker to the wood block, and tie the block to the lower hook of the balance. With the beaker empty, check that the sinker does not touch the bottom of the beaker and that the top of the wooden block is below the top of the beaker. Pour enough water into the beaker to cover the sinker, weigh, add more water until the wooden block is submerged, and then weigh again. Make certain that no air bubbles adhere to the objects during the submerged weighing procedures. The block is waxed so that it does not become waterlogged.

D. *Density of a Liquid* (ρ_ℓ)

6. Determine the specific gravity and density of the liquid provided, by the procedure described in the Theory section. Again, make certain that no air bubbles adhere to the object during the submerged weighing procedures.

7. Determine the specific gravity of the liquid using the hydrometer and cylinder. Compare this value with that found in procedure 6 by computing the percent difference.

EXPERIMENT 15

Archimedes' Principle: Buoyancy and Density

/TI/ *Laboratory Report*

A. *Direct Proof of Archimedes' Principle*

Type of metal _____

Mass of metal m_o in air _____

Mass of beaker m_b _____

Mass of metal m_o'
submerged in water _____

Mass of beaker and
displaced water $m_w + m_b$ _____

Mass of displaced
water m_w _____

Buoyant force
(in newtons) _____

Weight of displaced water
(in newtons) _____

Percent difference _____

Calculations
(show work)

Don't forget units

(continued)

B. *Density of a Heavy Solid* $(\rho_o > \rho_w)$

Calculations
(show work)

Specific gravity _____

Density _____

C. *Density of a Light Solid* $(\rho_o < \rho_w)$

Mass of block in air _____

Mass of block and sinker
with only sinker
submerged _____

Mass of block and sinker
with both submerged _____

Specific gravity _____

Density _____

Calculations
(show work)

EXPERIMENT 15   *Laboratory Report*

D. Density of a Liquid (ρ_ℓ)

Mass of object in air _____   Computed sp. gr. _____

Mass of object
submerged in liquid _____   Sp. gr. from hydrometer
                                                             measurement _____

Mass of object
submerged in water _____   Percent difference _____

Calculations
(show work)

/TI/ QUESTIONS

1. Look up the density of the metal of the object used in parts A and B of the procedure, and
    compare it with the experimental value. Comment on the purity of the metal of the object.
    (Archimedes developed his principle while working on a similar inquiry. His problem was
    to determine whether a crown alleged to be made of pure gold had actually been made with
    some content of cheaper metal.)

(continued)

2. In part B, the string will cause error. When does it lead to an experimental density that is too high? Too low?

3. Discuss the situation that occurs when an object is immersed in a fluid that has the same density as the object.

4. (a) Explain how a submarine is caused to submerge and surface without the use of its propulsion propeller and fins.

   (b) Which is heavier, a given volume of ice or the same volume of water? Justify your answer.

5. A block of wood floats in a beaker of water. According to Archimedes' principle, the block experiences an upward buoyant force. If the beaker with the water and floating block were weighed, would the measured weight be less than the sum of the weights of the individual components? Explain.

EXPERIMENT 15   *Laboratory Report*

6. A person can lift 45 kg ($\approx$ 100 lb). Using the experimental value of the specific gravity for the metal object in part B, how many cubic meters of the metal could the person lift (a) in air, (b) in water? How many actual kilograms of metal is this in air, and in water?

7. Explain the principle and construction of a hydrometer. What is the purpose of the common measurements of the specific gravities of an automobile's radiator coolant and battery electrolyte?

EXPERIMENT 15

Name _____ Section _____ Date _____

Lab Partner(s) _____

EXPERIMENT 16

Fields and Equipotentials

/TI/ *Advance Study Assignment*

Read the experiment and answer the following questions.

A. *Electric Field*

1. What is an electric field, and what does it tell you?

2. What are "lines of force," and what force is it?

3. What are equipotentials, and how are they experimentally determined? What is their relationship to the electric field lines?

Don't forget units

(continued)

B. *Magnetic Field*

4. What is a magnetic field, how is it defined, and what does it tell you?

5. Does the magnetic **B** field have the same relationship to electric charge as the electric **E** field? Explain.

6. How may a magnetic pole be moved in a magnetic field without doing work?

Fields and Equipotentials

INTRODUCTION AND OBJECTIVES

When buying groceries, we are often interested in the price per pound. Knowing this, we can determine the price for a given amount of an item. Analogously, it is convenient to know the electric force per unit charge at points in space due to an electric charge configuration, or the magnetic force per unit pole or "moving charge." Knowing these, we can easily calculate the electric force or magnetic force an interacting object would experience at different locations.

The electric force per unit charge is a vector quantity called the electric field intensity, or simply the **electric field** (**E**). By determining the electric force on a test charge at various points in the vicinity of a charge configuration, the electric field may be "mapped," or represented graphically, by lines of force. The English scientist Michael Faraday (1791–1867) introduced the concept of lines of force as an aid in visualizing the magnitude and direction of an electric field.

Similarly, the magnetic force per unit pole is a vector quantity called the magnetic field intensity, or **magnetic field (B)**. In this case, the field is mapped out by using the pole of a magnetic compass.

In this experiment, the concept of fields will be investigated and some electric and magnetic field configurations will be determined experimentally.

After performing this experiment and analyzing the data, you should be able to:

1. Describe clearly the concept of a force field.
2. Explain lines of force and the associated physical interpretations.
3. Distinguish between lines of force and equipotentials, and describe their relationships to work.

EQUIPMENT NEEDED

A. Electric Field

- Field mapping board and probes
- Conducting sheets with grids
- Conducting paint
- Connecting wires
- 1.5-V battery (or 10-V dc source)
- Galvanometer [or high-resistance voltmeter or multimeter, or vacuum-tube voltmeter (VTVM) with two-point contact field probe*]

- Single-throw switch
- 3 sheets of Cartesian graph paper

B. Magnetic Field

- 2 bar magnets and one horseshoe magnet
- Iron filings
- 3 sheets of paper or overhead transparency material
- Small compass
- 3 sheets of Cartesian graph paper or regular paper

* Leads from the dc input of an oscilloscope work nicely.

THEORY

A. Electric Field

The magnitude of the electrostatic force between two point charges q_1 and q_2 is given by Coulomb's law:

$$F = \frac{kq_1q_2}{r^2} \qquad (1)$$

where r is the distance between the charges and the constant $k = 9.0 \times 10^9$ N-m^2/C^2. The direction of the force on a charge may be determined by the **law of charges** or **charge-force law:**

Like-charges repel, and unlike charges attract.

The magnitude E of the **electric field** is defined as the electrical force per unit charge, or $E = F/q_0$ (N/C). By

convention, the electric field is determined by using a *positive* test charge q_0. In the case of the electric field associated with a single-source charge q, the magnitude of the electric field a distance r away from the charge is

$$E = \frac{F}{q_0} = \frac{kq_0q}{q_0r^2} = \frac{kq}{r^2} \qquad (2)$$

(electric field)

The direction of the electric field may be determined by the law of charges—that is, in the direction of the force experienced by the positive test charge.

The electric field vectors for several series of radial points from a positive source charge are illustrated in ● Fig. 1a. Notice that the lengths (magnitudes) of the vectors are smaller the greater the distance from the charge. (Why?)

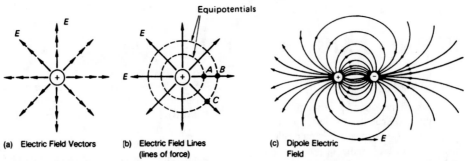

Figure 1   Electric field.   (a) Electric field vectors near a positive charge. (b) Lines of force with equipotentials for a positive charge. (c) An electric dipole and its electric field. The direction of the electric field at a particular location is tangent to the line of force through that point, as illustrated on the bottom line of force.

By drawing lines through the points in the direction of the field vectors, we form lines of force (Fig. 1b), which give a graphical representation of the electric field. The direction of the electric field at a particular location is tangent to the line of force through that point (Fig. 1c). The magnitudes of the electric field are not customarily listed, only the direction of the field lines. However, the closer together the lines of force, the stronger the field.

If a positive charge were released in the vicinity of a stationary positive source charge, it would move along a line of force in the direction indicated (away from the source charge). A negative charge would move along the line of force in the opposite direction. Once the electric field for a particular charge configuration is known, we tend to neglect the charge configuration itself, since the effect of the configuration is given by the field.

Since a free charge moves in an electric field by the action of the electric force, we say that work ($W = Fd$) is done by the field in moving charges from one point to another (e.g., from A to B in Fig. 1b).

To move a positive charge from B to A would require work supplied by an external force to move the charge against the electric field (force). The work W per charge q_o in moving the charge between two points in an electric field is called the **potential difference** ΔV between the points:

$$\Delta V_{BA} = V_B - V_A = \frac{W}{q_o} \qquad (3)$$

(It can be shown that the potential at a particular point a distance r from the source charge q is $V = kq/r$. See your textbook.)

If a charge is moved along a path at right angles or perpendicular to the field lines, no work is done ($W = 0$), since there is no force component along the path. Then along such a path (dashed-line paths in Fig. 1b), $\Delta V = V_B - V_C = W/q_o = 0$, and $V_C = V_B$. Hence, the potential

is constant along paths perpendicular to the field lines. Such paths are called **equipotentials.** (In three dimensions, the path is along an equipotential surface.)

An electric field may be mapped experimentally by determining either the field lines (of force) or the equipotential lines. Static electric fields are difficult to measure, and field lines are more easily determined by measuring small electric currents (flow of charges) maintained in a conducting medium between charge configurations in the form of metal electrodes.

The steady-state electric field lines closely resemble the static field that a like configuration of static charges would produce. The current is measured in terms of the voltage (potential) difference by a high-resistance voltmeter or multimeter (or VTVM).

In other instances, equipotentials are determined, and hence the field lines, using a simple galvanometer as a detector. When no current flows between two probe points, as indicated by a zero deflection on the galvanometer, there is no potential difference between the points ($\Delta V = 0$), and the points are on an equipotential.

B. Magnetic Field

Analogous to an electric field, a **magnetic field** was originally defined as the magnetic force per unit pole. The direction of the force at a particular location is that of the force experienced by a north magnetic pole.

Just as we may map the electric field around an electric charge, we may draw magnetic lines of force around a magnet. A single magnetic pole, or magnetic monopole, has never been observed, so the magnetic field is mapped using the north pole (by convention) of a magnetic dipole, for example, the magnetic needle of a compass. The torque on the compass needle resulting from the magnetic force causes the needle to line up with the field, and the north pole of the compass points in the direction of the field (● Fig. 2). If the compass is moved in the direction indicated by the north pole, the path of the compass traces out a field line.

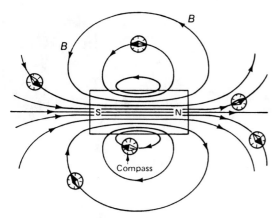

Figure 2 Magnetic field. The magnetic force causes a compass needle to line up with the field, and the north pole of the compass points in the direction of the field. If the compass is moved in the direction indicated by the north pole, the path of the compass needle traces out a field line.

Another observation is that an electric charge q moving nonparallel to a magnetic field experiences a force. For the special case in which the velocity vector **v** of the charge is perpendicular to the magnetic field **B**, the magnitude of the force is given by

$$F = qvB$$

This gives an expression for the strength (magnitude) of the magnetic field in terms of familiar quantities:

$$B = \frac{F}{qv} \qquad \textbf{(4)}$$

(magnetic field)

where the direction of **B** is perpendicular to the plane of **v** and **F**. Note that the SI unit of magnetic field is N/A-m, or tesla (T).*

The magnetic field may then be thought of as the magnetic force "per unit charge" per velocity. The **B** field has the same form as that mapped out using compass-needle poles.

It is instructive for comparative purposes to draw equipotential lines perpendicular to the field lines, as in the electric field case. No work would be done on a magnetic pole (or electric charge) when it was moved along these equipotential lines. (Why?)

A common method of demonstrating a magnetic field is to sprinkle iron filings over a paper or transparency material covering a magnet (● Fig. 3). The iron filings be-

---

* Other units of magnetic field are the weber/m² (Wb/m²) and the gauss (G). These units are named after early investigators of magnetic phenomena.

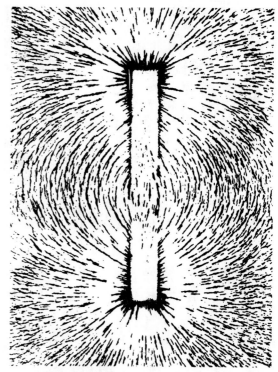

Figure 3 Iron filing pattern for a bar magnet. The iron filings become induced magnets and line up with the field, as would a compass needle. (Courtesy of *PSSC Physics,* D.C. Heath and Company with Educational Development Center, Inc., Newton, Massachusetts.)

come induced magnets and line up with the field as would a compass needle. This method allows one to visualize the magnetic field configuration quickly.

EXPERIMENTAL PROCEDURE

A. Electric Field

1. An electric field mapping setup with a galvanometer is shown in ● Fig. 4a. The apparatus consists of a flat board on which is placed a sheet of carbonized conducting paper imprinted with a grid. The sheet has an electrode configuration of conducting silver paint, which provides an electric field when connected to a voltage source (e.g., a battery).

    The common electrode configurations ordinarily provided are two dots representing point charges of an electric dipole configuration (Fig. 1c) and two parallel linear electrodes representing a two-dimensional cross section of a parallel-plate capacitor (on the board in the photo in Fig. 4a).

2. Draw the electric dipole configuration on a sheet of graph paper to the same scale and coordinates as those

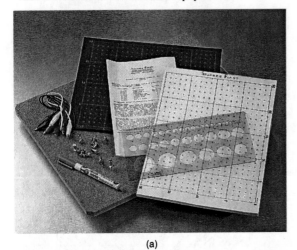

(a)

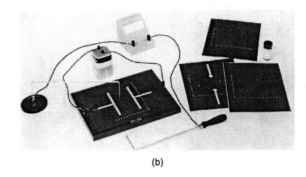

(b)

(c)

Figure 4   Electric field mapping equipment.   (a) Equipment for painting electrodes on conductive paper in preparation for measuring voltages to map equipotentials. (b) A parallel-plate capacitor configuration on the board and an electric dipole configuration to right. (c) This apparatus uses conductive plates, and the mapping is done on graph paper. [Courtesy of Sargent-Welch.]

of the painted dipole on the imprinted grid on the conducting sheet. Then place the dipole conducting sheet on the board, and set the contact terminals firmly on the painted electrode connections. If you are using a galvanometer, do procedures 3 through 7. If you are using a voltmeter, do procedures 8 through 12.

GALVANOMETER MEASUREMENTS

3. Connect the probes to the galvanometer as shown in Fig. 4b. The probes are used to locate points in the field that are at equipotential. Connect the voltage source (1.5-V battery) to the board terminals. Place a switch in the circuit (not shown in the figure) and leave it open until you are ready to take measurements.

Place the stationary probe on the electric dipole sheet at some general point near the edge of the grid area in the region between the electrodes. The potential at this point will serve as a reference potential. Mark the probe position on your graph-paper map.

The movable probe is then used to determine the location of a series of other points that have the same potential. When the movable probe is at a point with the same potential as that of the stationary reference probe, no deflection will be observed on the galvanometer.

4. Close the switch and place the movable probe on the conducting paper at some location an appreciable distance away from the stationary probe. Move the probe

until the galvanometer shows zero deflection (indicating a point of equipotential), and record this point on the graph-paper map.

Locate a series of 8 or 10 points of the same potential across the general field region, and draw a dashed-line curve through these points on the graph-paper map.

5. Choose a new location for the reference probe, 2 to 3 cm from the previous reference position, and locate another series of equipotential points. Continue this procedure until you have mapped the field region. Open the switch.

Draw curves perpendicular to the equipotential lines on the graph-paper map to represent the electric field lines. Do not forget to indicate the field direction on the field lines.

6. Repeat the procedure for the parallel linear (plate) electrode configuration. Be sure to investigate the regions around the ends of the plate electrodes.

7. (*Optional*) Your instructor may wish to have you map the electric field for a nonsymmetric electrode configuration or a configuration of your own choosing. These can be prepared by painting the desired electrode configuration on a conducting sheet with silver paint.

VOLTMETER MEASUREMENTS

8. For the high-resistance voltmeter (or VTVM), the field probe should have two contacts mounted about 2 cm apart. Connect the voltage source (10-V dc) to the board terminals. Place a switch in the circuit (not shown in Fig. 4b) and leave it open until you are ready to take measurements.

Close the switch, and with the zeroed voltmeter set on the 10-V scale, position the negative (−) contact of the field probe near the negative electrode. Using the negative probe point as a pivot, rotate the positive (+) contact around the fixed negative contact until the position with the maximum meter reading is found.

Record the positions of the probe contacts on the graph-paper map. (The sensitivity of the voltmeter may be increased by switching to a lower scale. A mid-scale reading is desirable.

9. Using the second probe point as a new negative probe point, repeat the procedure to determine another point of maximum meter reading, and record. Continue this procedure until the positive electrode is approached. Draw a smooth curve through these points on the graph-paper map.

Then, starting again at a new position near the negative electrode, repeat these procedures for another field line. Trace out four to six field lines in this man-

ner. Do not forget to indicate the field direction on the lines.

10. Place the negative probe near the center of the field region, and rotate the positive contact until a position is found that gives a *zero* meter reading. Record several of these points on the graph paper with a symbol different from that used for the field lines. Check the zero on the voltmeter frequently, particularly when changing scales.

Use the second point as a new pivot point, as before, and determine a series of null (zero) points. Draw a dashed-line curve through these equipotential points. Determine three to five equipotential lines in this manner.

11. Repeat this procedure for the parallel linear (plate) electrode configuration. Be sure to investigate the regions around the ends of the plate electrodes.

12. (*Optional*) Your instructor may wish to have you map the electric field for a nonsymmetric electrode configuration or a configuration of your own choosing. These can be prepared by painting the desired electrode configuration on a conducting sheet with silver paint.

B. Magnetic Field

13. Covering the magnets with sheets of paper or transparency material, sprinkle iron filings to obtain an iron filing pattern for each of the arrangements shown in ● Fig. 5.

For the bar magnet arrangements, the magnets should be separated by several centimeters, depending on the pole strengths of the magnets. Experiment with this distance so that there is enough space between the ends of the magnets to get a good pattern.

14. Sketch the observed magnetic field patterns on Fig. 5 in the Laboratory Report. After the patterns have been sketched, collect the iron filings on a piece of paper and return them to the filing container (recycling them for someone else's later use). Economy in the laboratory is important.

15. Place the magnets for each arrangement on a piece of graph paper or regular paper. Draw an outline of the magnets for each arrangement on the paper, and label the poles N and S. Using a small compass, trace out (marking on the paper) the magnetic field lines as smooth curves. Draw enough field lines so that the pattern of the magnetic field can be clearly seen. Do not forget to indicate the field direction on the lines.

16. Draw dashed-line curves perpendicular to the field lines.

Name _____ Section _____ Date _____

Lab Partner(s) _____

EXPERIMENT 16

Fields and Equipotentials
/TI/ *Laboratory Report*

Attach graphs to Laboratory Report.

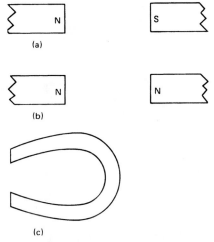

(a)

(b)

(c)

Figure 5   See Procedure Section B.

/TI/ QUESTIONS

1. Directions of the fields are indicated on field lines. Why are no directions indicated on equipotential lines?

2. For the dipole configuration, in what region(s) does the electric field have the greatest intensity? Explain how you know from your map, and justify.

Don't forget units

(continued)

3. Comment on the electric field of the parallel plates (a) between the plates, and (b) near the edges of the plates.

4. Sketch the electric field for (a) a negative point charge near a positively charged plate, and (b) two positive point charges.

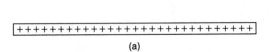

(a)

(b)

5. Compare the electric fields and magnetic fields of the experimental arrangements. Comment on any field similarities and differences.

6. Explain how a gravitational field might be mapped. Sketch the gravitational field for two point masses a short distance apart.

EXPERIMENT 17

Ohm's Law

TI *Advance Study Assignment*

Read the experiment and answer the following questions.

1. What is the definition of *electrical resistance*?

2. What is an "ohmic" resistance? Are all resistances ohmic in nature?

3. In what ways are liquid and electrical circuits analogous?

4. For a series circuit, what is the terminal voltage of a battery or power supply equal to in terms of the potential differences or voltage drops across circuit components?

(continued)

◆ *Advance Study Assignment*

Read the experiment and answer the following questions.

1. What is a triangle-wave voltage function?

2. What is a nonohmic resistance? How can we distinguish between an ohmic and a nonohmic resistance?

Ohm's Law

OVERVIEW

Experiment examines Ohm's law by TI and CI procedures. In the TI procedure, an experimental circuit makes it possible to investigate (1) the variation of current with voltage, and (2) the variation of current and resistance (constant voltage). The CI procedure looks not only at the voltage-current relationship for an ohmic resistance but also at a nonohmic resistance. Steadily increasing and decreasing voltages are obtained by using a signal generator to produce a triangle-wave voltage.

INTRODUCTION AND OBJECTIVES

One of the most frequently applied relationships in current electricity is that known as **Ohm's law.** This relationship, discovered by the German physicist Georg Ohm (1787–1854), is fundamental to the analysis of electrical circuits. Basically, it relates the voltage (V) and current (I) associated with a resistance (R).

Ohm's law applies to many, but not all, materials. Many materials show a constant resistance over a wide range of applied voltages and are said to be "ohmic." Those which do not are said to be "nonohmic." Common circuit resistors are ohmic, which allows Ohm's law to be used in simple circuit analysis. As we shall see in the theory section, Ohm's law is really a special case of the definition of resistance.

In this experiment, Ohm's law will be investigated as applied to components in a simple circuit.

After performing this experiment and analyzing the data, you should be able to:

TI/ OBJECTIVES

1. Distinguish between ohmic and nonohmic resistances.
2. Explain current-voltage relationships by Ohm's law.
3. Apply Ohm's law to obtain values of current or voltage in investigating a circuit resistance.

CI OBJECTIVES

1. Verify Ohm's law experimentally.
2. Study the behavior of the current in both an ohmic and a nonohmic resistance.

Ohm's Law

EQUIPMENT NEEDED*

- Ammeter (0 to 0.5 A)
- Voltmeter (0 to 10 V dc)
- (or multimeters)
- Decade resistance box (0.1 to 99.9 Ω)
- Rheostat (≈ 200 Ω)
- Unknown resistance

- Battery or power supply (6 V)
- Switch
- Connecting wires
- 2 sheets of Cartesian graph paper

* The ranges of the equipment are given as examples. These may be varied to apply to available equipment.

THEORY

When a voltage or potential difference (V) is applied across a material, the current (I) in the material is found to be proportional to the voltage, $I \propto V$. The resistance (R) of the material is defined as the ratio of the applied voltage and the resulting current—that is,

$$R = \frac{V}{I} \qquad \textbf{(TI 1)}$$

(definition of electrical resistance)

For many materials, the resistance is constant, or at least approximately so, over a range of voltages. A resistor that has constant resistance is said to obey Ohm's law or to be "ohmic." From TI Eq. 1, it can be seen that the unit of resistance is volt/ampere (V/A). However, the combined unit is called the ohm (Ω), in honor of Georg Ohm. Note that to avoid confusion with a zero, the ohm is abbreviated with a capital omega (Ω) instead of a capital "O".

A plot of V versus I for an ohmic resistance is a straight line (● TI Fig. 1). Materials that do not obey Ohm's law are said to be "nonohmic" and have a nonlinear voltage-

current relationship. Semiconductors and transistors are nonohmic.

In common practice, **Ohm's law** is written

$$V = IR \qquad \textbf{(TI 2)}$$

where it is understood that R is independent of V. Keep in mind that Ohm's law is not a fundamental law such as Newton's law of gravitation. It is a special case, there being no law that materials must have constant resistance.

To understand the relationships of the quantities in Ohm's law, it is often helpful to consider the analogy of a liquid circuit. (● TI Fig. 2).* In a liquid circuit, the force

* Keep in mind that an analogy only illustrates a resemblance. Liquid and electrical circuits are physically quite different.

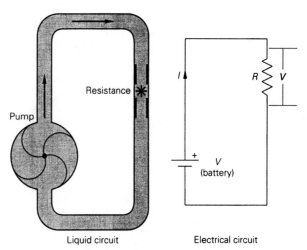

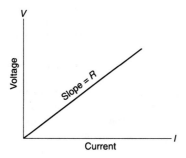

TI Figure 1 Ohmic resistance. A voltage-versus-current graph for an ohmic resistance is a straight line, the slope of which is equal to the value of the resistance ($R = V/I$).

TI Figure 2 Analogy to a liquid circuit. In the analogy between a simple electric circuit and a liquid circuit, the pump corresponds to a voltage source, the liquid flow corresponds to electric current, and the paddle wheel hindrance to the flow is analogous to a resistor.

to move the liquid is supplied by a pump. The rate of liquid flow depends on the resistance to the flow (e.g., due to some partial obstruction in the circuit pipe, here a paddle wheel)—the greater the resistance, the less liquid flow.

Analogously, in an electrical circuit, a voltage source (e.g., a battery or power supply) supplies the voltage (potential difference) for charge flow, and the magnitude of the current is determined by the resistance R in the circuit. For a given voltage, the greater the resistance, the less current through the resistance, as may be seen from Ohm's law, $I = V/R$. Notice that the voltage source supplies a voltage "rise" that is equal to the voltage "drop" across the resistance and is given by $V = IR$ (Ohm's law).

In an electrical circuit with two or more resistances and a single voltage source, Ohm's law may be applied to the entire circuit or to any portion of the circuit. When it is applied to the entire circuit, the voltage is the terminal input voltage supplied by the voltage source, and the resistance is the total resistance of the circuit. When Ohm's law is applied to a particular portion of the circuit, the individual voltage drops, currents, and resistances are used for that part of the circuit.

Consider the circuit diagram shown in ● TI Fig. 3. This is a series circuit. The applied voltage is supplied by a power supply or battery. R_h is a rheostat, a variable resistor that allows the voltage across the resistance R_s to be varied. (This combination is sometimes called a *voltage divider* because the rheostat divides the applied voltage across itself and R_s.)

An ammeter Ⓐ measures the current through the resistor R_s, and a voltmeter Ⓥ registers the voltage drop across both R_s and the ammeter Ⓐ. S is a switch for closing and opening (activating and deactivating) the circuit.

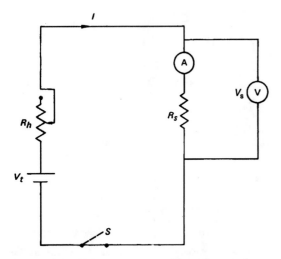

TI Figure 3   Circuit diagram.   The voltmeter is connected in parallel across the ammeter and the resistance R_s. The other resistance, R_h, is that of the rheostat (continuously variable resistor).

Any component in a circuit that does not generate or supply a voltage acts as a resistance in the circuit. This is true for the connecting wires, the ammeter, and the voltmeter. However, the metallic connecting wires and the ammeter have negligibly small resistances, so they do not greatly affect the current.

A voltmeter has a high resistance, so there is little current through the voltmeter. Hence, to good approximations, the ammeter registers the current in the resistor, and the voltmeter reads the voltage drop across the resistance. These approximations are adequate for most practical applications.

Applying Ohm's law to the portion of the circuit with R_s only, we have

$$V_s = IR_s \qquad \text{(TI 3)}$$

where V_s and I are the voltmeter and ammeter readings, respectively. Notice that the same current I flows through the rheostat R_h and the resistance R_s. The voltage drop across R_h is then

$$V_h = IR_h \qquad \text{(TI 4)}$$

To apply Ohm's law to the entire circuit, we use the fact that the applied voltage "rise" or the terminal voltage V_t of the voltage source must equal the voltage "drops" of the components around the circuit. Then

$$V_t = V_h + V_s$$

or

$$V_t = IR_h + IR_s = I(R_h + R_s) \qquad \text{(TI 5)}$$

From TI Eq. 5, we see that for a constant R_s, the current through this resistance, and hence its voltage drop V_s, can be varied by varying the rheostat resistance R_h. (The terminal voltage, V_t, is constant.) Similarly, when R_s is varied, the voltage V_s can be maintained constant by adjusting R_h.

/TI/ EXPERIMENTAL PROCEDURE

1. With the voltmeter, measure the terminal voltage of the power supply or battery, and record it in the Laboratory Report. Start with the voltmeter connection to the largest scale, and increase the sensitivity by changing to a smaller scale if necessary. Most common laboratory voltmeters and ammeters have three scale connections and one binding post common to all three scales.

   It is good practice to take measurements initially with the meters connected to the largest scales. This prevents the instruments from being "pegged" (the needle forced off scale in galvanometer-type meters) and possibly damaged, should the magnitude of the voltage or current exceed the smaller scale limits. A scale setting may be changed for greater sensitivity by moving the connection (or turning the switch on a multimeter) to a lower scale after the general magnitude and measurement are known.

Also, take care to ensure the proper polarity (+ and −); connect + to +, and − to −. Otherwise, the meter will be "pegged" in the opposite direction.

2. Set up the circuit shown in the circuit diagram (TI Fig. 3) with the switch open. A standard decade resistance box is used for R_s. Set the rheostat resistance R_h for maximum resistance and the value of the R_s to about 50 Ω. *Have the instructor check the circuit before closing the switch.*

A. Variation of Current with Voltage

3. After the instructor has checked the circuit, close the switch and read the voltage and current on the meters. Open the switch after the readings are taken, and record them in TI Data Table 1. Repeat this procedure for a series of four successively lower rheostat settings along the length of the rheostat.

It is convenient for data analysis to adjust the rheostat (after closing the switch) so that evenly spaced and convenient ammeter readings are obtained. The switch should be closed only long enough to obtain the necessary readings. This prevents unnecessary heating in the circuit and running the battery down.

4. Repeat procedure 3 for another value of R_s (about 30 Ω).

5. Repeat procedure 3 for the unknown resistance, and record data in TI Data Table 2. Relatively low values of voltage may be required. Your instructor will discuss this and the proper connection. *Do not proceed with this procedure without instructions.*

6. Plot the results for both decade box resistances on a single V_s-versus-I_s graph, and draw straight lines that best fit the sets of data. Determine the slopes of the

lines, and compare them with the constant values of R_s of the decade box by computing the percent errors. According to Ohm's law, the corresponding values should be equal.

7. Plot V_s versus I_s for the unknown resistance. What conclusions about the unknown resistance can you draw from the graphs?

B. Variation of Current and Resistance (V_s constant)

8. This portion of the experiment uses the same circuit arrangement as before. In this case, the voltage V_s is maintained constant by adjusting the rheostat resistance R_h when the R_s is varied.

Initially, set the rheostat near maximum resistance and the resistance R_s of the decade box to about 100 Ω. Record the value of R_s in TI Data Table 3.

Close the circuit and adjust the rheostat for a convenient voltmeter reading (about 4 V). Record the voltmeter reading as the constant voltage V_s in TI Data Table 3. Record the current and resistance in the table. Open the circuit after making the readings.

9. Repeat this procedure for four more successive steps of current by reducing the value of R_s of the decade box. Keep the voltage across R_s constant for each setting by adjusting the rheostat resistance R_h. Do not reduce R_s below 30 Ω.

10. Plot the results on an I_s-versus-$1/R_s$ graph and draw a straight line that best fits the data. (Reciprocal ohms, $1/R$, is commonly given the unit name "mhos.") Determine the slope of the line, and compare it with the constant value of V_s by computing the percent error. According to Ohm's law, these values should be equal.

T I **E X P E R I M E N T 17**

Ohm's Law

TI *Laboratory Report*

A. *Variation of Current with Voltage*

TI **DATA TABLE 1**   Terminal voltage V_t _____

Reading	Constant R_s _____		Constant R_s _____	
	Voltage V_s ( )	Current I_s ( )	Voltage V_s ( )	Current I_s ( )
1				
2				
3				
4				
5				

Calculations
(show work)

Slope of lines   Percent error from R_s

_____   _____

_____   _____

Don't forget units

(continued)

/TI/   **DATA TABLE 2**   Unknown Resistance

Reading	Voltage V_s ( )	Current I_s ( )
1		
2		
3		
4		
5		
6		
7		

Conclusions from graph:

EXPERIMENT 17

B. Variation of Current with Resistance (V_s constant)

/T1/ **DATA TABLE 3** Constant voltage V_s _____

Reading	Current I_s ( )	Resistance R_s ( )	$1/R_s$ ( )
1			
2			
3			
4			
5			

Calculations Slope of line _____
(show work)
Percent error from V_s _____

/T1/ QUESTIONS

1. If the switch were kept closed during the procedures and the circuit components heated up, how would this affect the measurements? (*Hint:* See Experiment 20.)

(continued)

2. Devise and draw a circuit using a long, straight wire resistor instead of a decade box that would allow the study of the variation of voltage with resistance (I_s constant). According to Ohm's law, what would a graph of the data from this circuit show?

3. Compute the values of R_h and the voltage drops across this resistance for the two situations in TI Data Table 1, reading 1.

CI EQUIPMENT NEEDED

This activity is designed for the Science Workshop 750 Interface, which has a built in function generator. It is easily adapted to use with an external wave function generator. Just substitute the available triangle-function generator for the signal generator in the procedure.

- 100-Ω resistor
- 2 cables with alligator clips
- 6-V light bulb
- Voltage sensor (PASCO CI-6503)
- Science Workshop 750 Interface

CI THEORY

As discussed in the TI Theory section, for many materials the resistance remains constant over a range of voltages. Such materials are called "ohmic" and they obey **Ohm's law:**

$$V = IR \qquad \textbf{(CI 1)}$$

For such a material, a graph of voltage versus current is a straight line, the slope of which is the value of the resistance, as shown in TI Fig. 1.

In this CI part of the experiment, we will investigate the relationship between current and voltage for both an ohmic and a nonohmic component of a circuit. The current will be measured as the voltage across a component is steadily increased and decreased. If the component is ohmic, the current should be directly proportional to the voltage.

To achieve a steadily increasing and decreasing voltage, we will use a signal generator, which can produce what is called a triangle-wave voltage. ● CI Fig. 1 shows how the voltage from such a source varies with time. Notice

that it increases up to a maximum value, then drops steadily back to zero, and then, with a change of polarity, increases in the opposite direction. This repeats with a certain fixed frequency.

SETTING UP DATA STUDIO

1. Open Data Studio and choose "Create Experiment."
2. From the sensor list, choose a Voltage Sensor. Connect the voltage sensor to the interface, as shown in the computer screen.
3. Right below the sensor list there is an icon for the Signal Output. Double-click on the icon. The signal generator window will open. (See ● CI Fig. 2.)
4. The default form of the signal generator function is a sine wave. Change it to a triangle wave by selecting from the drop menu.
5. Set the amplitude to 5.00 volts.
6. Set the frequency to 0.500 Hz. This will produce a triangle wave with a period of 2 seconds.
7. Click on the Measurement and Sample Rate button on the signal generator window. A list of measurements will open. Choose to measure the output current, and deselect all others.
8. Do not close the signal generator window. Move it toward the bottom of the screen.
9. The data list should now have two icons: one for the voltage reading of the sensor and one for the output current of the source.
10. Create a graph by dragging the "Voltage" icon from the Data list and dropping it on the "Graph" icon in the Displays list. A graph of voltage versus time will open. The graph window will be called Graph 1.
11. Drag the "Output Current" icon from the data list, and drop it on top of the X axis of the graph. The time axis should change to a current axis. Graph 1 is now a graph of voltage versus current.
12. ● CI Fig. 3 shows what the screen should look like once the setup is complete. The size of the graph window can be changed, if needed. The signal generator window will need to stay visible for procedure B of the experiment, where the output voltage will be manually controlled.

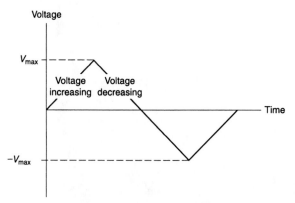

CI Figure 1   A triangle-wave voltage function.   With a triangle-wave voltage function, the voltage will increase up to a maximum value, drop steadily back to zero, and then change the polarity and increase in the opposite direction. This will repeat with a certain fixed frequency.

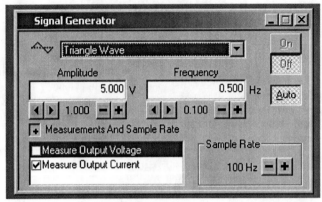

CI Figure 2   The signal generator window.   Choose a triangle-wave function, adjust the amplitude and the frequency as specified in the setup procedure, and choose to measure the output current. (Data displayed using DataStudio Software. Reprinted courtesy of PASCO scientific.)

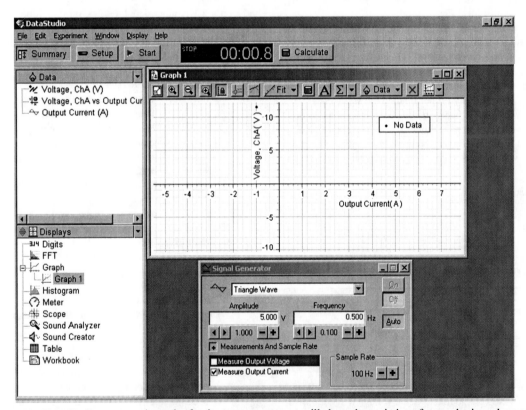

CI Figure   3   Data Studio setup.   A graph of voltage versus current will show the variations for an ohmic and a nonohmic resistor. The signal generator window remains active to manually control the output during experimental procedure B. (Data displayed using DataStudio Software. Reprinted courtesy of PASCO scientific.)

CI EXPERIMENTAL PROCEDURE

A. *Ohmic Component*

1. Connect the signal generator to the 100-Ω resistor. The voltage sensor will measure the voltage drop across the resistor, as shown in the circuit of ● CI Fig. 4.

2. Press the START button and click on the Scale-to-Fit button of the graph toolbar. (That is the leftmost button of the graph toolbar.) After a few seconds, press the STOP button. A cycle is complete after 2 seconds, but it will not affect the experiment if it runs longer than that. In fact, let it run longer and follow the plot on the screen as it appears. What is happening to the current as the voltage changes?

3. Print a copy of the graph and paste it to the Laboratory Report.

4. As expected, the graph for the ohmic resistor is a straight line. Use the Fit drop menu (on the graph toolbar) to select a "Linear Fit" for the data. Record the slope of the line, and compare it to the known value of the resistance by calculating a percent error.

B. *Nonohmic Component*

1. Change the 100-Ω resistor for a small 6-V light bulb.

2. Click on the button labeled "Auto" in the signal generator window. This will cancel the automatic ON/OFF feature of the generator and give manual control of the signal.

3. Press the "On" button of the signal generator.

4. Press the START button and collect data for a few seconds, enough to observe the pattern on the screen. Press the Scale-to-Fit button if needed to see the data better. What is happening to the current now, as the voltage changes?

5. Press the STOP button to end the data collection.

6. Press the "Off" button of the signal generator to turn off the output voltage.

7. Print a copy of the graph and paste it to the Laboratory Report.

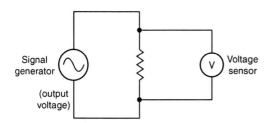

CI Figure 4   The experimental setup.   The resistor (ohmic or nonohmic) is connected to the signal generator. The voltage sensor measures the voltage drop across the resistor.

   **C I**   **E X P E R I M E N T 17**

Ohm's Law
CI *Laboratory Report*

A. *Ohmic Component*

Don't forget to attach the graph to the Laboratory Report.

Calculations
(show work)

Slope of line _____

Percent error from R _____

B. *Nonohmic Component*

Don't forget to attach the graph to the Laboratory Report.

CI QUESTIONS

1. The graph of voltage versus current for the nonohmic resistor was not a straight line. Describe what happened to the current as the voltage increased, compared to what happened for the ohmic resistor.

2. Why does the graph for the nonohmic resistor "loop"? *Hint:* What happens to the light bulb filament as the current increases?

Don't forget units

(continued)

3. Describe what is happening to the resistance of the light bulb as the voltage increases.
   (*Hint:* Look at the graph in segments, and treat each segment as though it were a straight
   line with slope equal to the resistance.)

Name _____ Section _____ Date _____

Lab Partner(s) _____

The Measurement of Resistance: Ammeter-Voltmeter Methods and Wheatstone Bridge Method

$\boxed{\text{TI}}$ *Advance Study Assignment*

Read the experiment and answer the following questions.

A. Ammeter-Voltmeter Method

1. When one is measuring a resistance with an ammeter and voltmeter, is the resistance given exactly by $R = V/I$? Explain.

2. Comment on the relative magnitudes of the resistances of an ammeter and a voltmeter.

3. Is (a) an ammeter and (b) a voltmeter connected in series or parallel with a circuit component (a resistance)? Explain.

(continued)

B. *Wheatstone Bridge Method*

4. Why is the Wheatstone bridge called a "null" instrument?

5. When the galvanometer in a Wheatstone bridge circuit shows no deflection, why are the voltages across opposite branches on each side of the galvanometer necessarily equal?

6. For a slide-wire Wheatstone bridge, why should the sliding key not be moved with the key depressed?

The Measurement of Resistance:
Ammeter-Voltmeter Methods
and Wheatstone Bridge Method

INTRODUCTION AND OBJECTIVES

The magnitude of a resistance can be measured by several methods. One common method is to measure the voltage drop V across a resistance in circuit with a voltmeter and the current I through the resistance with an ammeter. By Ohm's law, then, $R = V/I$. However, the ratio of the measured voltage and current does not give an exact value of the resistance because of the resistances of the meters.

This problem is eliminated when one measures a resistance, or, more properly, compares a resistance with a standard resistance in a Wheatstone bridge circuit [named after the British physicist Sir Charles Wheatstone (1802–1875)].* In this experiment, the ammeter-voltmeter and the Wheatstone bridge methods of measuring resistances will be investigated.

After performing this experiment and analyzing the data, you should be able to:

1. Describe the two ways to measure resistance with an ammeter and voltmeter, and explain how they differ.
2. Describe the basic principle and operation of the Wheatstone bridge.
3. Discuss the relative accuracy of the ammeter-voltmeter methods and the Wheatstone bridge method of measuring resistance.

* The Wheatstone bridge was popularized and promoted by Sir Charles Wheatstone; however, the British mathematician Samuel Christie invented it.

EQUIPMENT NEEDED[†]

A. Ammeter-Voltmeter Methods

- Ammeter (0 to 0.5 A)
- Voltmeter (0 to 3 V)
- Rheostat (10 Ω)
- Resistors (e.g., 10 Ω and 25 Ω)
- Battery or power supply (3 V)
- Connecting wires

B. Wheatstone Bridge Method

- Slide-wire Wheatstone bridge
- Galvanometer
- Standard decade resistance box (0.1 to 99.9 Ω)
- Single-pole, single-throw switch

† The ranges of the equipment are given as examples. These may be varied to apply to available equipment.

THEORY

A. Ammeter-Voltmeter Methods

There are two basic arrangements by which resistance is measured with an ammeter and a voltmeter. One circuit is shown in ● Fig. 1. The current I through the resistance R is measured with an ammeter, and the potential difference or voltage drop V across the resistance is measured with a voltmeter. Then, by Ohm's law, $R = V/I$.

Strictly speaking, however, this value of the resistance is not altogether correct, since the current registered on the ammeter divides between the resistance R and the voltmeter in parallel. A voltmeter is a high-resistance instru-

ment and draws relatively little current, provided that voltmeter resistance R_v is much greater than R. Hence, it is more appropriate to write

$$R \simeq \frac{V}{I} \quad \text{if} \quad R_v \gg R \quad (1)$$

For more accurate resistance measurement, one must take the resistance of the voltmeter into account. The current drawn by the voltmeter is $I_v = V/R_v$. Since the total current I divides between the resistance and the voltmeter in the parallel branch, we have

$$I = I_R + I_v$$

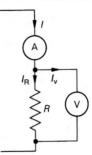

Figure 1 Resistance measurement. One of the basic arrangements for measuring resistance with an ammeter and a voltmeter. The ammeter measures the sum of the currents through the resistance and the voltmeter. Therefore, the true value of R is *greater* than the measured value, if the measured value is taken to be V/I.

or

$$I_R = I - I_v \tag{2}$$

where I_R is the true current through the resistance. Then, by Ohm's law,

$$R = \frac{V}{I_R} = \frac{V}{I - I_v} = \frac{V}{I - V/R_v} \tag{3}$$

Another possible arrangement for measuring R is shown in the circuit diagram in ● Fig. 2. In this case, the ammeter measures the current through R alone, but now the voltmeter reads the voltage drop across *both* the ammeter *and* the resistance. Since the ammeter is a low-resistance instrument, to a good approximation

$$R \simeq \frac{V}{I} \qquad \text{if} \qquad R_a \ll R \tag{4}$$

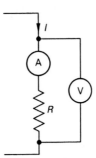

Figure 2 Resistance measurement. Another basic arrangement for measuring resistance with an ammeter and a voltmeter. The ammeter measures the current through R, but the voltmeter is across R *and* the ammeter. Therefore, the true value of R is *less* than the measured value, if the measured value is taken to be V/I.

where R_a is the resistance of the ammeter. When $R_a \ll R$, the voltage drop across R_a—that is, $V_a = IR_a$—is small compared to that across R, which is $V_R = IR$.

Taking the voltage drop or the resistance of the ammeter into account, we have

$$V = V_R + V_a = IR + IR_a$$
$$= I(R + R_a) = IR'$$

and

$$R' = R + R_a \tag{5}$$

Or, from Eq. 5, we may write

$$R = \frac{V}{I} - R_a$$

B. Wheatstone Bridge Method

The basic diagram of a Wheatstone bridge circuit is shown in ● Fig. 3. In its simplest form, the bridge circuit consists of four resistors, a battery or voltage source, and a sensitive galvanometer. The values of R_1, R_2, and R_s are all known, and R_x is the unknown resistance.

Switch S is closed, and the bridge is balanced by adjusting the standard resistance R_s until the galvanometer shows no deflection (indicating no current flow through the galvanometer branch). As a result, the Wheatstone bridge is called a "null" instrument. This is analogous to an ordinary double-pan beam balance, which shows a null or zero reading when there are equal masses on its pans.

Assume that the Wheatstone bridge is balanced so that the galvanometer registers no current. Then points b and c in the circuit are at the same potential; current I_1 flows through both R_s and R_x, and current I_2 flows through both R_1 and R_2.

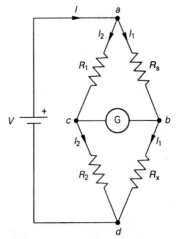

Figure 3 Wheatstone bridge circuit diagram. See text for description.

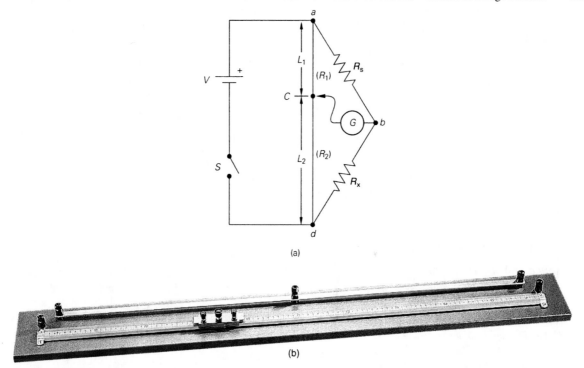

(a)

(b)

Figure 4 Slide-wire Wheatstone bridge.   (a) Circuit diagram for resistance measurements. The resistances R_1 and R_2 are varied by sliding the contact C along the wire. The galvanometer G is used to indicate when the bridge is "balanced." (b) The contact slides over the wire on a meterstick. When contact is made, the lengths L_1 and L_2 on either side of C are easily read. (Courtesy of Sargent-Welch.)

Also, the voltage drop V_{ab} across R_s is equal to the voltage drop across R_1, V_{ac}, for a zero galvanometer deflection:

$$V_{ab} = V_{ac}$$

Similarly,

$$V_{bd} = V_{cd} \qquad (6)$$

(Why?)

Writing these equations in terms of currents and resistances, by Ohm's law,

$$I_1 R_x = I_2 R_2$$
$$I_1 R_s = I_2 R_1 \qquad (7)$$

Then, dividing one equation by the other and solving for R_x yields

$$R_x = \left(\frac{R_2}{R_1}\right) R_s \qquad (8)$$

Hence, when the bridge is balanced, the unknown resistance R_x can be found in terms of the standard resistance R_s and the ratio R_2/R_1.

Notice that the difficulties of the preceding method are eliminated. The Wheatstone bridge in effect compares the

unknown resistance R with a standard resistance R_s. Should $R_1 = R_2$, then $R_x = R_s$.

The circuit diagram for a slide-wire form of the Wheatstone bridge is shown in ● Fig. 4 along with a photo of an actual bridge. The line from a to d represents a wire, and C is a contact key that slides along the wire so as to divide the wire into different-length segments.

The resistances of the segments are proportional to their lengths, so the resistance ratio may be expressed in terms of a length ratio:

$$\frac{R_2}{R_1} = \frac{L_2}{L_1} \qquad (9)$$

Equation 8 can then be written in terms of the length   ratio:

$$R_x = \left(\frac{L_2}{L_1}\right) R_s \qquad (10)$$

This type of bridge is convenient since the length segments can be measured easily. The resistances R_1 and R_2 of the length segments may be quite small relative to R_x and R_s because the bridge equation depends only on the ratio R_2/R_1 or L_2/L_1. This fact makes it possible to use a wire as one side of the bridge.

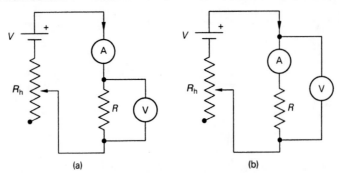

(a)   (b)

Figure 5   Resistance measurement.   Circuit diagrams for experimental procedures for ammeter-voltmeter methods of measuring resistance.

EXPERIMENTAL PROCEDURE

A. Ammeter-Voltmeter Methods

1. Set up a circuit as shown in ● Fig. 5a, where R is a small known resistance and R_h is the rheostat. However, *do not connect the wire to the positive side of the battery until the instructor has checked it.* Record the value of R in the first of the spaces provided for this purpose in part A of the Laboratory Report.

   Most common meters have three scale connections, with a binding post common to all three scales. It is good practice initially to make connections to the largest scale. This prevents the instruments from being "pegged" (and possibly damaged) should the magnitudes of the current and voltage exceed the smaller scales' limits.

   The scale setting may be changed for greater sensitivity by moving the connection to a lower scale after the general magnitude of a measurement is known.

   Also, attention should be given to the proper polarity (+ and −). Otherwise, the meter will be "pegged" in the opposite direction. Connect + to + and − to −. However, *do not activate the circuit until your laboratory instructor has checked it.*

2. The current in the circuit is varied by varying the rheostat resistance R_h. Activate the circuit and take three different readings of the ammeter and voltmeter for three different currents. Adjust R_h so that the three currents differ as much as possible. Record the data in Data Table 1 and deactivate the circuit after each of the three readings until the rheostat is set for the next reading.

   Also, record the resistance of the voltmeter. The resistance of the meter will be found on the meter face or will be supplied by the instructor. The voltmeter resistance is commonly given as so many ohms per volt, which is the total resistance of the meter divided by the full-scale reading.

For example, if the meter has a resistance of 1000 Ω/V and the full-scale reading of a particular range is 3 V, then $R_v = 3$ V$(1000\ \Omega$/V$) = 3000\ \Omega$. The resistance in ohms/volt applies to any range setting of the meter. (*Note:* If voltmeter scales are changed during readings, R_v will be different for different sets of V and I measurements. Be sure to record this if it occurs.)

3. Using Eq. 3, compute the value of R for each current setting and find the average value. Compare this with the accepted value by finding the percent error.

4. Set up a circuit as shown in Fig. 5b. This is accomplished by changing only one wire in the previous circuit. Repeat the measurements as in procedure 2 for this circuit, recording your findings in Data Table 2.

5. (a) Compute the resistance $R' = V/I$ directly from each set of current and voltage measurements, and find the average value.
   (b) When one is not taking into account the ammeter resistance, R' is taken to be the value of the resistance R. Compare the average experimental value of R' with the accepted value of R by finding the percent error.
   (c) Using the values of R and R', compute R_a (Eq. 5). Mentally compare the magnitudes of the ammeter and voltmeter resistances.

6. Repeat the previous procedures with a large known resistance. Record its accepted value in the space provided above Data Table 4, and use Data Tables 3 and 4 for your findings.

B. Wheatstone Bridge Method

7. Set up a slide-wire Wheatstone bridge circuit as in Fig. 4a, using the previous small known resistance R as R_x. *Leave the switch open until the instructor checks*

the circuit. The wires connecting the resistances and the bridge should be as short as practically possible. The decade resistance box is used for R_s. This should be initially set for a value about equal to R_x.

Contact is made to the wire by sliding contact key C. *Do not slide the key along the wire while it is pressed down.* This will scrape the wire, causing it to be nonuniform. *Have the instructor check your setup before activating the circuit.*

8. Activate the circuit by closing the switch *S*, and balance the bridge by moving the slide-wire contact. Open the switch and record R_s, L_1, and L_2 in Data

Table 5. Leave the switch open except when you are actually making measurements.

9. Repeat procedures 7 and 8 for R_s settings of (a) $R_s \approx 3R_x$, and (b) $R_s \approx 0.3R_x$.

10. Compute the value of R_x for each case and find the average value. Compare this value to the accepted value of R by finding the percent error.

11. Repeat the previous procedures with a large known resistance. Record your findings in Data Table 6.

EXPERIMENT 18

The Measurement of Resistance: Ammeter-Voltmeter Methods and Wheatstone Bridge Method

|TI| *Laboratory Report*

A. Ammeter-Voltmeter Methods

DATA TABLE 1
Purpose: To measure resistance values.

Accepted value of R _____

Voltmeter resistance R_v _____

Rheostat setting R_h	Current I ( )	Voltage V ( )	Resistance R ( )
1			
2			
3			
		Average R	

DATA TABLE 2
Purpose: To measure resistance values.

Rheostat setting R_h	Current I ( )	Voltage V ( )	$R' = V/I$ ( )
1			
2			
3			
		Average R'	

Percent error of R _____

Percent error of R' _____

Ammeter resistance R_a _____

Calculations
(show work)

Don't forget units

(continued)

DATA TABLE 3

Purpose: To measure resistance values.

Voltmeter resistance R_v _____

Rheostat setting R_h	Current I ( )	Voltage V ( )	Resistance R ( )
1			
2			
3			
		Average R	

Percent error of R _____

Calculations
(show work)

Accepted value of R _____

DATA TABLE 4

Purpose: To measure resistance values.

Rheostat setting R_h	Current I ( )	Voltage V ( )	$R' = V/I$ ( )
1			
2			
3			
		Average R'	

Percent error of R' _____

Ammeter resistance R_a _____

EXPERIMENT 18   *Laboratory Report*

B. Wheatstone Bridge Method

DATA TABLE 5

Purpose: To measure resistance values.

Accepted value of R _____

R_s ( )	L_1 ( )	L_2 ( )	R ( )
		Average R	

Percent error _____

DATA TABLE 6

Purpose: To measure resistance values.

Accepted value of R _____

R_s ( )	L_1 ( )	L_2 ( )	R ( )
		Average R	

Percent error _____

Calculations
(show work)

/TI/ QUESTIONS

A. Ammeter-Voltmeter Methods

1. An ideal ammeter would have zero resistance, and an ideal voltmeter would have an infinite resistance. Explain why we would desire these ideal cases when using the meters.

(continued)

2. If in general R were calculated as $R = V/I$, which circuit arrangement in part A of the experiment would have the smallest error? Explain.

3. (a) Prove that the true resistance R is given by

$$R = R'\left(1 - \frac{R_a}{R'}\right)$$

where $R' = V/I$ is the measured resistance as given by the voltmeter and ammeter readings for measurements done by the arrangement in Fig. 2 or Fig. 5b. Is the true resistance larger or smaller than the apparent resistance?

(b) Prove that the true resistance R is given approximately by

$$R = R'\left(1 + \frac{R'}{R_v}\right)$$

where $R' = V/I$ is the measured resistance as given by the voltmeter and ammeter readings for measurements done by the arrangement in Fig. 1 or Fig. 5a.
Hint: Use the binomial theorem,

$$\frac{1}{1 - \dfrac{R'}{R_v}} \simeq 1 + \frac{R'}{R_v}$$

Is the true resistance larger or smaller than the apparent resistance? Explain.

4. For each of the circuits used in the preceding question, for what values of R (large or small) does the error in taking R as equal to V/I become large enough to be important?

B. Wheatstone Bridge Method

5. Why should the wires connecting the resistances and the bridge be as short as possible?

6. Suppose that the slide-wire on the bridge did not have a uniform cross section. How would this affect your measurements? Was there any experimental evidence of this?

EXPERIMENT 18

The Temperature Dependence of Resistance

$\boxed{\text{TI}}$ *Advance Study Assignment*

Read the experiment and answer the following questions.

1. Does the resistance of all substances increase with temperature? Explain.

2. What is the temperature coefficient of resistance, and what are its units?

3. Distinguish between a positive and a negative temperature coefficient of resistance.

4. Are the α of a metal conductor and the β of a thermistor the same? Explain.

(continued)

5. What are the circuit conditions when a Wheatstone bridge is "balanced"?

6. In using the equations to determine the temperature dependence of resistance, what temperature scale is used?

The Temperature Dependence of Resistance

INTRODUCTION AND OBJECTIVES

The electrical resistance of all substances varies somewhat with temperature. For pure metals and most alloys, the resistance increases with increasing temperature. However, for some substances, such as carbon and many electrolytes (conducting solutions), the resistance decreases with increasing temperature. Then, too, for some special alloys [for example, constantan (55% Cu–45% Ni)] the resistance is virtually independent of temperature over a limited range.

The temperature dependence of resistance for a substance is commonly expressed in terms of its **temperature coefficient of resistance,** which is the fractional change in the resistance per degree change in temperature. For many electrical applications, it is important to know temperature coefficients and to take into account the temperature dependence of resistances. In this experiment, this temperature dependence will be investigated and the temperature coefficients of some materials will be determined.

After performing this experiment and analyzing the data, you should be able to:

1. Explain how the resistances of common metallic conductors vary with temperature.
2. Discuss the temperature coefficient of resistance for various materials and the differences among them.
3. Describe what is meant by the exponential temperature coefficient of a thermistor.

EQUIPMENT NEEDED

- Slide-wire Wheatstone bridge assembly (with a 3-V battery and a single-pole, single-throw switch)
- Standard decade resistance box
- Copper coil (and optional constantan or manganese coil)

- Thermistor
- Immersion vessel and stirrer
- Thermometer
- Immersion heater and power source (or Bunsen burner and stand or hot plate)
- 2 sheets of Cartesian graph paper

THEORY

The change in resistance ΔR of a substance is proportional to the change in temperature ΔT. This change in resistance is commonly expressed in terms of the fractional change $\Delta R/R_0$, where R_0 is the initial resistance. For many substances—for example, metals—the change in resistance is to a good approximation a linear function of temperature:

$$\frac{\Delta R}{R_0} = \alpha \Delta T \qquad (1)$$

where the constant of proportionality α is called the **temperature coefficient of resistance** and has the units of inverse temperature ($1/\text{C}°$ or $\text{C}^{°-1}$, that is, per Celsius degree).

For the change in temperature $\Delta T = T - T_0$, it is convenient to take the initial temperature T_0 as 0°C and with $\Delta R = R - R_0$. Equation 1 can be written

$$\frac{R - R_0}{R_0} = \alpha T$$

or

$$R = R_0 + R_0\alpha T = R_0(1 + \alpha T) \qquad (2)$$

where R is then the resistance of the conductor at some temperature T (°C), and R_0 is the resistance at $T_0 = 0$°C. The linearity of the temperature dependence is only approximate, but Eq. 2 can be used over moderate temperature ranges for all but the most accurate work.

In contrast to pure metals, which have positive temperature coefficients of resistance (increase in resistance with increase in temperature), some materials have negative temperature coefficients (decrease in resistance with an increase in temperature). Carbon is an example, and negative temperature coefficients of resistance generally occur in materials of intermediate conductivity, or semiconducting materials.

Carbon has a relatively small negative temperature coefficient of resistance compared to other semiconducting materials. Such materials with large negative temperature coefficients are used in commercial components called **thermistors.** A thermistor is a thermally sensitive resistor made of semiconducting materials such as oxides of manganese, nickel, and cobalt.

Because of relatively large (negative) temperature coefficients, thermistors are very sensitive to small temperature changes and are used for temperature measurement and in a variety of temperature-sensing applications such as voltage regulation and time-delay switches.

Unlike common metal conductors, for a thermistor, the change of resistance with a change of temperature is nonlinear, and the α in Eq. 1 is not constant. The temperature dependence of a thermistor is given by an exponential function,

$$R = R_a e^{\beta(1/T - 1/T_a)} \qquad (3)$$

where
R = resistance at a temperature T (in kelvins, K)
R_a = resistance at temperature T_a (K)
T_a = initial temperature (K), near ambient room temperature in the experiment
e = 2.718, the base of natural logarithms
β = *exponential* temperature coefficient of resistance, which has Kelvin temperature units (K).

In this case, as T increases, the exponential function and hence the resistance R become smaller. This expression can be written in terms of the natural logarithm (to the base e) as

$$\ln\left(\frac{R}{R_a}\right) = \beta\left(\frac{1}{T} - \frac{1}{T_a}\right) \qquad (4)$$

Hence, when we plot $y = \ln(R/R_a)$ versus $x = (1/T - 1/T_a)$ on a Cartesian graph, β is the slope of the line. This, too, is an approximation, but β is reasonably constant for moderate temperature ranges.

The temperature coefficient of resistance of a material can be determined by using an experimental arrangement with a slide-wire Wheatstone bridge circuit, as illustrated in ● Fig. 1. The resistance R_c of a material (coil of wire) when the bridge circuit is balanced is given by

$$R_c = \left(\frac{R_2}{R_1}\right)R_s$$

or

$$R_c = \left(\frac{L_2}{L_1}\right)R_s \qquad (5)$$

where R_s is a standard resistance and R_2/R_1 and L_2/L_1 are the ratios of the resistances and lengths of the slide-wire segments, respectively. (See Experiment 19 for the theory of the Wheatstone bridge.) By measuring the resistance of a material at various temperatures, we can determine the temperature coefficient.

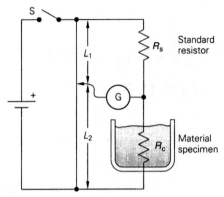

Figure 1   Temperature dependence of resistance.   The circuit diagram for the experimental procedure to measure the temperature dependence of resistance. See text for description.

Experimental Procedure

A. Metal Conductor(s)

1. Set up the circuit as in Fig. 1 with the copper coil in the container of water (near room temperature) and the heating arrangement for the water (immersion heater or other heat source). Place the thermometer in the water. *Have the instructor check your setup.*

2. After your setup has been checked, close the switch and balance the bridge circuit to measure the resistance R_c of the coil at the initial water temperature. The value of the standard resistance R_s should be selected so that the bridge is balanced with the contact key C near the center of the slide-wire. Then with $L_1 \simeq L_2$, we have $R_c \simeq R_s$ (Eq. 5).
   Record in Data Table 1 the initial temperature of the water, the magnitude of R_s, and the lengths of the wire segments of the bridge.

3. *Slowly,* raise the temperature of the water by about $10\,\mathrm{C}°$. Stir the water while heating, and discontinue heating when the temperature is about $2\,\mathrm{C}°$ below the desired temperature. Continue stirring until a maximum steady temperature is reached. Balance the bridge and record the measurements. Adjust R_s if necessary. Record the measurements of temperature and bridge length in the data table.

4. Repeat procedure 3, taking a series of measurements at approximately $10\,\mathrm{C}°$ temperature intervals until a final temperature of about 90°C is reached.

5. (*Optional*) Repeat the foregoing procedures using the constantan wire coil, starting near room temperature. (Use Data Table 1A.)

6. Compute R_c of the coil(s) at the various temperatures and plot a graph of R_c versus T with a temperature range of 0 to 100°C. Draw the straight line(s) that best fit(s) the data, and extrapolate the line(s) to the y axis. Determine the slope and y intercept of the line(s).

   From the slope, find the temperature coefficient of resistance for the specimen(s) and compare with the accepted value found in Appendix A, Table A6, by computing the percent error.

B. Thermistor

7. Replace the coil with the thermistor in the bridge circuit and repeat the previous measurement procedures 1–4 starting at a temperature near room temperature.

In this portion of the experiment, exercise great care in order to have temperatures as constant as possible when making resistance measurements, since a thermistor shows considerable variation in resistance with temperature.

8. (a) Find the quantities listed in the second part of Data Table 2.
   (b) Plot a graph of $y = \ln (R/R_a)$ versus $x = (1/T - 1/T_a) \, \mathrm{K}^{-1}$, and draw the straight line that best fits the data.
   (c) Determine the slope of the line, which is the value of β. Compare this to the accepted value provided by the instructor by computing the percent error.

Name _____ Section _____ Date _____

Lab Partner(s) _____

The Temperature Dependence of Resistance

/TI/ *Laboratory Report*

A. Metal Conductor(s)

DATA TABLE 1

Purpose: To determine the temperature coefficient of resistance.

Material _____

Temperature ( )	Decade box resistance R_s ( )	L_1 ( )	L_2 ( )	Coil resistance $R = (L_2/L_1)R_s$ ( )

Calculations
(show work)

Slope $R_0\alpha$ _____

Intercept R_0 _____

Experimental α _____

Accepted α _____

Percent error _____

Don't forget units

(continued)

DATA TABLE 1A (optional)
Purpose: To determine the temperature coefficient of resistance.

Material _____

Temperature ( )	Decade box resistance R_s ( )	L_1 ( )	L_2 ( )	Coil resistance R_c ( )

Calculations
(show work)

Slope $R_0\alpha$ _____

Intercept R_0 _____

Experimental α _____

Accepted α _____

Percent error _____

EXPERIMENT 19   *Laboratory Report*

B. Thermistor

DATA TABLE 2

Purpose: To determine the exponential temperature coefficient of resistance.

Calculations
(show work)

Temperature T ( )	Decade box resistance R_s ( )	L_1 ( )	L_2 ( )	Thermistor resistance $R = (L_2/L_1)R_s$ ( )
(T_a)				(R_a)

Temperature ( ) $T_K = T_C + 273$	$1/T$	$1/T - 1/T_a$	R/R_a	$\ln(R/R_a)$

Slope β _____

Accepted β _____

Percent error _____

(continued)

/TI/ QUESTIONS

A. *Metal Conductor*

1. What is the value of α for copper in terms of Fahrenheit degrees $(1/F°)$? If the resistance is a linear function on the Celsius scale, will it be a linear function on the Fahrenheit scale? Explain.

2. Replot the copper data for R_c versus T with a smaller temperature scale extending to $-300°C$, and extrapolate the line to the temperature axis. At what temperature would the resistance go to zero? What are the practical electrical implications for a conductor with zero resistance?

   [It is interesting to note that the value of α is roughly the same for many pure metals: approximately $\frac{1}{273}$, or 0.004 C$°^{-1}$. This is the same as the value of the coefficient of expansion of an ideal gas. Also, some metals and alloys do become "superconductors," or have zero resistance at low temperatures. Recently discovered "high-temperature" ceramic materials show superconductivity at liquid nitrogen temperatures (77 K, or $-196°C$, or $-321°F$).]

3. A coil of copper wire has a resistance of 10.0 Ω, and a coil of silver wire has a resistance of 10.1 Ω, both at 0°C. At what temperature would the resistance of the coils be equal?

EXPERIMENT 19   *Laboratory Report*

B. Thermistor

4. Explain why the ambient temperature T_a for the thermistor cannot be taken as $T_a = 0°C$ and why the expression for the resistance is written $R = R_o e^{\beta/T}$, where T is in degrees Celsius.

5. Assuming that β remained constant, what would be the resistance of the thermistor in the experiment as the temperature approached absolute zero?

6. Assume the temperature coefficient of resistance α to be defined over the temperature range $\Delta T = T - T_a$, where $T_a > 273K$ ($0°C$), by $R - R_a = -R_a \alpha(T - T_a)$. Show that for a thermistor, α is a function of temperature given by

$$\alpha = \frac{1 - e^{\beta(1/T - 1/T_a)}}{T - T_a}$$

EXPERIMENT 19

EXPERIMENT 20

Resistances in Series and Parallel

/TI/ *Advance Study Assignment*

Read the experiment and answer the following questions.

1. Explain the difference between series and parallel connections.

2. Consider resistors are connected in series.
   a. How are the voltage drops across the individual resistors related to the voltage supplied by the battery?

   b. How are the currents through the individual resistors related to the current supplied by the battery?

3. Consider resistors are connected in parallel.
   a. How are the voltage drops across the individual resistors related to the voltage supplied by the battery?

   b. How are the currents through the individual resistors related to the current supplied by the battery?

(continued)

4. Give (draw and explain) an analogy to liquid flow for the series-parallel circuit in part C of the experiment.

5. How would the current divide in a parallel branch of a circuit containing two resistors R_1 and R_2 if (a) $R_1 = R_2$, and (b) $R_1 = 4R_2$?

EXPERIMENT 20   *Advance Study Assignment*

CI *Advance Study Assignment*

Read the experiment and answer the following questions.

1. Explain the difference between series and parallel connections.

2. Consider resistors are connected in series.
   a. How are the voltage drops across the individual resistors related to the voltage supplied by the battery?

   b. How are the currents through the individual resistors related to the current supplied by the battery?

3. Consider resistors are connected in parallel.
   a. How are the voltage drops across the individual resistors related to the voltage supplied by the battery?

   b. How are the currents through the individual resistors related to the current supplied by the battery?

(continued)

4. In a plot of voltage versus current, what physical quantity is represented by the slope of the graph?

Resistances in Series and Parallel

OVERVIEW

Experiment examines resistances in parallel and series combinations with both TI and CI procedures. In the TI procedure, the resistances are measured using a voltmeter and ammeter. In the CI procedure, measurements are made with a voltage (and current) sensor, and graphs of V versus I are plotted, from which the resistances are given by the slopes.

INTRODUCTION AND OBJECTIVES

The components of simple circuits are connected in series and/or parallel arrangements. Each component may be represented as a resistance to the current in the circuit. In computing the voltage and current requirements of the circuit (or part of the circuit), it is necessary to know the equivalent resistances of the series and parallel arrangements.

In this experiment, the circuit characteristics of resistors in series and parallel will be investigated. A particular circuit will first be analyzed theoretically, and then those predictions will be checked experimentally.

After performing this experiment and analyzing the data, you should be able to:

/TI/ OBJECTIVES

1. Describe the current-voltage relationships for resistances in series.

2. Describe the current-voltage relationships for resistances in parallel.

3. Reduce a simple series–parallel resistance circuit to a single equivalent resistance, and compute the voltage drops across and the currents through each resistance in the circuit.

CI OBJECTIVES

1. Describe the current-voltage relationships for resistances in series.

2. Describe the current-voltage relationships for resistances in parallel.

3. Describe the changes in the slopes of V-versus-I graphs as more resistors are connected in (a) series and (b) parallel.

Resistances in Series and Parallel

T I EQUIPMENT NEEDED*

- Battery or power supply (3 V)
- Ammeter (0 to 500 mA)
- Voltmeter (0 to 3 V)
- Single-pole, single-throw (SPST) switch

- Four resistors (10 Ω, 20 Ω, 100 Ω, and 10 kΩ, composition type, 1 W)
- Connecting wires

* The ranges of the equipment are given as examples. These may be varied to apply to available equipment.

T I THEORY

A. Resistances in Series

Resistors are said to be **connected in series** when they are connected as in ● TI Fig. 1. (The resistors are connected in line or "head to tail," so to speak, although there is no distinction between the connecting ends of a resistor.) When they are connected to a voltage source V and the switch is closed, the source supplies a current I to the circuit.

By the conservation of charge, this current I flows through each resistor. The voltage drop across each resistor is not equal to V, but the *sum* of the voltage drops is:

$$V = V_1 + V_2 + V_3 \qquad \textbf{(TI 1)}$$

In an analogous liquid-gravity circuit (TI Fig. 1), a pump, corresponding to the voltage source, raises the liquid a distance h. The liquid then falls or "drops" through three paddle wheel "resistors" and the distances h_1, h_2, and h_3. The liquid "rise" supplied by the pump is equal to the sum of the liquid "drops," $h = h_1 + h_2 + h_3$. Analogously, the voltage "rise" supplied by the source is equal to the sum of the voltage drops across the resistors (TI Eq. 1).*

The voltage drop across each resistor is given by Ohm's law (e.g., $V_1 = IR_1$). TI Eq. 1 may be written

$$
\begin{aligned}
V &= V_1 + V_2 + V_3 \\
&= IR_1 + IR_2 + IR_3 \\
&= I(R_1 + R_2 + R_3) \qquad \textbf{(TI 2)}
\end{aligned}
$$

For a voltage across a single resistance R_s in a circuit, $V = IR_s$, and by comparison,

$$\boxed{R_s = R_1 + R_2 + R_3} \qquad \textbf{(TI 3)}$$

(resistances in series)

where R_s *is the equivalent resistance of the resistors in series.* That is, the three resistors in series could be replaced by a single resistor with a value of R_s, and the same current I would be drawn from the battery.

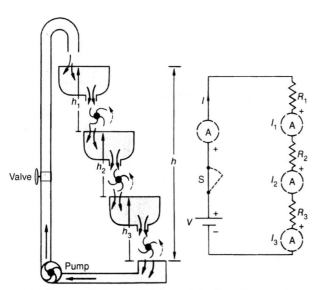

TI Figure 1 Series connection. A liquid analogy on the left for the circuit diagram of resistors in series on the right. The analogies are: pump–voltage source, valve–switch, liquid flow–current, and paddle wheels–resistors. See text for more description.

* Keep in mind that an analogy represents only a resemblance. Liquid and electrical circuits are quite different physically.

B. Resistances in Parallel

Resistors are said to be **connected in parallel** when connected as in ● TI Fig. 2. (In this arrangement, all the "heads" are connected, as are all of the "tails.") The voltage drops across all the resistors are the same and equal to the voltage V of the source. However, the current I from the source divides among the resistors such that

$$I = I_1 + I_2 + I_3 \qquad \textbf{(TI 4)}$$

In the liquid circuit analogy (TI Fig. 2), the height h the pump raises the liquid is equal to the distance the liquid "drops" through each paddle wheel "resistor." The liquid flow coming into the junction of the parallel arrangement divides among the three pipe paths, analogously to the current dividing in the electrical circuit.

The current in an electrical parallel circuit divides according to the magnitudes of the resistances in the parallel branches—the smaller the resistance of a given branch, the greater the current through that branch. The current through each resistor is given by Ohm's law (e.g., $I_1 = V/R_1$), and TI Eq. 4 may be written

$$I = I_1 + I_2 + I_3 = \frac{V}{R_1} + \frac{V}{R_2} + \frac{V}{R_3}$$

$$= V\left(\frac{1}{R_1} + \frac{1}{R_2} + \frac{1}{R_3}\right) \qquad \textbf{(TI 5)}$$

For the current through a single resistance R_p in a circuit, $I = V/R_p$, and by comparison,

$$\boxed{\frac{1}{R_p} = \frac{1}{R_1} + \frac{1}{R_2} + \frac{1}{R_3}} \qquad \textbf{(TI 6)}$$

(resistances in parallel)

where R_p *is the equivalent resistance of the resistors in parallel.* That is, the three resistors in parallel could be replaced by a single resistor with a value of R_p, and the same current I would be drawn from the battery.

The previous developments for equivalent resistances may be extended to any number of resistors (i.e., $R_s = R_1 + R_2 + R_3 + R_4 + \cdots$ and $1/R_p = 1/R_1 + 1/R_2 + 1/R_3 + 1/R_4 + \cdots$).

In many instances, two resistors are connected in parallel in a circuit, and

$$\frac{1}{R_p} = \frac{1}{R_1} + \frac{1}{R_2}$$

or

$$\boxed{R_p = \frac{R_1 R_2}{R_1 + R_2}} \qquad \textbf{(TI 7)}$$

(two resistances in parallel)

This particular form of R_p for two resistors may be more convenient for calculations than the reciprocal form.

Also, in a circuit with three resistors in parallel, the equivalent resistance of two of the resistors can be found by TI Eq. 7, and then the equation may be applied again to the equivalent resistance and the other resistance in parallel to find the total equivalent resistance of the three parallel resistors. However, if your calculator has a $1/x$ function, the reciprocal form may be easier to use.

Note that the voltage drops across R_1 and R_2 in parallel are the same, and by Ohm's law,

$$I_1 R_1 = I_2 R_2$$

or

$$\frac{I_1}{I_2} = \frac{R_2}{R_1} \qquad \textbf{(TI 8)}$$

TI Example 1   Given two resistors R_1 and R_2, with $R_2 = 2R_1$, in parallel in a circuit, what fraction of the current I from the voltage source goes through each resistor?

Solution   With $R_2 = 2R_1$, or $R_2/R_1 = 2$, by TI Eq. 8

$$I_1 = \left(\frac{R_2}{R_1}\right)I_2 = 2I_2$$

Since $I = I_1 + I_2$, we have

$$I = I_1 + I_2 = 2I_2 + I_2 = 3I_2$$

or

$$I_2 = \frac{I}{3}$$

Hence, the current divides with one-third going through R_2 and two-thirds going through R_1.

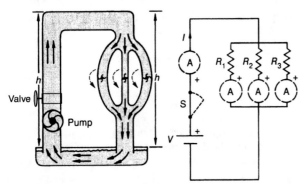

TI Figure 2   Parallel connection.   A liquid analogy on the left for the circuit diagram of resistors in parallel on the right. See text for description.

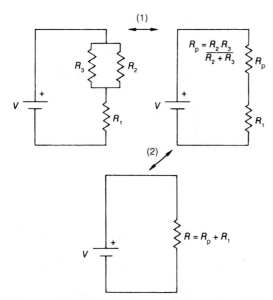

TI Figure 3 Circuit eduction. Series and parallel resistances are combined to find the equivalent resistance of a series-parallel circuit. See text for description.

Thus, the ratio of the resistances gives the relative magnitudes of the currents in the resistors.

Consider the circuit in ● TI Fig. 3. To find the equivalent resistance of this series–parallel circuit, one first "collapses" the parallel branch into a single equivalent resistance, which is given by TI Eq. 7. This equivalent resistance is in series with R_1, and the total equivalent resistance R of the circuit is $R = R_1 + R_p$.

/TI/ EXPERIMENTAL PROCEDURE

1. Examine the resistors. The colored bands conform to a color code that gives the value of a resistor. Look up the color code in Appendix A, Table A5, read the value of each resistor, and record in the Laboratory Report. Designate the smallest resistance as R_1 and consecutively larger values as R_2, R_3, and R_4.

2. In the following procedures, you will be asked to compute theoretically various quantities for a given circuit arrangement. The quantities are then determined by actual circuit measurements, and the calculated and experimental results are compared. Before initially activating each circuit arrangement, *have the circuit checked by the instructor, unless otherwise instructed.*

A. Resistors in Series

3. Set up a series circuit with R_1, R_2, and R_3, as in TI Fig. 1, with a switch and only one ammeter in the circuit next to the voltage source. A convenient way to check

a circuit to see whether it is properly connected is to trace the path of the current (with your finger) through the circuit.

Do this for the circuit under consideration to make sure that the current goes through each circuit component in series. Remember, an ammeter is *always* connected in series, and for proper polarity, + is connected to +.

Connect the voltmeter across (in parallel with) the voltage source. After having the circuit checked by the instructor, close the switch. If using a variable power supply, adjust the voltage, if necessary, to the suggested value (3.0 V). Read and record the voltmeter value (V). This is the voltage "rise" of the source.

(*Note:* If the needle of the ammeter goes in the wrong direction, reverse the polarity, i.e., reverse the hook-up of the leads of the ammeter.)

Open the circuit after completing the reading.

4. Using the resistor values and the measured voltage, *compute* (a) the equivalent resistance R_s of the circuit, (b) the current in the circuit, and (c) the voltage drop across each resistor. Show your calculations in the Laboratory Report.

5. Returning to the experimental circuit, close the switch and read the current I. Compare this with the computed value by finding the percent difference. Open the switch and move the ammeter in the circuit to the position "after" the first resistor [i.e., on the opposite side of the resistor from the voltage source so as to measure the current through (coming from) the resistor]. Record this as I_1.

Carry out this procedure for each resistor and record the currents in the Laboratory Report. The ammeter positions are shown in TI Fig. 1. Leave the switch closed only while readings are being taken.

6. Remove the ammeter from the circuit, and with the voltmeter, measure and record the voltage drop across each resistor and across all three resistors as a group. Remember, a voltmeter is *always* connected in parallel or "across" a circuit element to measure its voltage drop.

7. Compare the experimentally measured values with the theoretically computed values by finding the percent error. (Use the theoretical values as the accepted values.)

B. Resistors in Parallel

8. Set up a parallel circuit with R_1, R_2, and R_3, as in TI Fig. 2, with the ammeter and voltmeter connected as before in procedure 3. Check the circuit arrangement by tracing the current from the source through the circuit to see that it divides into three parallel

branches at the junction of the resistors and comes together again at the opposite junction.

Close the circuit (after it's been checked) and record the voltage and current readings in the Laboratory Report. (If using a variable power supply, adjust the voltage if necessary.)

Open the circuit after reading the voltage.

9. Using the resistor values and the measured voltage, *compute* (a) the equivalent resistance R_p of the circuit, (b) the current supplied by the source, and (c) the current through each resistor. Show your calculations in the Laboratory Report.

10. Returning to the experimental circuit, measure and record the voltage drops across each resistor and across all three resistors as a group.

Remove the voltmeter and connect the ammeter so as to measure the current I supplied by the source. Then move the ammeter to measure the current through each resistor by connecting the meter between a given resistor and one of the common junctions. The ammeter positions are shown in TI Fig. 2. Leave the switch closed only while readings are being taken.

11. Compare the theoretical and experimental values by computing the percent errors.

12. (*Optional*) Repeat procedures 8 through 11 with R_2 replaced by R_4.

C. Resistors in Series–Parallel

13. (Compute the following and record in the Laboratory Report.) If R_1 were connected in series with R_2 and R_3 in parallel (TI Fig. 3):
    (a) What would be the equivalent resistance R_{sp} of the resistors?
    (b) How much current would be supplied by the source?
    (c) What would be the voltage drop across R_1?
    (d) What would be the voltage drop across R_2 and R_3?
    (e) What would be the voltage drop across all three resistors?
    (f) What would be the currents through R_2 and R_3?

14. Set up the actual circuit and trace the current flow to check the circuit. With the voltmeter and ammeter, measure and record the calculated quantities.

You need not compute the percent errors in this case. However, make a mental comparison to satisfy yourself that the measured quantities agree with the computed values within experimental error.

Name _____ Section _____ Date _____

Lab Partner(s) _____

EXPERIMENT 20

Resistances in Series and Parallel

TI/ *Laboratory Report*

Resistor values R_1 _____ R_3 _____

R_2 _____ R_4 _____

A. Resistors in Series

Calculations
(show work)

Source voltage V _____

Equivalent resistance R_s _____

Current I _____

Voltage drops
across resistors V_1 _____

V_2 _____

V_3 _____

Don't forget units

(continued)

Experimental measurements

Percent error

I _____   _____

I_1 _____ V_1 _____   _____

I_2 _____ V_2 _____   _____

I_3 _____ V_3 _____   _____

$V_1 + V_2 + V_3$ _____

V across resistors as a group _____

B. Resistors in Parallel

Calculations
(show work)

Source voltage V _____

Equivalent resistance R_p _____

Current I _____

Current through resistors I_1 _____

I_2 _____

I_3 _____

Experimental measurements

Percent error

I _____   _____

V_1 _____ I_1 _____   _____

V_2 _____ I_2 _____   _____

V_3 _____ I_3 _____   _____

$I_1 + I_2 + I_3$ _____

EXPERIMENT 20

(Optional Procedure)

Calculations
(show work)

Source voltage V _____

Equivalent resistance R_p _____

Current I _____

Current through
resistors I_1 _____

I_3 _____

I_4 _____

Experimental measurements

Percent error

I _____ _____

V_1 _____ I_1 _____ _____

V_3 _____ I_3 _____ _____

V_4 _____ I_4 _____ _____

(continued)

C. Resistors in Series–Parallel

Calculations
(show work)

Source voltage V _____

Equivalent resistance R_{sp} _____

Current I _____

Voltage drops

V_1 _____

$V_2 = V_3$ _____

Experimental measurements

Currents

I_2 _____

I_3 _____

I _____

V_1 _____

$V_2 = V_3$ _____

I_2 _____

I_3 _____

/TI/ QUESTIONS

1. Discuss the sources of error in the experiment.

2. Suppose that the resistors in the various circuit diagrams represented the resistances of light bulbs. When a light bulb "burns out," the circuit is open through that particular component, i.e., R is infinite. Would the remaining bulbs continue to burn for the following conditions? If so, would the bulbs burn more brightly (draw more current) or burn more dimly (draw less current), if:

(a) R_2 burned out in the circuit in part A?

(b) R_1 burned out in the circuit in part B?

(c) Then R_3 also burned out in the circuit in part B?

(d) R_3 burned out in the circuit in part C?

(continued)

(e) Then R_1 also burned out in the circuit in part C?

3. Explain the effect of replacing R_2 with R_4 in procedure 12. (Explain theoretically even if procedure 12 of the experiment was not done.)

4. For the circuit in Fig. 3, $V = 12$ V, $R_1 = 4$ Ω, $R_2 = 6$ Ω, and $R_3 = 3$ Ω. Show that the power supplied by the battery is equal to that dissipated in the resistors. What principle does this illustrate? Use the accompanying table. (Consider values significant to two decimal places.)

(Show calculations)

Circuit element	Current I	Power dissipated p
$R_1 = 4$ Ω		
$R_2 = 6$ Ω		
$R_3 = 3$ Ω		
		(total)
		Power supplied
Battery $V = 12$ V		

5. Given three resistors of different values, how many possible resistance values could be obtained by using one or more of the resistors? (List the specific combinations—for example, R_1 and R_2 in series.)

EQUIPMENT NEEDED

This activity is designed for the Science Workshop 750 Interface, which has a built-in function generator.

- Voltage sensor (PASCO CI-6503)
- Science Workshop 750 Interface
- Cables and alligator clips
- Three 1000-Ω resistors

THEORY

According to Ohm's law, the current through a resistor is proportional to the voltage but inversely proportional to the resistance:

$$I = \frac{V}{R} \qquad \textbf{(CI 1)}$$

Thus if the resistance of a circuit increases, the current decreases, and if the resistance of a circuit decreases, the current increases. On the other hand, the larger the voltage, the larger the current. The overall current in a circuit thus depends on the interplay between the amount of voltage and the amount of resistance.

In this experiment, the total amount of resistance in a circuit will be varied by connecting resistors in series and then in parallel. An increasing voltage will be applied, and the overall current in the circuit (through the voltage source) will be measured.

Rewriting Ohm's law as $V = IR$, notice that a plot of voltage (in the y axis) versus current (in the x axis) must result in a straight line, with the slope equal to the overall resistance in the circuit:

$$\begin{array}{ccc} V & = & R\,I \\ \downarrow & & \downarrow\downarrow \\ y & = & m\,x \end{array}$$

Using voltage (and current) sensors, we will find the resistances of the circuits by measuring the slope of a voltage versus current plot.

SETTING UP DATA STUDIO

1. Open Data Studio and choose "Create Experiment."
2. From the sensor list, choose a Voltage Sensor. Connect the voltage sensor to the interface, as shown in the computer screen.
3. Double-click on the "Voltage Sensor" icon. The sensor properties window will open. Set the sample rate to 20 Hz, then click OK.
4. Press the Options button on the experiment setup window. The "Sampling Options" window will open. Under "Automatic Stop," set the time to 4.5 seconds. Click OK.
5. Directly below the sensor list there is an icon for Signal Output. Double-click on this icon. The signal generator window will open. (See ● CI Figure 1.)
6. The default form of the signal generator function is a sine wave. Change it to a "Positive Up Ramp Wave" by selecting from the drop menu.
7. Set the amplitude to 2.0 volts and the frequency to 0.20 Hz.
8. Click on the Measurements and Sample Rate button on the signal generator window. A list of measurements will open. Choose to measure the output current. Deselect the measurement of the output voltage.

9. Click on the Calculate button on the main toolbar. The calculator will open. Follow the next steps:
   a. Clear the definition box at the top, and enter the following formula in it:

   $$\text{Voltage} = \text{smooth}\,(20,\,x)$$

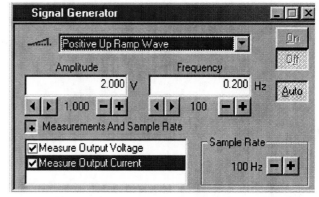

CI Figure 1   The Signal Generator Window.   Choose a positive up ramp wave function, adjust the amplitude and the frequency as specified in the setup procedure, and choose to measure the output current. (Data displayed using DataStudio Software. Reprinted courtesy of PASCO scientific.)

b. Press the top <u>Accept</u> button after entering the formula. Notice that the variable x will appear, waiting to be defined.

c. To define the variable, click on the drop menu button on the side of the variable. Define x as a Data Measurement and when prompted choose Voltage (ChA).

d. Press the <u>Accept</u> button.

e. Click on the <u>New</u> button again to define another calculation.

f. Clear the definition box and enter the following formula in it:

$$\text{Current} = \text{smooth}\,(20,\,x)$$

g. Press the <u>Accept</u> button after entering the formula. Notice that the variable x will again appear, waiting to be defined.

h. This time define x as a Data Measurement and, when prompted, choose Output Current.

i. Press <u>Accept</u> again.

10. The data list on the top left of the screen should now have the following items: Voltage ChA, Output Current, Voltage, and Current, where a small calculator icon identifies the quantities that are calculated, not measured.

11. Drag the "Voltage" (calculator) icon from the data list and drop it on the "Graph" icon of the displays list. A graph of voltage versus time will open, in a window called Graph 1.

12. Drag the "Current" (calculator) icon from the data list and drop it on top of the time axis of Graph 1. The time axis will change into a current axis. The graph should now be of voltage versus current. ● CI Figure 2 shows what the screen should look like at this point.

13. Double-click anywhere on the graph. The graph settings window will open. Make the following selections:

Under the tab Appearance:

Data:

Connect data points in bold

Deselect the buttons marked "Show Data Points" and "Show Legend Symbols"

Click OK to accept the changes and exit the graph settings window.

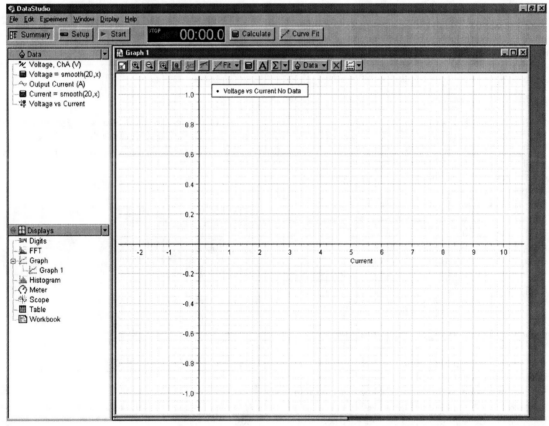

CI Figure 2   Data Studio setup.   A graph of voltage versus current will be used to examine different simple circuits. The slope of the graph will represent the resistance of the circuit. (Data displayed using DataStudio Software. Reprinted courtesy of PASCO scientific.)

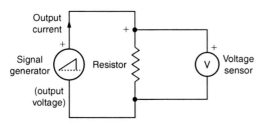

CI Figure 3 The experimental setup. A single resistor is connected to the source, with the voltage sensor connected across the resistor. The positive (red) lead of the voltage sensor must connect to the positive lead of the resistor.

CI EXPERIMENTAL PROCEDURE

A. Measuring Resistance

1. Get three resistors and label them R_1, R_2, and R_3.
2. Connect R_1 to the output source of the 750 Interface, using cables and alligator clips, if needed. A circuit diagram for this setup is shown in ● CI Figure 3.
3. Put alligator clips on the prongs of the voltage sensor, and connect the voltage sensor across the resistor. Make sure that the positive of the voltage sensor (red lead) is connected to the positive lead of the resistor.
4. Press the START button. Data collection will stop automatically after 4.5 seconds.
5. Press the Scale-to-Fit button of the graph toolbar. The Scale-to-Fit button is the leftmost button of the graph toolbar. This will scale all data to fit the full screen.
6. Use the Fit menu (on the graph toolbar) to do a "Linear Fit" of the data. A box with information about the fit will appear. Report the slope of the line in CI Data Table 1 as the value of R_1. Do not forget units.
7. Repeat the experiment two more times and determine an average value for R_1.
8. Repeat the process individually with R_2 and R_3.
9. If the graph window gets too crowded, go to "Experiment" (in the main menu, top of the screen) and choose "Delete all Data Runs." This will com-

pletely erase the data already collected. The fits can also be removed by deselecting them in the Fit menu.

B. Resistances in Series

1. Delete all the data to clear the graph. Also clear all the fits.
2. **Run 1:** Connect R_1 alone to the voltage source and take data, as before. Do a linear fit, and report the measured resistance in CI Data Table 2.
3. Introduce R_2 to the circuit by connecting it in series with R_1.
4. Connect the voltage sensor across both R_2 and R_1. (See ● CI Figure 4.)
5. **Run 2:** Press START and collect the data. Do a linear fit, and report the measured resistance in CI Data Table 2.
6. Now introduce R_3 to the circuit by connecting it in series with both R_2 and R_1.
7. Connect the voltage sensor across all three resistors. (See CI Figure 4.)
8. **Run 3:** Press START and collect the data. Do a linear fit, and report the measured resistance in CI Data Table 2.
9. Remove the fit information boxes and print the graph. Label it "Series Circuits" and attach it to the Laboratory Report.
10. Calculate the theoretical (expected) value of the equivalent resistance of each circuit. Compare the theoretical values with the measured ones by taking a percent difference.
11. Using the print-out of the graph or the Smart-Tool of the graph toolbar, determine the maximum value of the voltage and the maximum value of the current for each run. Report them in CI Data Table 2.

C. Resistances in Parallel

1. Delete all the data to clear the graph. Also clear all the fits.
2. **Run 1:** Connect R_1 alone to the voltage source and take data, as before. Do a linear fit, and report the measured resistance in CI Data Table 3.
3. Introduce R_2 to the circuit by connecting it in parallel with R_1.

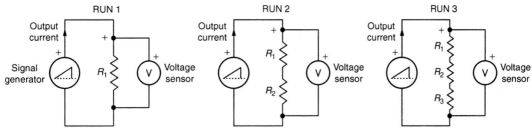

CI Figure 4 Resistors connected in series. Three different series circuits will be analyzed, each time adding an extra resistor to the series.

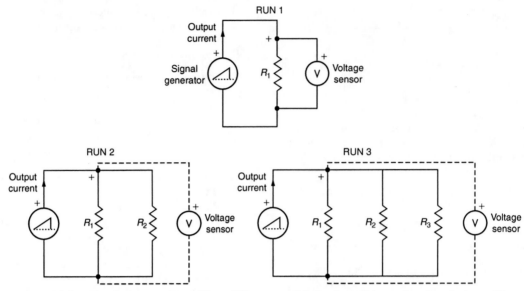

CI Figure 5   Resistors connected in parallel.   Three different parallel circuits will be analyzed, each time adding an extra branch to the circuit.

4. Connect the voltage sensor across both R_2 and R_1. (See ● CI Figure 5.)

5. **Run 2:** Press START and collect the data. Do a linear fit, and report the measured resistance in CI Data Table 3.

6. Now introduce R_3 to the circuit by connecting it in parallel with both R_2 and R_1.

7. Connect the voltage sensor across the three resistors. (See CI Figure 5.)

8. **Run 3:** Press START and collect the data. Do a linear fit, and report the measured resistance in CI Data Table 3.

9. Remove the fit information boxes and print the graph. Label it "Parallel Circuits" and attach it to the Laboratory Report.

10. Calculate the theoretical (expected) value of the equivalent resistance of each circuit. Compare the theoretical values with the measured ones by taking a percent difference.

11. Using the print-out of the graph or the Smart-Tool of the graph toolbar, determine the maximum value of the voltage and the maximum value of the current for each run. Report them in CI Data Table 3.

EXPERIMENT 20

Resistances in Series and Parallel

CI *Laboratory Report*

A. Measuring Resistance

CI DATA TABLE 1

Purpose: To measure the actual resistance of each of the three resistors.

Resistor	Slope measurements	Average resistance
R_1	1.	
	2.	
	3.	
R_2	1.	
	2.	
	3.	
R_3	1.	
	2.	
	3.	

Don't forget units

(continued)

B. Resistances in Series

CI **DATA TABLE 2**

Purpose: To experimentally measure the equivalent resistance of series circuits.

	Run 1 R_1 alone	Run 2 R_1 and R_2 in series	Run 3 R_1, R_2, and R_3 in series
Measured equivalent resistance			
Theoretical equivalent resistance $R_s = R_1 + R_2 + \cdots$			
Percent difference			
Maximum voltage			
Maximum current			

C. Resistances in Parallel

CI **DATA TABLE 3**

Purpose: To experimentally measure the equivalent resistance of parallel circuits.

	Run 1 R_1 alone	Run 2 R_1 and R_2 in parallel	Run 3 R_1, R_2, and R_3 in parallel
Measured equivalent resistance			
Theoretical equivalent resistance $R_p = \left(\dfrac{1}{R_1} + \dfrac{1}{R_2} + \cdots \right)^{-1}$			
Percent difference			
Maximum voltage			
Maximum current			

Don't forget units

(continued)

CI QUESTIONS

1. As more resistors were added to the series circuit, what happened to the total resistance of the circuit?

2. For approximately the same maximum voltage, what happened to the maximum current as more resistors were added to the series circuit?

3. As more resistors were added to the parallel circuit, what happened to the total resistance of the circuit?

4. For approximately the same maximum voltage, what happened to the maximum current as more resistors were added to the parallel circuit?

Name _____ Section _____ Date _____

Lab Partner(s) _____

EXPERIMENT 21

The *RC* Time Constant

(Electronic Timing)

/TI/ *Advance Study Assignment*

Read the experiment and answer the following questions.

1. Compare the voltages across a capacitor in dc and ac *RC* circuits.

2. How is the time base of the horizontal oscilloscope trace determined?

3. What is the significance of the *RC* time constant for the circuit?

4. Explain how the time constant of an *RC* circuit is determined from a stationary oscilloscope pattern.

CI *Advance Study Assignment*

Read the experiment and answer the following questions.

1. What is the time constant of an *RC* circuit and what are the units of measurement?

2. How many time constants will you have to wait before you can consider the capacitor "fully charged"?

The *RC* Time Constant

(Electronic Timing)

OVERVIEW

Experiment   examines the *RC* time constant using complementary electronic TI and CI approaches. In the TI procedure, the time constant of an *RC* circuit is determined from an oscilloscope trace of voltage versus time, using the $0.63V_o$ position. This is done for combinations of *RC* values.

In the CI procedure, a voltage sensor that feeds data into the computer monitors voltage changes for charging and discharging. From computer-drawn graphs of voltage versus time, the time constant is determined—the point of 63% of maximum voltage for charging and 37% of the maximum voltage for discharging. The procedure is done for two resistances.

INTRODUCTION AND OBJECTIVES

The oscilloscope can be used to study many ac circuit characteristics. The screen display of voltage versus time makes it possible to observe a variety of measurements. In particular, in an *RC* **(resistance-capacitance) circuit,** the charging of the capacitor can be visually observed. And using the horizontal time scale, the time constant of the charging process can be readily determined.

In this experiment, the oscilloscope will be used to determine the time constant of an *RC* circuit as the capacitor is continually charged and discharged by an ac signal voltage.

TI OBJECTIVES

After performing this experiment and analyzing the data, you should be able to:

1. Explain the charging characteristics of a capacitor with ac voltage.
2. Appreciate how the oscilloscope can be used to monitor electrical characteristics and to make electrical measurements.

3. Describe how an *RC* time constant may be measured from an oscilloscope trace.

CI OBJECTIVES

The purpose of this experiment is to investigate the charging and discharging of a capacitor in a series *RC* circuit. The time constant of the circuit will be determined experimentally and compared to the theoretical value. After performing this experiment and analyzing the data, you should be able to:

1. Describe the charging and discharging of a capacitor through a resistor.
2. Explain how the time constant can be measured experimentally.
3. Explain what the *RC* time constant means in terms of circuit characteristics.

The *RC* Time Constant

(Electronic Timing)

/TI/ EQUIPMENT NEEDED

- Function generator (square wave)
- Oscilloscope
- Three capacitors (0.05 μF, 0.1 μF, and 0.2 μF, or capacitor substitution box)
- Three resistors (5 kΩ, 10 kΩ, and 20 kΩ, or resistance box)

- Connecting wires
- 2 sheets of Cartesian graph paper
- (*Optional*) Unknown resistor wrapped in masking tape to conceal value

/TI/ THEORY

When an *RC* circuit is connected to a dc voltage source, charge must flow into the capacitor before the voltage across the capacitor can change. This takes time. As the voltage across the capacitor becomes closer to that of the source, the flow of charge becomes slower and slower. The capacitor voltage approaches the supply voltage as an asymptote—coming ever closer, but never getting there.

When the capacitor starts with no voltage across it, $V = 0$ at $t = 0$, the changing voltage is given by the equation

$$V = V_0(1 - e^{-t/RC})$$
$$= V_0(1 - e^{-t/\tau}) \qquad \textbf{(TI 1)}$$

where e is the base of the natural logarithms ($e = 2.718\ldots$), V_0 is the voltage of the dc source, R is the resistance in the circuit, and C is the capacitance. The quantity $\tau = RC$ is the **time constant** of the circuit. (See the Theory section in Experiment   .)

After a time of one time constant, $t = \tau = RC$, the voltage is

$$V = V_0(1 - e^{-RC/RC}) = V_0(1 - e^{-1}) = V_0(0.63)$$

or

$$\frac{V}{V_0} = 0.63 \qquad \textbf{(TI 2)}$$

That is, the voltage across the capacitor is 0.63 (or 63%) of its maximum value (● TI Fig. 1). For a dc voltage source, the capacitor voltage further increases to V_0 and maintains this voltage unless discharged.

However, for an ac voltage source, the capacitor voltage increases and decreases as the voltage of the applied signal alternately increases and decreases. For example,

suppose that a square-wave ac signal as illustrated in ● TI Fig. 2 is applied to the circuit. This has the effect of continuously charging and discharging the capacitor.* The voltage across the capacitor increases according to TI Eq. and then decreases according to the relationship[†]

$$V = V_0 e^{-t/RC} \qquad \textbf{(TI 3)}$$

* The square-wave generator actually is constantly reversing the charge on the capacitor, but the trace has the same rise time as though it were charging and discharging.

[†] It should be noted that the high point on the charging curve and the low point on the decay curve in Fig. 2 are not $V = V_0$ and $V = 0$, respectively, since it takes infinite times for the capacitor to charge and discharge to these values. However, if the time constant is several times smaller than one-half the period T of the square wave, $T = 1/f$, then to a good approximation the high and low points of the curve may be taken to correspond to $V = V_0$ and $V = 0$, respectively.

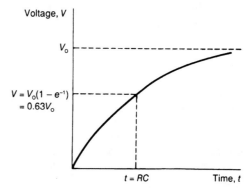

TI Figure 1 Voltage rise. A typical graph of voltage versus time for a capacitor charging in an *RC* circuit. In a time $t = RC$, the capacitor charges to 63% of its maximum value.

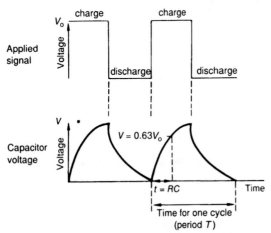

TI Figure 2 Charging and discharging. When a square-wave signal is applied to a capacitor in an *RC* circuit, the capacitor periodically charges and discharges, as shown here on a voltage-versus-time graph.

On an oscilloscope, the time base or the magnitude of the horizontal time axis is determined by the SWEEP TIME/DIV. From this control setting, you can determine time functions for traces on the screen. For example, suppose two complete wave cycles of a stationary sinusoidal pattern cover 6.66 horizontal divisions with a SWEEP TIME/DIV setting of 5 ms/div.

Then, the time for these two cycles is time = ST/div × div = 5 ms/div × 6.66 div = 33.3 ms, so the time for one cycle or the period of the wave is T = 33.3 ms/2 = 16.7 ms. (What is the frequency of the wave?)

The time constant of an *RC* circuit can be determined from a stationary oscilloscope pattern of the capacitor voltage versus time. This is done by finding the horizontal distance (time) needed for the trace to reach $0.63V_o$.

On an oscilloscope, time is measured as a horizontal distance. The scale is set by the knob marked SWEEP TIME/DIV.

TI Example 1 If the horizontal distance from the starting point to the point where the trace reaches 63% of the maximum voltage V_o, as shown in TI Fig. 1, is 6.5 divisions (1 division ≈ 1 cm), the time for 6.5 horizontal divisions is equal to one time constant (τ).

With the SWEEP TIME/DIV set at 5 ms/div, the value of the *RC* time constant would be (6.5 div) × (5 ms/div) = 32.5 ms.

/TI/ **EXPERIMENTAL PROCEDURE**

1. Turn on the oscilloscope and function generator. Set the function generator frequency to 100 Hz and the wave amplitude near maximum. Connect the square-wave output of the function generator directly to the vertical input terminals of the oscilloscope.

   Set the oscilloscope as follows. (*Note:* Different oscilloscopes differ somewhat in the names and locations of controls. **Vertical:** CH A DC, VOLTS/DIV 0.5, MODE CH A, POSITION Center Trace, CH B GND. **Horizontal:** TIME/DIV 2 mSEC, POSITION Center the trace. **Triggering:** LEVEL 12:00 position, COUPLING/SYNC AC SLOW, SOURCE INT, SLOPE +). The Vertical VOLTS/DIV and Horizontal TIME/DIV will be used here.

   Check that the small red knobs in the center of the VOLTS/DIV and TIME/DIV controls are in the Calibrated position. Adjust the FOCUS and INTENSITY controls for a sharp, clear trace. *Caution:* Intensity should be kept low to protect the phosphor on the screen. If time permits, experiment with the controls to see how they affect the display.

   Obtain a stationary trace of one or two cycles of the square-wave pattern on the screen. Adjust the vertical VOLTS/DIV and the function generator amplitude until the pattern is exactly 8 divisions high. (This is about 8 cm high.) If V_o is 8 divisions, the 5-division horizontal line will be very close to the $0.63V_o$ criterion for measuring the time constant (since 5/8 = 0.625, actually $0.625V_o$).

2. Then set up the circuit as shown in ● TI Fig. 3, with $R = R_1 = 10$ kΩ and $C = C_1 = 0.1$ μF. *Have the instructor check the circuit before attaching the final lead to the oscilloscope.*

3. Close the oscilloscope circuit by connecting the wire to the circuit and note the pattern. Carefully adjust the trigger controls so that the curve starts upward at the left end of the trace. The exponential rise time can be observed in greater detail by increasing the sweep rate (decreasing the TIME/DIV).

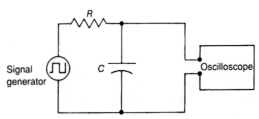

TI Figure 3 *RC* circuit. Circuit diagram for the experimental procedure for studying *RC* circuits. See text for description.

Adjust the time (TIME/DIV) until the rising curve extends well across the screen. Be sure that the variable TIME/DIV remains in the calibrated position.

4. With the total pattern 8 divisions high, the time constant is represented by the horizontal distance from the point where the trace starts to move up to the point where it crosses the horizontal line 5 divisions up. The time is found by multiplying the horizontal distance by the TIME/DIV setting (see TI Example 1). Record in TI Data Table 1.

5. Open the circuit and repeat procedures 3 and 4 with $R = R_2 = 5$ kΩ and $R = R_3 = 20$ kΩ. Record in TI Data Table 2.

6. On a Cartesian graph, plot the experimental τ versus R. Determine the slope of the straight line that best fits the data. To what does the value of the slope correspond?

7. Replace R with $R_1 = 10$ kΩ, and repeat procedures 3 and 4 with $C = C_2 = 0.05$ μF and $C = C_3 = 0.2$ μF.

8. On a Cartesian graph, plot the experimental τ versus C. (You should have three data points for τ with R_1. Why?) Determine the slope of the straight line that best fits the data. To what does the value of the slope correspond?

9. Compute the time constants for each of the RC combinations using the known R and C values, and compare with the experimentally determined values by finding the percent errors.

10. *(Optional)* Use your knowledge gained in this experiment to determine experimentally the value of the unknown resistor. Remove the masking tape after doing so and compute the percent error.

Name _____ Section _____ Date _____

Lab Partner(s) _____

EXPERIMENT 21

The *RC* Time Constant

(Electronic Timing)

TI/ *Laboratory Report*

TI/ DATA TABLE 1

Purpose: To determine the effect of R on the time constant.

	R ( )	C ( )	Divisions for 0.63 rise	Sweep time = div	Exp. time constant	Computed RC	Percent error
Case 1 R_1C_1							
Case 2 R_2C_1							
Case 3 R_3C_1							

Calculations
(show work)

Slope of the τ-versus-R plot _____

Percent difference between slope and C_1 _____

Don't forget units

(continued)

/TI/ DATA TABLE 2

Purpose: To determine the effect of C on the time constant.

	R ( )	C ( )	Divisions for 0.63 rise	Sweep time = div	Exp. time constant	Computed RC	Percent error
Case 4 R_1C_2							
Case 5 R_1C_3							

Slope of the τ-versus-C plot _____

Percent difference between slope and R_1 _____

Experimental RC time constant _____

Capacitance C _____

Computed R _____

Marked value of R _____

Percent error _____

EXPERIMENT 21   *Laboratory Report*

TI/ QUESTIONS

1. Judging on the basis of your experimental results, under what conditions are the charging times of different RC circuits the same?

2. In the form $V = V_0(1 - e^{-t/\tau})$, the $\tau = RC$ in the exponential must have units of time. (Why?) Show that this is the case.

3. How could the value of an unknown capacitance be determined using the experimental procedures? Show explicitly by assuming a value for an experimentally determined time constant.

EXPERIMENT 21

This activity is designed for the Science Workshop 750 Interface, which has a built-in function generator.

- 1000-Ω resistor
- 330-μF capacitor

- Voltage sensor (PASCO CI-6503)
- Cables and alligator clips
- Multimeter (that can measure resistance and capacitance)
- Second resistor of different value

CI T H E O R Y

A. Charging a Capacitor

● CI Fig. 1 shows a series RC circuit: a resistor connected in series with a capacitor and a power source of voltage V_0. As soon as the voltage source is turned on, the capacitor starts charging. As the charge in the capacitor increases exponentially with time, so does the voltage across its plates. The voltage across the capacitor at any time t is given by

$$V = V_0(1 - e^{-t/RC}) \quad \textbf{(CI 1)}$$

The quantity RC is called the time constant τ of the circuit—that is,

$$\tau = RC \quad \textbf{(CI 2)}$$

With the resistance measured in ohms and the capacitance in farads, it is easy to show that the time constant has units of seconds. (See Question 1.) In terms of the time constant, CI Eq. 1 can be written as

$$V = V_0(1 - e^{-t/\tau}) \quad \textbf{(CI 3)}$$

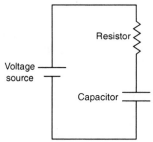

CI Figure 1 A series RC circuit. A capacitor and a resistor are connected in series to a voltage source.

311

The voltage across the capacitor will increase exponentially with time until it matches the voltage of the source. The capacitor is fully charged when $V = V_0$, which theoretically requires an infinite amount of time, $t \to \infty$. In practice, however, it is said the capacitor is fully charged if we wait long enough. But how long is "long enough"? Let's say until the voltage across the capacitor is 99.9% of the voltage of the source. The time it takes for this to happen can be calculated as follows:

$$V = V_0(1 - e^{-t/\tau})$$
$$0.999\,V_0 = V_0(1 - e^{-t/\tau})$$
$$0.999 = 1 - e^{-t/\tau}$$
$$e^{-t/\tau} = 1 - 0.999$$
$$e^{-t/\tau} = 0.001$$
$$\frac{-t}{\tau} = \ln(0.001)$$

Thus the time needed is

$$t = -\tau \ln(0.001) = 6.9\tau \approx 7\tau \quad \textbf{(CI 4)}$$

For experimental purposes, for a time of about seven time constants, the capacitor is considered to be fully charged.

Another time that is of special interest is the time constant itself. Notice that at a time $t = \tau = RC$, one time constant after starting the charging process, the voltage across the capacitor has increased to 63% of the voltage of the source, as shown here:

$$V = V_0(1 - e^{-t/\tau})$$
$$= V_0(1 - e^{-\tau/\tau})$$
$$= V_0(1 - e^{-1})$$
$$= 0.63V_0 \quad \textbf{(CI 5)}$$

Notice that if you experimentally find at what time the voltage is 63% of the maximum, you are finding the time constant of the circuit.

In this experiment, the voltage source will be a signal generator that will produce a positive square wave. The circuit is shown in ● CI Fig. 2. The voltage source is the signal generator of the PASCO Science Workshop 750

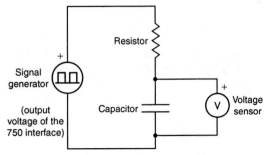

CI Figure 2 The experimental setup. The signal generator of the 750 Interface will be the voltage source for this experiment. A positive-square-wave voltage function will be used to periodically charge and discharge the capacitor. A voltage sensor will keep track of the voltage across the capacitor.

Interface. A voltage sensor will keep track of the voltage across the capacitor. A positive square wave is shown in ● CI Fig. 3. The voltage source will periodically turn on and off, charging and discharging the capacitor. To make sure that the capacitor gets fully charged before the source turns off, it will be necessary to set up the square wave so that the time it remains "ON" is at least seven time constants, as explained by CI Eq. 4. The experimental procedure contains detailed instructions on how to do this.

B. Discharging a Capacitor

When the voltage source is turned off, the charge in the capacitor flows back through the resistor. As the charge in the capacitor decreases, the voltage across the capacitor also decreases. The decrease is exponential, and as a function of time, it is described by the equation

$$V = V_o e^{-t/\tau} \qquad \textbf{(CI 6)}$$

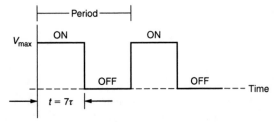

CI Figure 3 A positive square wave. The voltage periodically turns ON and OFF. To make sure the time it remains ON is enough to charge the capacitor fully, the time needed will be approximated to seven time constants (7τ), and the frequency of the signal will be adjusted accordingly.

In this case, notice that one time constant after the discharge begins, the voltage across the capacitor will be 37% of the original fully charged voltage of V_o:

$$
\begin{aligned}
V &= V_o e^{-t/\tau} \\
&= V_o e^{-\tau/\tau} \\
&= V_o e^{-1} \\
&= 0.37 V_o \qquad \textbf{(CI 7)}
\end{aligned}
$$

Thus the discharging of the capacitor can also be used to find the time constant experimentally, by determining how long it takes for the voltage to decrease to 37% of the initial maximum value.

In this experiment, the charging and discharging of the capacitor will be observed in a plot of voltage versus time. The time constant of the circuit will be directly measured from the plot.

SETTING UP DATA STUDIO

1. Open Data Studio and choose "Create Experiment."
2. From the sensor list, choose a voltage sensor. Connect the sensor to the interface, as shown in the computer screen.
3. Create a graph by dragging the "Voltage" icon from the data list and dropping it on the "Graph" icon on the displays list. A graph of voltage versus time will open in a window called Graph 1.
4. Directly below the sensor list there is an icon for the Signal Output. Double-click on the icon. The signal generator window will open.
5. The default form of the signal generator function is a sine wave. Change it to a positive square wave of amplitude 3.0 V. (*Note:* Be sure to choose the "Positive Square Wave," not the one that says just "Square Wave." Scrolling down the list may be needed.) The frequency of the signal will depend on the values of R and C and will be entered later on.

CI EXPERIMENTAL PROCEDURE

1. Measure the resistance of the resistor using a multimeter, and record the value in CI Data Table 1.
2. Measure the capacitance of the capacitor using a multimeter, and record the value in CI Data Table 1. If the available multimeter does not measure capacitance, then use the manufacturer's value as the capacitance.
3. Calculate the theoretical time constant, and enter the value in CI Data Table 1.
4. Calculate the approximate time needed to consider the capacitor fully charged. (See CI Eq. 4.) Enter the value in CI Data Table 1.

5. As explained in the CI Theory section, the frequency of the square wave needs adjusting so that the voltage source remains "ON" for enough time to charge the capacitor fully before it automatically turns "OFF" and discharges, as shown in CI Fig. 3. This is accomplished by following these steps:

a. The time to charge, calculated in Step 4, is half the required period of the square wave. (See CI Fig. 3.) Calculate the required period and enter it in CI Data Table 1.

b. Calculate the frequency, remembering that the frequency is the inverse of the period. Report the frequency in Data Table 1.

c. Enter the required frequency in the signal generator window, and set the generator to AUTO.

6. Set up the circuit shown in CI Fig. 2. The resistor, the capacitor, and the voltage source are connected in series. The voltage source is the output source of the 750 Interface, set to 3 V.*

7. Connect the voltage sensor across the capacitor, as shown in CI Fig. 2.

8. Press the START button. The capacitor will begin to charge and discharge. Press the STOP button after two cycles have been completed. Press the Scale-to-Fit button (left-most button of the graph toolbar) to scale all data to fit on the screen.

* The voltage value of 3 V is suggested for the values of *R* and *C* specified before because it produces an easy-to-read plot. The voltage sensor can measure a high range of voltages, and you may use a different value.

9. Print the graph. If no printer is available, make a careful drawing of the graph. Paste the graph to the Laboratory Report.

10. Record the maximum voltage across the capacitor. Then calculate 63% of this value. Report these values in CI Data Table 1.

A. *Charging*

11. Look at the charging part of the graph. Use the graph tools to find the time at which the voltage reached 63% of the maximum. This is the experimental time constant of the circuit. (Refer to CI Eq. 5.) Enter the value in the table, and compare it to the theoretical value with a percent error.

B. *Discharging*

12. Determine 37% of the maximum voltage, and record this value in the table.

13. From the graph, determine how long *after the start of the discharge* the voltage was only 37% of the maximum. This is again the time constant of the circuit. (Refer to CI Eq. 7.) Enter this value in the laboratory report and compare it to the theoretical value by calculating the percent error.

14. Repeat the experiment with a different value of resistance, keeping the capacitor and the voltage source constant. Do not forget to recalculate and adjust the required frequency of the positive square-wave function. Report the results as Trial 2 in CI Data Table 1.

Name _____ Section _____ Date _____

Lab Partner(s) _____

EXPERIMENT 21

The *RC* Time Constant

(Electronic Timing)

CI *Laboratory Report*

CI **DATA TABLE 1**

Purpose: To experimentally determine the time constant of the *RC* circuit.

			Trial 1	Trial 2
Theoretical Values	R			
	C			
	τ_{theo}			
	Time to fully charge $\approx 7\tau_{theo}$			
Output Signal	Period, T			
	Frequency, $f = \frac{1}{T}$			
Experimental Values	V_{max}			
	Charging	0.63 of V_{max}		
		τ_{exp}		
		Percent error		
	Discharging	0.37 of V_{max}		
		τ_{exp}		
		Percent error		

Don't forget units

(continued)

CI QUESTIONS

1. Show, by dimensional analysis, that the time constant $\tau = RC$ has units of time.

2. Compare the charging and discharging of the capacitors from Trial 1 and Trial 2. What things were similar and what things were different? Be specific.

3. Suppose that a particular RC series circuit has a time constant of 5.0 seconds. What does that mean in terms of the charging and discharging? How would this circuit compare to the ones you tried? Explain qualitatively and quantitatively.

4. What could be a practical application of an RC circuit?

EXPERIMENT 22

Reflection and Refraction

TI/ *Advance Study Assignment*

Read the experiment and answer the following questions.

1. What is the law of reflection, and does it apply to all reflecting surfaces?

2. Distinguish between regular and irregular reflection. Give an example of each.

3. Why is light refracted when it passes from one medium into an optically different medium?

(continued)

4. Show by Snell's law that if the speed of light is less in a particular medium, then a light ray is bent toward the normal when entering that medium. What happens if the speed of light is greater in the medium?

5. What is the difference between the relative index of refraction and the absolute index of refraction? Explain why we can experimentally determine the absolute index of refraction fairly accurately using air as a medium.

Reflection and Refraction

INTRODUCTION AND OBJECTIVES

Reflection and refraction are two commonly observed properties of light. The reflection of light from smooth and polished surfaces, such as ponds of water and mirrors, enables us to view the images of objects. Also, when light passes from one medium into another, it is bent, or refracted. As a result, a stick in a pond or a spoon in a glass of water appears to be bent.

As part of geometrical optics, these phenomena are explained by the behavior of light rays. Through ray tracing, we can conveniently investigate the physical laws of reflection and refraction in the laboratory. In this experiment, we employ a plane mirror and a glass plate to study these laws and the parameters used in describing the reflection and refraction of light.

After performing this experiment and analyzing the data, you should be able to:

1. Describe the law of reflection and explain how it can be verified experimentally.
2. Explain Snell's law and its application to transparent materials.
3. Explain what the index of refraction tells you about a transparent material and how it can be measured experimentally.

EQUIPMENT NEEDED

- Pins
- Pin board (cardboard or poster board suffices)
- Sheets of white paper ($8\frac{1}{2} \times 11$ in.)
- Ruler and protractor
- Short candle (less than 5 cm) or some similar light source
- Rectangular mirror (and holder if available)
- Thick glass plate (approximately 8×10 cm)

Note: Ray boxes may be used if available.

THEORY

A. Reflection

When light strikes the surface of a material, some light is usually reflected. The reflection of light rays from a plane surface such as a glass plate or a plane mirror is described by the **law of reflection:**

The angle of incidence is equal to the angle of reflection (that is, $\theta_i = \theta_r$).

These angles are measured from a line perpendicular or *normal* to the reflecting surface at the point of incidence (● Fig. 1). Also, the incident and reflected rays and the normal lie in the same plane.

The rays from an object reflected by a smooth plane surface appear to come from an image behind the surface, as shown in the figure. From congruent triangles it can be seen that the image distance d_i from the reflecting surface is the same as the object distance d_o. Such reflection is called **regular** or **specular reflection.**

The law of reflection applies to any reflecting surface. If the surface is relatively rough, like the paper of this page, the reflection becomes diffused or mixed, and no image of the source or object will be produced. This type of reflection is called **irregular** or **diffuse reflection.**

B. Refraction

When light passes from one medium into an optically different medium at an angle other than normal to the surface,

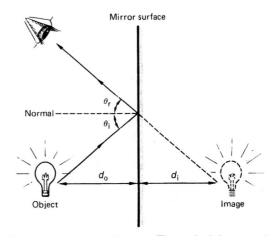

Figure 1 Law of reflection. The angle θ_i between the incident ray and the normal to the surface is equal to the angle θ_r between the reflected ray and the normal; that is, $\theta_i = \theta_r$. (Only a single ray is shown.) The object distance d_o is also equal to the image distance d_i for a plane mirror.

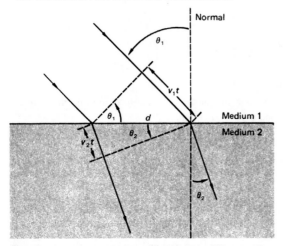

Figure 2 Refraction of two parallel rays. When medium 2 is more optically dense than medium 1, then $v_2 < v_1$ and the rays are bent toward the normal as shown here. If $v_2 > v_1$, the rays are bent away from the normal (as though the ray arrows were reversed in reverse ray tracing here).

it is "bent," or undergoes a change in direction, as illustrated in ● Fig. 2 for two parallel rays in a beam of light. This is due to the different velocities of light in the two media. In the case of refraction, θ_1 is the angle of incidence and θ_2 is the angle of refraction.

From the geometry of Fig. 2, where d is the distance between the parallel rays at the boundary, we have

$$\sin \theta_1 = \frac{v_1 t}{d} \quad \text{and} \quad \sin \theta_2 = \frac{v_2 t}{d}$$

or

$$\frac{\sin \theta_1}{\sin \theta_2} = \frac{v_1}{v_2} = n_{12} \qquad (1)$$

where the ratio of the velocities n_{12} is called the **relative index of refraction.** Equation 1 is known as **Snell's law.** If $v_2 < v_1$ (as in Fig. 2), the rays are bent toward the normal in the second medium. And if $v_2 > v_1$, the rays are bent away from the normal (e.g., reversed rays in Fig. 2 with medium 2 taken as medium 1).

For light traveling initially in vacuum (or approximately for light traveling initially in air), the relative index of refraction is called the **absolute index of refraction** or simply the **index of refraction,** and

$$n = \frac{c}{v} \qquad (2)$$

where c is the speed of light in vacuum and v is the speed of light in the medium. Hence, the index of refraction of vacuum is $n = c/c = 1$, and for air $n \simeq c/c = 1$. For water, $n = 1.33$.

Snell's law can then be written

$$\frac{\sin \theta_1}{\sin \theta_2} = \frac{v_1}{v_2} = \frac{c/n_1}{c/n_2} = \frac{n_2}{n_1}$$

or

$$n_1 \sin \theta_1 = n_2 \sin \theta_2 \qquad (3)$$

where n_1 and n_2 are in indices of refraction of the first and second media, respectively.

We see from Eq. 2 that the index of refraction is a measure of the speed of light in a transparent material, or a measure of what is called the **optical density** of a material.* For example, the speed of light in water is less than that in air, so water is said to have a greater optical density than air. Thus the greater the index of refraction of a material, the greater its optical density and the lesser the speed of light in the material.

In terms of the indices of refraction and Snell's law (Eq. 3), we have the following relationships for refraction:

- If the second medium is more optically dense than the first medium ($n_2 > n_1$), the refracted ray is bent *toward* the normal ($\theta_2 < \theta_1$), as in Fig. 2.
- If the second medium is less optically dense than the first medium ($n_2 < n_1$), the refracted ray is bent *away from* the normal ($\theta_2 > \theta_1$), as for reverse ray tracing in Fig. 2.

EXPERIMENTAL PROCEDURE

A. Reflection

GLASS PLATE AS A MIRROR
1. Place a sheet of white paper on the table. As illustrated in ● Fig. 3, draw a line where the candle (or object) will be placed. The line should be drawn parallel to the shorter edge of the page and about 3 to 4 cm from that edge. Make a mark near the center of the line, and place the candle on the mark.

---

* Optical density does not correlate directly with mass density. In some instances, a material with a greater optical density than another will have a lower mass density.

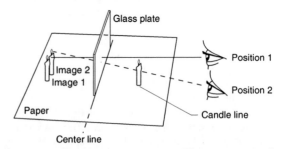

Figure 3 Glass plate as a mirror. The arrangement for the experimental procedure using a glass plate as a mirror. See text for description. (Images are displaced for illustration.)

Put the glass plate near the center of the paper, as shown in the figure. With the length of the plate parallel to the candle line, draw a line along the edge of the glass plate (side toward the candle). Light the candle.

Caution: Take care not to burn yourself during the experimental procedure.

Looking *directly over the candle* with your eye as in position 1 in Fig. 3, you will observe an image of the candle (image 1) in the glass plate. The glass plate reflects light and serves as a mirror. (Observing should be done with only one eye open.)

2. Observing the top of the flame from a side position (position 2 in Fig. 3), you will see a double image, one nearer than the other. Can you explain why?

Place a pin in the pin board near the glass plate so that it is aligned (in the line of sight) with the front or nearer image of the candle (image 2 in Fig 3; double image not shown in figure). Place another pin closer to you or to the edge of the paper so that both pins and the candle image are aligned. Mark the locations of the pins.

Repeat this procedure, viewing from a position on the other side of the candle.

3. Remove the equipment from the paper. Draw straight lines through the pair of pin points extending from the candle line through the glass-plate line. (Extend the candle line if necessary.) The lines will intersect on the opposite side of the plate line at the location of the candle image.

Draw lines from the actual candle position or mark to the points of intersection of the previously drawn lines *and* the plate line. These lines from the candle (mark) to the glass-plate line and back to the observation positions are ray tracings of light rays.

4. Draw normal lines to the glass-plate line at the points of intersection of the ray lines. Label and measure the angles of incidence θ_i and reflection θ_r. Record the data in the Laboratory Report.

Also, measure the perpendicular distances from the glass-plate line to the candle mark (the object distance d_o) and to the candle image position (the image distance d_i). Compute the percent differences of the quantities, as indicated in the Laboratory Report.

PLANE MIRROR

5. (a) Place the mirror near the center of a sheet of paper as with the glass plate used previously. (The mirror may be propped up by some means, or a holder may be used if available.) Draw a line along the silvered side of the mirror. Then lay an object pin about 10 cm in front of the mirror and parallel to its length (● Fig. 4).

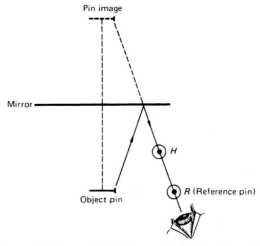

Figure 4 Plane mirror. The arrangement for the experimental procedure for a plane mirror. See text for description.

Mark the locations of the ends of the object pin on the paper with a pencil.

(b) Stick a reference pin R in the board to one side of the object pin and near the edge of the paper, as illustrated in Fig. 4, and mark its location.

(c) Placing another pin nearer the mirror so that it is visually aligned with the reference pin and the head of the object pin's image in the mirror. Mark the position of this pin, and label it with an H. Then move this pin over so that it aligns with the reference pin and the "tail" of the image pin. Mark this location, and label it with a T.

(d) Repeat this procedure on the opposite side of the object pin with another reference pin.

6. Remove the equipment from the paper, and draw straight lines from the reference points through each of the H and T locations and the mirror line. The H lines and T lines will intersect and define the locations of the head and tail of the pin image, respectively.

Draw a line between the line intersections (the length of the pin image). Measure the length of this line and the length of the object pin, and record. Also, measure the object distance d_o and the image distance d_i from the mirror line, and record.

Compute the percent differences of the respective measured quantities.

ROTATION OF A MIRROR

7. Place the mirror near the center of a sheet of paper (as described above) and draw a line along the length of the silvered side of the mirror. Measure so as to find the center of the line, and mark that location.

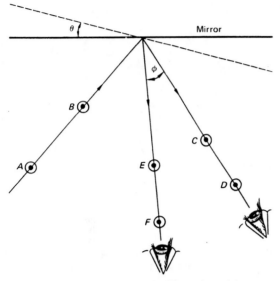

Figure 5 Mirror rotation. An illustration of the experimental arrangement and procedure for the rotation of a mirror. See text for description.

Stick two pins (*A* and *B*) in the board to one side and in front of and in line with the center of the mirror, as in ● Fig. 5. Viewing the aligned images of these pins from the other side of the page, place two more pins (*C* and *D*) in alignment. Label the locations of the pins.

8. Leaving pins *A* and *B* in place, rotate the mirror a small but measurable angle θ (approximately 10 to 15°) about its center point, and draw a line along the silvered side of the mirror.

Align two pins (*E* and *F*) with the aligned images of *A* and *B*, and mark and label the locations of *E* and *F*.

9. Remove the equipment from the paper and draw the incident ray and the two reflected rays. Measure the angle of rotation θ of the mirror and the angle of deflection ϕ between the two reflected rays, and record in the Laboratory Report.

Double θ, and compute the percent difference between 2θ and ϕ. Make a conclusion about the relationship between the angle of rotation of a mirror and the angle of deflection of a ray.

B. Refraction

INDEX OF REFRACTION OF A GLASS PLATE
10. Lay the glass plate in the center of a sheet of paper, and outline its shape with a pencil (● Fig. 6). Draw a

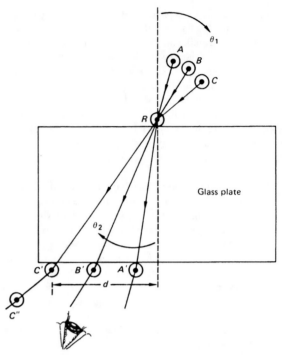

Figure 6 Index of refraction. An illustration (top view) of the experimental arrangement and procedure for determining the index of refraction of a glass plate. See text for description.

line normal to one of the sides of the plate, and place a pin (*R*) at the intersection of this line and the face of the plate. Measure an angle θ_1 of 15° relative to this line, and place a pin (*A*) about 6 to 8 cm from the plate at this angle.

Then, sighting through the edge of the plate from the eye position shown in Fig. 6, place a pin (*A'*) adjacent to the face of the plate so that it is aligned with *R* and *A*. Mark and label the locations of the pins.

Repeat with pins *B* and *C* at angles of 30° and 45°, respectively. For the 45°-angle case, align an additional pin (*C''* Fig. 6).

11. Trace the various rays, and measure and record θ_1 and θ_2 for each case. Also measure and record the displacement *d* of ray *C'C''* from the normal and the thickness of the plate. Using Eq. 3, compute the index of refraction of the glass.

Compare the average experimental value of the index of refraction with the general range of the index of refraction of glass ($n = 1.5-1.7$, depending on type).

Reflection and Refraction
TI/ *Laboratory Report*

A. Reflection

Glass Plate as a Mirror

	θ_i	θ_r	
Ray 1	_____	_____	d_o _____
Ray 2	_____	_____	d_i _____

Percent differences between θ_i and θ_r

Ray 1 _____

Ray 2 _____

Percent difference
between d_o and d_i _____

Plane Mirror

Length of pin _____ d_o _____

Length of image _____ d_i _____

Percent difference
between pin length
and image length _____

Percent difference
between d_o and d_i _____

Rotation of a Mirror

Angle of rotation, θ _____ 2θ _____

Angle of deflection of ray, ϕ _____

Percent difference between ϕ and 2θ _____

Calculations
(show work)

Don't forget units

(continued)

B. Refraction

Index of Refraction of a Glass Plate

	θ_1	θ_2	Computed n
Ray *ARA'*	_____	_____	_____
Ray *BRB'*	_____	_____	_____
Ray *CRC'*	_____	_____	_____

Average n _____

General range of the index of
refraction of glass _____

Displacement d of ray $C'C''$ _____

Thickness of glass plate _____

Calculations
(show work)

/TI/ QUESTIONS

1. (a) Why are two images seen in the glass plate when it is viewed from position 2 in part A
of the experiment? Why is only one image seen when it is viewed from position 1?

(b) Explain why reflection images are easily seen at night in a window pane from inside the house, whereas during the day they are not.

2. Judging on the basis of your experimental data, draw conclusions about (a) the relationship of the distance of the object in front of a plane mirror and the distance of its image "behind" the mirror; and (b) the image magnification (i.e., how much bigger the image is than the object).

3. Explain the situation shown in ● Fig. 7. How can this be done without hurting one's hand? (*Hint:* The author's hand extends inside the sliding glass-windowed door of a laboratory cabinet.)

Figure 7 See Question 3.

(continued)

4. Prove mathematically that when a plane mirror is rotated an angle θ about an axis through its center (part A of the experiment), the angle of deflection ϕ of a light ray is equal to 2θ. Draw a diagram and show the work involved in your proof. Attach an additional sheet if necessary.

5. Referring to the situation in Fig. 6, show theoretically that ray $C'C''$ is parallel to ray CR. Compute the displacement d of the ray passing through the glass plate. Compare this with the measured experimental displacement.

6. Using the experimentally determined n for the glass plate, compute the speed of light in the glass plate.

EXPERIMENT 23

Spherical Mirrors and Lenses
/TI/ *Advance Study Assignment*

Read the experiment and answer the following questions.

1. Distinguish between concave and convex spherical mirrors.

2. What is the difference between a real image and a virtual image?

3. Distinguish between diverging and converging lenses.

4. What does the word *focal* mean with regard to the focal point of spherical mirrors and lenses?

(continued)

5. If an object is placed 15 cm in front of a concave mirror with a radius of curvature of 20 cm, what are the image characteristics? (Show your work.)

Spherical Mirrors and Lenses

INTRODUCTION AND OBJECTIVES

Mirrors and lenses are familiar objects that are used daily. The most common mirror is a plane mirror, the type we look into every morning to see our image. Spherical mirrors also have many common applications. For example, convex spherical mirrors are used in stores to monitor aisles and merchandise, and concave spherical mirrors are used as flashlight reflectors and as cosmetic mirrors that magnify.

Mirrors reflect light, whereas **lenses** transmit light. Spherical lenses are used to cause light rays to converge and hence focus them (biconvex spherical lenses) and to cause light rays to diverge (biconcave spherical lenses). Many of us wear lenses in the form of eyeglasses. Cameras and projectors use lens systems to form images. Cameras form reduced-size images on film or a chip (digital), and projectors form magnified images on a screen.

In this experiment, we will investigate the fundamental properties of spherical mirrors and lenses to learn the parameters that govern their use.

After performing this experiment and analyzing the data, you should be able to:

1. Distinguish among converging and diverging spherical mirrors and lenses.
2. Determine the image characteristics for spherical mirrors graphically using ray diagrams and analytically using the mirror equation and magnification factor.
3. Determine the image characteristics for spherical lenses graphically using ray diagrams and analytically using the thin-lens equation and magnification factor.

EQUIPMENT NEEDED

- Concave and convex spherical mirrors
- Convex lens (focal length 10 to 20 cm)
- Concave lens (focal length at least 5 cm longer than convex lens)
- Meterstick optical bench (or precision bench) with lens holder, screen, and screen holder (white cardboard can serve as the screen)
- Light source: candle and candle holder, or electric light source with object arrow

THEORY

A. Spherical Mirrors

A **spherical mirror** is a section of a sphere and is characterized by a center of curvature C (● Fig. 1). The distance from the center of curvature to the vertex of the mirror along the optic axis is called the **radius of curvature** R. This also may be measured to any point on the surface of the mirror. (Why?)

The focal point F is midway between C and the vertex, and the **focal length** f is one-half the radius of curvature:

$$f = \frac{R}{2} \qquad (1)$$

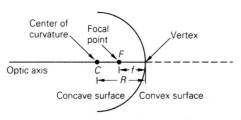

If the reflecting surface is on the inside of the spherical section, the mirror is said to be **concave.** For a **convex** mirror, the reflecting surface is on the outside of the spherical section.*

The characteristics of the images formed by spherical mirrors can be determined either graphically or analytically. Examples of the graphical ray method are shown in the ray diagrams in ● Fig. 2.

As illustrated for a concave mirror (Fig. 2a):

1. A **chief ray** from the object goes through the center of curvature C and is reflected back through C.
2. A **parallel ray** from the object is parallel to the optic axis and is reflected through the focal point F.
3. A **focal ray** from the object passes through the focal point F and is reflected parallel to the optic axis.

The intersection of these rays defines the location of the tip of the **image arrow,** which extends to the optic axis. The focal ray is a "mirror" image of the parallel ray and is not needed to locate the tip of the image. It is often

Figure 1 Spherical mirrors. The parameters used to describe spherical mirror surfaces. See text for description.

* To help remember the difference, note that a con*cave* mirror is recessed, as though one were looking into a *cave.*

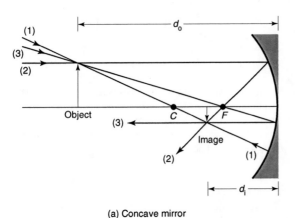

(a) Concave mirror

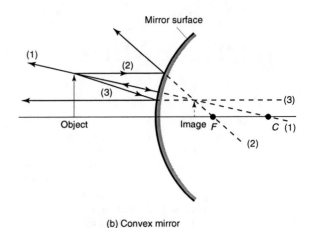

(b) Convex mirror

Figure 2 Mirror ray diagrams. Examples of the ray diagram method for determining the image characteristics for (a) a concave, or converging, spherical mirror and (b) a convex, or diverging, spherical mirror.

omitted but is helpful when the object is inside the center of curvature.

For a convex mirror, the chief and parallel rays appear to go through C and F, as illustrated in Fig. 2b.

A concave mirror is called a **converging mirror** because rays parallel to the optic axis converge at the focal point. Similarly, a convex mirror is called a **diverging mirror** because the rays parallel to the optic axis appear to diverge from the focal point.

If the image is formed on the same side of the mirror as the object, the image is said to be a **real image.** In this case, the light rays converge and are concentrated, and an image can be observed on a screen placed at the image distance. An image that is formed "behind" or "inside" the mirror is called a **virtual image.** Here, the rays appear to diverge from the image, and no image can be formed on a screen.

In general, an image is described in terms of whether it is:

1. Real or virtual,
2. Upright (erect) or inverted (relative to the object orientation), and
3. Magnified or reduced (or smaller)

In Fig. 2a the image is real, inverted, and reduced; in Fig. 2b the image is virtual, upright, and reduced.

The distance from the object to the vertex along the optic axis is called the **object distance** d_o, and the distance from the vertex to the image is the **image distance** d_i. Knowing the focal length f of the mirror, we can find the position of the image d_i from the **spherical mirror equation,**

$$\frac{1}{d_o} + \frac{1}{d_i} = \frac{1}{f} \qquad \textbf{(2a)}$$

Another convenient form of this equation is

$$d_i = \frac{d_o f}{d_o - f} \qquad \textbf{(2b)}$$

In the case of a concave mirror, the focal length is taken to be positive $(+)$; for a convex mirror, the focal length is taken to be negative $(-)$. The object distance d_o is taken to be positive in either case. The sign convention is as follows: If d_i is positive, the image is real, and if d_i is negative, the image is virtual. The **magnification factor** M is given by

$$M = -\frac{d_i}{d_o} \qquad \textbf{(3)}$$

If M is positive (d_i negative), the image is upright; if M is negative (d_i positive), the image is inverted. The sign convention is summarized in Table 1.

TABLE 1 Sign Convention for Spherical Mirrors and Lenses

Quantity	Conditions	Sign
Focal length f	Concave mirror	+
	Convex mirror	−
	Convex lens	+
	Concave lens	−
Object distance d_o	Usually* (Always in this experiment)	+
Image distance d_i	Image real	+
	Image virtual	−
Magnification M	Image upright	+
	Image inverted	−

* In some cases of lens combinations, d_o may be negative when the image of one lens is used as the object for the next lens.

Example 1 An object is placed 45 cm in front of a concave mirror with a focal length of 15 cm (corresponding to the case in Fig. 2a). Determine the image characteristics analytically. (Neglect significant figures.)

Solution With $d_o = 45$ cm and $f = 15$ cm, from Eq. .2a,

$$\frac{1}{45} + \frac{1}{d_i} = \frac{1}{15} = \frac{3}{45}$$

Then

$$\frac{1}{d_i} = \frac{2}{45} \quad \text{or} \quad d_i = \frac{45}{2} = 22.5 \text{ cm}$$

Then

$$M = -\frac{d_i}{d_o} = -\frac{22.5 \text{ cm}}{45 \text{ cm}} = -\frac{1}{2}$$

Thus the image is real (positive d_i), inverted (negative M), and reduced by a factor of $\frac{1}{2}$ (i.e., one-half as tall as the object).

B. Spherical Lenses

The shapes of biconvex and biconcave spherical lenses are illustrated in ● Fig. 3. A radius of curvature is defined for each spherical surface, but only the focal points (one for each spherical surface) are needed for ray diagrams.

A convex lens is called a **converging lens** because rays parallel to the principal axis converge at the focal point. A concave lens is called a **diverging lens** because rays parallel to the principal axis appear to diverge from the focal point.

As with spherical mirrors, the characteristics of the images formed by spherical lenses can be determined graphically or analytically. The chief (1) and parallel (2) rays for the graphical method are illustrated in the ray diagrams in ● Fig. 4. In the case of a convex lens (Fig. 4a), the chief ray (1) through the center of the lens passes straight through. A ray parallel (2) to the principal axis is refracted in such a way that it goes through the focal point on the far side of the lens. Also, a focal ray (3) through the near focal point is refracted by the lens so it leaves parallel to the axis. In the case of a concave lens (Fig. 4b), the chief ray (1) still goes straight through the centre of the lens. The ray parallel (2) to the principal axis is refracted upward so that it appears to have passed through the focal point on the object side of the lens. The focal ray (3), which is headed for the focal point on the far side of the lens, is refracted so that it leaves parallel to the principal axis.

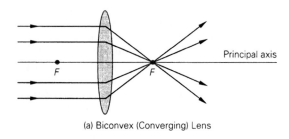

(a) Biconvex (Converging) Lens

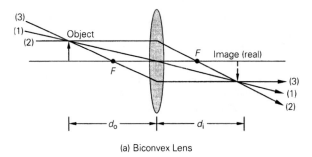

(a) Biconvex Lens

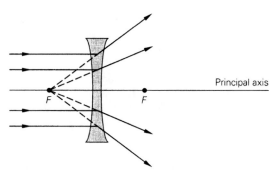

(b) Biconcave (Diverging) Lens

Figure 3   Spherical lenses. (a) A biconvex, or converging, lens and (b) a biconcave, or diverging, lens showing the refraction of parallel incident rays.

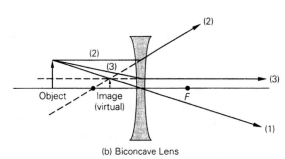

(b) Biconcave Lens

Figure 4   Lens ray diagrams. Examples of the ray diagram method for determining the image characteristics for (a) a biconvex, or converging, lens and (b) a biconcave, or diverging, lens.

If the image is formed on the side of the lens opposite to the object, it is real and can be observed on a screen. However, if the image is on the same side of the lens as the object, it is virtual and cannot be seen on a screen.

The spherical **thin-lens equation** and **magnification factor** for analytically determining the image characteristics are identical to the equations for spherical mirrors (Eqs. 2 and 3). The sign convention is also similar (see Table 1). It should be noted that this lens equation applies only to *thin* lenses.

Example 2 An object is placed 30 cm from a biconcave lens with a focal length of 10 cm (corresponding to the case in Fig. 4b). Determine the image characteristics analytically.

Solution With $d_0 = 30$ cm and $f = -10$ cm (negative by convention for a concave lens), using Eq. 2b yields

$$d_i = \frac{d_0 f}{d_0 - f} = \frac{(30 \text{ cm})(-10 \text{ cm})}{30 \text{ cm} - (-10 \text{ cm})}$$

$$= \frac{-300 \text{ cm}}{40 \text{ cm}} = \frac{-30 \text{ cm}}{4} = -7.5 \text{ cm}$$

Then

$$M = -\frac{d_i}{d_0} = \frac{-(-30/4)}{30} = +\frac{1}{4}$$

Thus the image is virtual (negative d_i), upright (positive M), and reduced by a factor of $\frac{1}{4}$.

However, the relationship between the focal length and the radius of curvature for a spherical lens is not as simple as for a spherical mirror (Eq. 1). For a lens, the focal length is given by what is known as the *lensmaker's equation:*

$$\frac{1}{f} = (n - 1)\left(\frac{1}{R_1} + \frac{1}{R_2}\right) \tag{4}$$

where n is the index of refraction for the lens material and the R's are taken as positive for *convex* surfaces. (See your textbook.)

The index of refraction of glass varies, $n = 1.5$–1.7. For example, for glass with $n = 1.5$ and symmetric converging lenses ($R_1 = R$ and $R_2 = R$), Eq. 4 yields $f = R$.* Keep in mind, however, that the focal length of a lens depends in general on the R values, which can be different, as well as on n. In computations, the experimentally determined value of f will be used.

---

* For f to be equal to $R/2$ for a symmetric lens, as for a spherical mirror, requires $n = 2$, which is greater than the index of refraction of glass.

EXPERIMENTAL PROCEDURE

A. *Spherical Mirrors*

CONCAVE MIRROR

1. **(a)** Construct a ray diagram for a concave mirror with an object located at its focal point. (Drawing provided in the Laboratory Report.) It should be observed from the diagram that the reflected rays are parallel. In this case we say that the rays "converge" at infinity or that the image is formed at infinity.

    Inversely, rays coming from an object at infinity converge to form an image at the focal point or in the focal plane (the plane perpendicular to the optic axis).

    (b) In the open area at the lower right corner of the Laboratory Report sheet, construct a ray diagram with several rays parallel to the optic axis to show they converge at f.

    (c) Using the spherical-mirror equation, determine the image distance for an object at infinity (∞).

2. This focal property makes possible the experimental determination of the focal length of the mirror. An object a great distance from the mirror is essentially at infinity relative to the dimensions of the mirror.

    Take the mirror and screen to a window. Holding the mirror in one hand and the screen in the other, adjust the distance of the screen from the mirror until the image of some outside distant object is observed on the screen (hence a real image).[†]

    Measure the distance f from the mirror vertex to the screen, and record it in the Laboratory Report. Repeat this procedure twice, and take the average of the three measurements as the focal length of the mirror.

3. *Case 1: $d_0 > R$.*

    (a) Sketch a ray diagram for an object at a distance slightly beyond R (that is, $d_0 > R$) and note the image characteristics.

    (b) Set this situation up on the optical bench as illustrated in ● Fig. 5, with the object placed several centimeters beyond the radius of curvature (known from f determination in procedure 2, with $R = 2f$). Measure the object distance d_0, and record it in Data Table 1.

    It is usually convenient to hold the mirror manually and adjust the object distance by moving the mirror rather than the object light source. Move the screen along the side of the optical bench until an image is observed on the screen.

[†] If a window is not available or it is a dark day, use procedure 4 to determine f experimentally. In this case, show first that $d_i = d_0 = R$ and $M = 1$. Then, d_i having been measured, the focal length is $f = d_i/2 = R/2$.

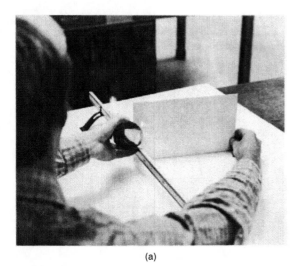

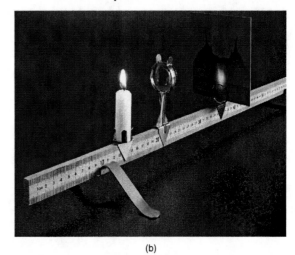

(a) (b)

Figure 5 Experimental arrangements. Arrangements for experimental procedures for (a) spherical mirrors and (b) spherical lenses.

This is best observed in a darkened room. The mirror may have to be turned slightly to direct the rays toward the screen.

(c) Estimate the magnification factor M, and measure and record the image distance d_i.

(d) Using the mirror equation, compute the image distance and the magnification factor.

(e) Compare the computed value of d_i with the experimental value by computing the percent difference.

4. *Case 2: $d_o = R$.* Repeat procedure 3 for this case.

5. *Case 3: $f < d_o < R$.* Repeat procedure 3 for this case.

6. *Case 4: $d_o < f$.* Repeat procedure 3 for this case.

CONVEX MIRROR

7. Sketch ray diagrams for objects at (1) $d_o > R$, (2) $f < d_o < R$, and (3) $d_o < f$, and draw conclusions about the characteristics of the image of a convex mirror. Experimentally verify that the image of a convex mirror is virtual (i.e., try to locate the image on the screen).

B. Spherical Lenses

CONVEX LENS

8. (a) Sketch a ray diagram for a convex lens with the object at its focal point. As with the concave mirror (procedure 1), the image is formed at infinity.

(b) Using the lens equation, determine the image characteristics for an object at infinity.

(c) Experimentally determine the focal length of the lens by a procedure similar to that used for the concave mirror. (The lens may be placed in a lens holder and mounted on a meterstick.)*

9. Repeat the four cases for the lens as was done for the concave mirror in procedures 3 to 6, with R replaced by $2f$ (see Fig. 5). It is initially instructive to move the lens continuously toward the object light source (decreasing d_o) from a $d_o > 2f$ and to observe the image on the screen, which also must be moved continuously to obtain a sharp image. In particular, notice the change in the size of the image as d_o approaches f.

CONCAVE LENS

10. Repeat the procedures carried out for the convex mirror in procedure 7 for the concave lens, with R replaced by $2f$.

11. It is possible to determine the focal length of a concave lens experimentally by placing it in contact with a convex lens so as to form a lens combination. The combination forms a real image. If two lenses of focal lengths f_1 and f_2 are placed in contact, the lens combination has focal length f_c given by

$$\frac{1}{f_c} = \frac{1}{f_1} + \frac{1}{f_2} \tag{5}$$

* In general for a lens, $f \neq R/2$. However, it can be shown for the case of $d_i = d_o$ that $d_o = 2f$. See Question 4 at the end of the experiment.

Place the concave lens in contact with the convex lens (convex surface to concave surface) in a lens holder, and determine the focal length of the lens combination f_c by finding the image of a distant object as in procedure 8. Record in the Laboratory Report.

Using Eq. 5 with the focal length of the convex lens determined in procedure 8, compute the focal length of the concave lens.

EXPERIMENT 23

Spherical Mirrors and Lenses

$\boxed{\text{TI}}$ *Laboratory Report*

A. Spherical Mirrors

Concave Mirror: Ray diagrams

$d_o = f$

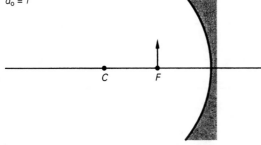

$f < d_o < R$

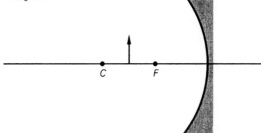

$d_o > R$

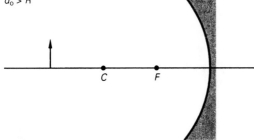

$d_o < f$

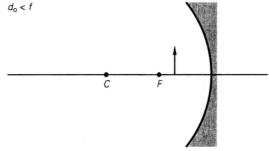

$d_o = R$

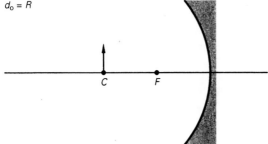

Don't forget units

(continued)

Calculation of d_i for object at ∞　　　　　　Experimental focal length f _____

_____

_____

Average _____

DATA TABLE 1

Purpose: To determine the image distance and magnification.

	Experimental			Computed		d_i percent difference
	d_o ( )	d_i ( )	M factor (estimated)	d_i ( )	M	
$d_o > R$						
$d_o = R$						
$f < d_o < R$						
$d_o < f$						

Calculations
(show work)

EXPERIMENT 23

Convex mirror: Ray diagrams

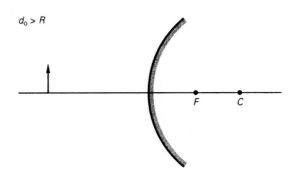

$d_o > R$

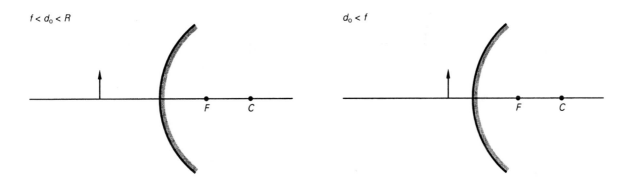

$f < d_o < R$

$d_o < f$

Conclusions

(continued)

B. Spherical Lenses

Convex lens: Ray diagrams

$d_o = f$

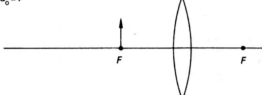

$f < d_o < 2f$

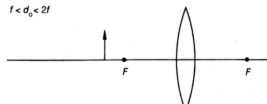

$d_o > 2f$

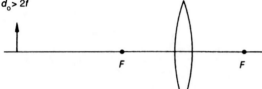

$d_o < f$

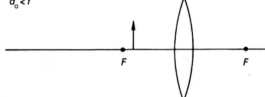

$d_o = 2f$

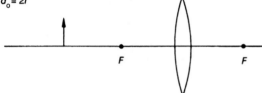

Calculation of d_i for object at ∞

Experimental focal length f _____

_____

_____

Average _____

EXPERIMENT 23   *Laboratory Report*

DATA TABLE 2

Purpose: To determine the image distance and magnification.

	Experimental			Computed		d_i percent difference
	d_o ( )	d_i ( )	M factor (estimated)	d_i ( )	M	
$d_o > 2f$						
$d_o = 2f$						
$f < d_o < 2f$						
$d_o < f$						

Calculations
(show work)

(continued)

Concave lens: Ray diagrams

$d_o > 2f$

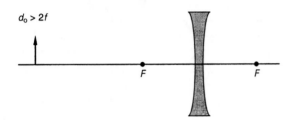

Conclusions

$f < d_o < 2f$

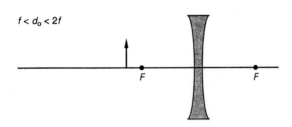

$d_o < f$

Focal length determination:

f_c, focal length of the combination _____

f, focal length of convex lens _____

f, focal length of concave lens _____

/TI/ QUESTIONS

1. A plane mirror essentially has a radius of curvature of infinity. Using the mirror equation, show that (a) the image of a plane mirror is always virtual, (b) the image is "behind" the mirror the same distance as the object is in front of the mirror, and (c) the image is always upright.

2. Show that the magnification factor for a mirror or lens $M = d_i/d_o$ (sign convention omitted) is the lateral magnification, or the ratio of the height (lateral size) of the image to that of the object. (*Hint:* Draw a ray diagram.)

3. Explain what characteristics make convex spherical mirrors applicable for store monitoring and concave spherical mirrors applicable as flashlight reflectors.

4. Prove that for a converging lens, for the case $d_i = d_o$, it is true that $d_i = d_o = 2f$.

(continued)

5. Using the thin-lens equation and the magnification factor, show that for a spherical diverging lens the image of a real object is always virtual, upright, and reduced. Does the same apply for a spherical diverging mirror?

6. (*Optional*) (a) Using the experimental value of f for the biconvex converging lens and $n = 1.5$, compute the radius of curvature of the lens's surfaces using the lensmaker's equation. (The radius of curvature for each surface is the same.)

(b) A student incorrectly assumes that $f = R/2$ for the lens and computes f using the value of R found in part (a). Compare this computed value of f with the experimental value.

(c) The index of refraction of the lens could have a different value (n of glass varies generally from 1.5 to 1.7). Would this make a difference? Explain.

EXPERIMENT 24

Line Spectra and the Rydberg Constant

/TI/ *Advance Study Assignment*

Read the experiment and answer the following questions.

1. Distinguish between continuous spectra and line spectra, and describe their causes.

2. Why does a gas discharge tube (e.g., a neon light) have a certain color?

3. What are (a) the Balmer series and (b) the Rydberg constant?

(continued)

4. Explain briefly how the prism spectrometer is calibrated.

5. Explain briefly how the Rydberg constant is determined experimentally.

Line Spectra and the Rydberg Constant

INTRODUCTION AND OBJECTIVES

In spectroscopic analysis, two types of spectra are observed: continuous spectra and line or discrete spectra. The spectrum of visible light from an incandescent source is found to consist of a **continuous spectrum** or band of merging colors and contains all the wavelengths of the visible spectrum.

However, when the light from a gas discharge tube (for example, mercury or helium) is observed through a spectroscope, only a few colors, or wavelengths, are observed. The colored images of the spectroscope slit appear as bright lines separated by dark regions; hence the name **line** or **discrete spectra.**

Each gas emits a particular set of spectral lines and hence has a characteristic spectrum. Thus, spectroscopy (the study of spectra) provides a method of identifying elements. The discrete lines of a given spectrum depend on the atomic structure of the atoms and are due to electron transitions.

The line spectrum of hydrogen was explained by Bohr's theory of the hydrogen atom. However, before this, the line spectrum of hydrogen was described by an empirical relationship involving the Rydberg constant. In this experiment, line spectra will be observed and the relationship of the Rydberg constant to the theoretical quantities of the Bohr theory will be investigated.

After performing this experiment and analyzing the data, you should be able to:

1. Clearly distinguish between continuous and line (discrete) spectra.
2. Explain why gas discharge tubes emit line spectra.
3. Tell what is meant by the Balmer series and the Rydberg constant.

EQUIPMENT NEEDED

- Prism spectrometer
- Incandescent light source
- Mercury or helium discharge tube
- Hydrogen discharge tube
- Discharge-tube power supply
- 2 sheets of Cartesian graph paper

THEORY

The electrons in an incandescent light source undergo thermal agitation and emit electromagnetic radiation (light) of many different wavelengths, hence producing a continuous spectrum. However, when light emitted from excited gases or vaporized liquids or solids is analyzed, line spectra such as those illustrated in ● Fig. 1 are observed.

Modern theory explains spectra in terms of photons of light of discrete wavelengths being emitted as the result of electron transitions between atomic energy levels. Different substances have characteristic spectra—that is, they have a characteristic set of lines at specific wavelengths. In a manner of speaking, the spectrum of a substance acts as a "fingerprint" by which the substance can be identified.

The characteristic color of light from a gas discharge tube is often indicative of the most intense spectral line(s) in the visible region. For example, light from a hydrogen discharge tube has a characteristic red glow resulting from an intense emission line with a wavelength of 656.1 nm. Similarly, when table salt is vaporized in a flame, yellow light is observed because of the intense yellow discharge line in the spectrum of sodium. It is the presence of sodium

that gives candles and wood fires their yellow glow. You may have noticed that many highway and parking lot lights are bright yellow. They are sodium lights, used because sodium discharge is a very efficient way to produce light. Wavelengths are commonly measured in nanometers (nm):

$$1 \text{ nm} = 10^{-9} \text{ m} = 10^{-7} \text{ cm}$$

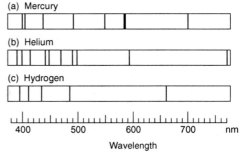

Figure 1 Line spectra. Illustrations of visible line spectra for (a) mercury, (b) helium, and (c) hydrogen. From Wilson/Buffa, *College Physics,* Fifth Edition. Copyright © 2003. Reprinted by permission of Pearson Education.

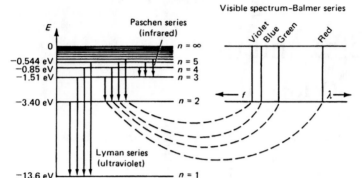

Figure 2 Energy level transitions. The energy level transmissions for the hydrogen atom. The Balmer series, $n_f = 2$, produces a line spectrum in the visible region.

The systematic spacing of the spectral lines in the hydrogen spectrum was empirically described by spectroscopists in the late 1800s. For example, the wavelengths of spectral lines in the visible region, called the Balmer series, were found to fit the formula

$$\frac{1}{\lambda} = R\left(\frac{1}{2^2} - \frac{1}{n^2}\right) \qquad n = 3, 4, 5, \ldots \qquad (1)$$

where R is the **Rydberg constant,*** with a value of 1.097×10^{-2} nm^{-1}.

The hydrogen spectrum is of particular theoretical interest because hydrogen, having only one proton and one electron, is the simplest of all atoms. Niels Bohr (1885–1962), a Danish physicist, developed a theory for the hydrogen atom that explains the spectral lines as resulting from electron transitions between energy levels, or discrete electron orbits (● Fig. 2), with the wavelengths of the spectral lines being given by the theoretical equation

$$\lambda = \frac{hc}{\Delta E} \qquad (2)$$

where

$$\Delta E = 13.6\left(\frac{1}{n_f^2} - \frac{1}{n_i^2}\right) \text{eV}$$

is the energy difference between the initial and final states, n_i and n_f, $h = 6.63 \times 10^{-34}$ J-s $= 4.14 \times 10^{-15}$ eV-s (Planck's constant), and $c = 3.00 \times 10^8$ m/s (speed of light in vacuum). The values $n = 1, 2, 3, 4, \ldots$ are called the **principal quantum numbers.** Different final states account for the different series.[†]

---

* After J. R. Rydberg (1854–1919), the Swedish physicist who developed the series relationship.

[†] The three such spectral series (shown in Fig. 2) are named after 19th-century scientists: the Swiss mathematician Johann Balmer and the German Physicists Theodore Lyman and Friedrich Paschen.

For spectral lines in the visible region, the final state is $n_f = 2$, and

$$\lambda = \frac{hc}{\Delta E} = \frac{hc}{13.6[(1/2^2) - 1/n^2]} \qquad n = 3, 4, 5, \ldots$$

or

$$\frac{1}{\lambda} = \frac{13.6}{hc}\left(\frac{1}{2^2} - \frac{1}{n_i^2}\right) \qquad n = 3, 4, 5, \ldots \qquad (3)$$

Comparing this theoretical equation with the empirical equation, Eq. 1 reveals that the forms are identical, with the prediction that $R = (13.6 \text{ eV})/hc$.

EXPERIMENTAL PROCEDURE

1. A prism spectrometer will be used to analyze and study spectra in this experiment. The prism spectrometer is illustrated and its use described in Experiment . Review the operation of this instrument. Place the incandescent source in front of the collimator slit, and observe the continuous spectrum that results from the prism dispersion (see Experiment). List the colors of the spectrum in the Laboratory Report, beginning with red.

2. A convenient type of discharge tube and power supply is shown in ● Fig. 3.
   Caution: Great care should be taken, because the discharge tube operates at high voltage and you could receive an electrical shock.

   Mount a mercury (or helium) discharge tube in the power supply holder, and place in front of the collimator slit. *Caution: If a larger mercury source is used, it should be properly shielded because of the ultraviolet radiation that may be emitted. Consult your instructor.*

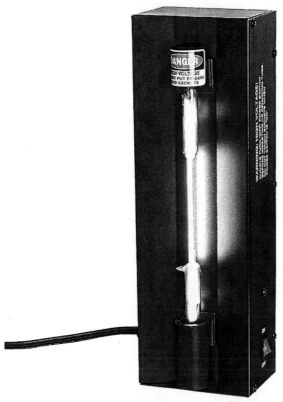

Figure 3 Experimental apparatus. A gas discharge tube and power supply. (Photo courtesy of Sargent-Welch.)

Turn on the power supply, observe the mercury (or helium) spectrum through the telescope, and note its line nature.

With the slit as narrow as possible, rotate the prism slightly back and forth and notice the reversal of direction of the motion of the spectrum when the prism is rotated in one direction. Focusing on the yellow line (for mercury, the brighter yellow line), stop rotating the prism at the position of the reversal of motion of this line.

This sets the prism for minimum deviation for the yellow line (Experiment   ), which will be taken as an average for the spectrum. (The other lines have slightly different minimum deviations.)

3. **(a)** Without disturbing the prism, starting at the red end of the spectrum, set the cross hairs of the telescope on the extreme red line, and record the color and divided circle reading in the Laboratory Report. Repeat this procedure for each spectral line in order. (Turn off the discharge tube as soon as possible to conserve the life of the tube.)

   (b) Find the wavelengths of the spectral lines for the discharge tube gas in Appendix A, Table A8, and match them to the line readings.

   (c) Using these data, plot the wavelength λ versus the divided circle reading θ. This calibrates the spectrometer, and unknown wavelengths can be determined from divided circle readings from the calibration curve.

4. With the discharge tube power supply off, replace the mercury (or helium) discharge tube with a hydrogen discharge tube. Turn on the power supply, and starting with the red line of the hydrogen spectrum, determine the divided circle reading for each spectral line with the cross hairs of the telescope positioned on the center of the line. Record in the Laboratory Report.

   The red line is referred to as H_α in spectroscopic notation. The other sequential lines are referred to as H_β, etc., with subscripts in Greek alphabetical order.

5. Determine the wavelengths of the hydrogen lines from the calibration curve, and plot the reciprocal of the wavelength $1/\lambda$ versus $1/n^2$. (Begin the abscissa scale with zero.) Draw the best straight line that fits the data points and determine the slope of the line.

   Note that Eq. 1,

$$\frac{1}{\lambda} = R\left(\frac{1}{2^2} - \frac{1}{n^2}\right) = \frac{R}{4} - \frac{R}{n^2}$$

has the form of a straight line, $y = mx + b$, with the negative slope equal to the Rydberg constant. Compare the slope of the line with the accepted value of the Rydberg constant by computing the percent error. Compare the intercept of this line with $R/4$.

Line Spectra and the Rydberg Constant

$\overline{\text{TI}}$ *Laboratory Report*

Colors of the Continuous Spectrum

DATA TABLE 1   Mercury (or Helium) Spectrum

Color	Divided circle reading	Wavelength ( ) (from Table A8)

Sequence of colors

red
_____

_____

_____

_____

_____

_____

_____

_____

_____

Don't forget units

(continued)

DATA TABLE 2 Hydrogen Spectrum

Line	Color	Divided circle reading	Wavelength ( )	$1/\lambda$ ( )	$1/n^2$
H_α, $n = 3$					
H_β, $n = 4$					
H_γ, $n = 5$					
H_δ, $n = 6$					

Calculations (Slope of Graph)
(show work)

R (experimental) _____

Accepted value _____

Percent error _____

$R/4$ from graph _____

Accepted value _____

Percent error _____

/TI/ QUESTIONS

1. Compute the value of the Rydberg constant from the Bohr theory, and compare it with the accepted empirical value.

EXPERIMENT 24   *Laboratory Report*

2. Why are only four lines seen in the Balmer series? (Transitions for $n_i > 6$ also exist.) Justify your answer mathematically.

3. As n becomes very large, the wavelengths of the Balmer (and other) series approach a minimum wavelength, or series limit (Eq. 1).   What is the wavelength of the series limit for the Balmer series?

EXPERIMENT 24

Name _____ Section _____ Date _____

Lab Partner(s) _____

  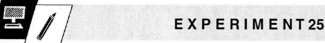

EXPERIMENT 25

/TI/ The Transmission Diffraction Grating: Measuring the Wavelengths of Light

/CI/ Single-Slit and Double-Slit Diffraction

/TI/ *Advance Study Assignment*

Read the experiment and answer the following questions.

1. What is a diffraction grating? Distinguish between the two types of gratings.

2. What is the grating constant? What would be the grating constant for a grating with 300 lines/mm? (Express the constant in nanometers.)

(continued)

3. Explain why there is a spectrum for each diffraction order when multicolored light is analyzed.

4. Will the red or the violet end of the first-order spectrum be nearer the central maximum? Justify your answer.

5. It will be observed that the second-order spectrum is "spread out" more than the first-order spectrum. Why?

CI *Advance Study Assignment*

Read the experiment and answer the following questions.

1. What is diffraction?

EXPERIMENT 25   *Advance Study Assignment*

2. What type of pattern is produced by a double slit? By a single slit?

3. What causes dark and bright fringes?

EXPERIMENT 25

TI The Transmission Diffraction Grating: Measuring the Wavelengths of Light

CI Single-Slit and Double-Slit Diffraction

OVERVIEW

Experiment   examines diffraction, but the TI and CI procedures differ in focus. The TI procedure uses a transmission diffraction grating to measure the wavelengths of light from an incandescent source and the Hg line spectrum. The CI procedure complements this by investigating single-slit and double-slit diffraction. By using the diffraction patterns formed by laser light, it examines the conditions for single-slit dark fringes and double-slit bright fringes.

INTRODUCTION AND OBJECTIVES

In Experiment   the prism spectrometer had to be calibrated in terms of known wavelengths before we were able to determine unknown wavelengths of light experimentally. How, then, are the wavelengths of spectral lines or colors initially determined? This is most commonly done with a diffraction grating, a simple device that allows for the study of spectra and the measurement of wavelengths.

By replacing the prism with a diffraction grating, a prism spectrometer (Experiments   and   becomes a grating spectrometer. When a diffraction grating is used, the angle(s) at which the incident beam is defracted relate simply to the wavelength(s) of the light. In this experiment the properties of a transmission grating will be investigated and the wavelengths of several spectral lines will be determined.

TI OBJECTIVES

After performing this experiment and analyzing the data, you should be able to:

1. Describe the principle of a diffraction grating.
2. Explain the operation of a grating spectrometer.
3. Tell how the wavelength of light can be measured with a grating spectrometer.

CI OBJECTIVES

A diffraction grating has many slits. But what about using a single slit or a double slit? When illuminated with monochromatic light, these slits produce interference and diffraction patterns with bright and dark fringes. Geometric analyses give equations for the positions of single-slit dark fringes and double-slit bright fringes. The CI portion of this experiment investigates these fringe relationships. After performing this experiment and analyzing the data, you should be able to:

1. Verify that the positions of the minima in a diffraction pattern match the positions predicted by theory.
2. Use a diffraction and interference pattern to determine the wavelength of light.
3. Compare the patterns formed by single slits to those formed by double slits.
4. Investigate the effects of changing slit width and slit separation.

The Transmission Diffraction Grating: Measuring the Wavelengths of Light

/TI/ EQUIPMENT NEEDED

- Spectrometer*
- Diffraction grating and holder
- Mercury discharge tube

- Power supply for discharge tube
- Incandescent light source

* *Instructor's note:* If a spectrometer is not available, an alternative, inexpensive method is described in the Instructor's Manual.

/TI/ THEORY

A **diffraction grating** consists of a piece of metal or glass with a very large number of evenly spaced parallel lines or grooves. This gives two types of gratings: reflection gratings and transmission gratings.

Reflection gratings are ruled on polished metal surfaces; light is reflected from the unruled areas, which act as a row of "slits." **Transmission gratings** are ruled on glass, and the unruled slit areas transmit incident light.

The transmission type is used in this experiment. Common laboratory gratings have 300 grooves per mm and 600 grooves per mm (about 7500 grooves per in. and 15,000 grooves per in.), and are pressed plastic *replicas* mounted on glass. Glass originals are very expensive.

Diffraction consists of the "bending," or deviation, of waves around sharp edges or corners. The slits of a grating give rise to diffraction, and the diffracted light interferes so as to set up interference patterns (● TI Fig. 1).

Complete constructive interference of the waves occurs when the phase or path difference is equal to one wavelength, and the first-order maximum occurs for

$$d \sin \theta_1 = \lambda \qquad \text{(TI 1)}$$

where d is the grating constant, or distance between the grating lines, θ_1 is the angle the rays are diffracted from the incident direction, and $d \sin \theta_1$ is the path difference between adjacent rays. The **grating constant** is given by

$$d = 1/N \qquad \text{(TI 2)}$$

where N is the number of lines or grooves per unit length (usually per millimeter or per inch) of the grating.

A second-order maximum occurs for $d \sin \theta_2 = 2\lambda$, and so on, so that in general we may write

$$d \sin \theta_n = n\lambda \qquad n = 1, 2, 3, \ldots \qquad \text{(TI 3)}$$

where n is the order of the image maximum. The interference is symmetric on either side of an undeviated and undiffracted central maximum of the slit image, so the angle between symmetric image orders is $2\theta_n$ (● TI Fig. 2).

In practice, only the first few orders are easily observed, with the number of orders depending on the grating constant. If the incident light is other than monochromatic, each order corresponds to a spectrum. (That is, the grating spreads the light out into a spectrum.)

As can be seen from Eq. TI 1, since d is constant, each wavelength (color) deviates by a slightly different angle, so that the component wavelengths are separated into a spectrum. Each diffraction order in this case corresponds to a spectrum order. (The colorful displays seen on compact disks [CDs] result from diffraction.)

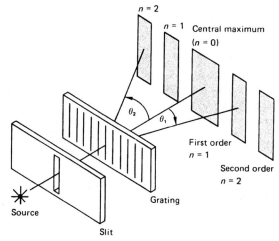

TI Figure 1 Diffraction pattern. A simplistic view of the diffraction pattern (two orders) produced by a diffraction grating. (Patterns and angles are exaggerated for illustration.)

(a)

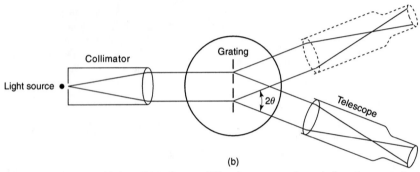

(b)

TI Figure 2 Grating spectrometer.   (a) A student views a diffraction pattern through the telescope of a grating spectrometer. The light source and collimator are on the right. (b) A diagram of a top view of a grating spectrometer. When the symmetric images of a particular order n are viewed from both sides of the central maximum, the angle between the two viewing positions is $2\theta_n$. (Photo courtesy of Sargent-Welch.)

TI Example 1   In an experiment using a diffraction grating with 600 lines/mm, the angle between the corresponding lines of a particular component of the first-order spectrum on either side of the incident beam is 41.30°. What is the wavelength of the spectral line?

Solution   Given $2\theta_1 = 41.30°$, or $\theta_1 = 20.65°$, and with a grating ruling of $N = 600$ lines/mm, the grating constant d is (TI Eq. 2)

$$d = \frac{1}{N} = \frac{1}{600/\text{mm}} = 1.67 \times 10^{-3} \text{ mm}$$

When doing several calculations, it is convenient to express the grating constant in nanometers (nm). Converting to nanometers (1 mm = 10^{-6} nm) yields

$$d = 1.67 \times 10^{-3} \text{ mm } (10^{-6} \text{ nm/mm}) = 1.67 \times 10^{3} \text{ nm}$$

Then for first-order ($n = 1$) interference, by Eq. TI 3,

$$d \sin \theta_1 = \lambda$$

or

$$\lambda = d \sin \theta_1 = (1.67 \times 10^{-3} \text{ nm})(\sin 20.65°) = 589 \text{ nm}$$

TI EXPERIMENTAL PROCEDURE

1. Review the general operation of a spectrometer if necessary (Experiment   ). Record the number of lines per mm of your diffraction grating in the Laboratory Report. Mount the grating on the spectrometer table with the grating ruling parallel to the collimator slit and the plane of the grating perpendicular to the collimator axis.

DETERMINATION OF THE WAVELENGTH RANGE OF THE VISIBLE SPECTRUM

2. Mount an incandescent light source in front of the collimator slit. Move the spectrometer telescope into the line of the slit of the collimator, and focus the cross hairs on the central slit image.

   Notice that this central maximum, or "zeroth"-order image, does not depend on the wavelength of light, so that a white image is observed. Then move the telescope to either side of the incident beam and observe the first- and second-order spectra. Note which is spread out more.

3. (a) Focus the cross hairs on the blue (violet) end of the first-order spectrum at the position where you judge the spectrum just becomes visible. Record the divided circle reading (to the nearest minute of arc) in TI Data Table 1.

   (b) Move the telescope to the other (red) end of the spectrum, and record the divided circle reading of its visible limit.

   (c) Repeat this procedure for the first-order spectrum on the opposite side of the central maximum. The angular difference between the respective readings corresponds to an angle of 2θ (TI Fig. 2b).

4. Compute the grating constant d in millimeters, and with the experimentally measured θ's, compute the range of the wavelengths of the visible spectrum in nanometers.

DETERMINATION OF THE WAVELENGTHS OF SPECTRAL LINES

5. Mount the mercury discharge tube in its power supply holder, and place in front of the collimator slit.

   Caution: Work very carefully, as the discharge tube operates at high voltage and you could receive an elec-

trical shock. Make certain the power supply is turned off before inserting the tube. If a large mercury source is used, it should be properly shielded because of the ultraviolet radiation that may be emitted. Consult with your instructor.

Turn on the power supply and observe the first- and second-order mercury line spectra on both sides of the central image.

6. Because some of the lines are brighter than others and the weaker lines are difficult to observe in the second-order spectra, the wavelengths of only the brightest lines will be determined. Find the listing of the mercury spectral lines in Appendix A, Table A8, and record the color and wavelength in TI Data Table 2.

   Then, beginning with either first-order spectra, set the telescope cross hairs on each of the four brightest spectral lines, and record the divided circle readings (read to the nearest minute of arc). Repeat the readings for the first-order spectrum on the opposite side of the central image.

7. Repeat the measurement procedure for the four lines in the second-order spectra and, using TI Eq. 2, compute the wavelength of each of the lines for both orders of spectra. Compare with the accepted values by computing the percent error of your measurements in each case.

   Note: In the second-order spectra, two yellow lines—a *doublet*—may be observed. Make certain that you choose the appropriate line. (*Hint:* See the wavelengths of the yellow lines in Appendix A, Table A8. Which is closer to the red end of the spectrum?)

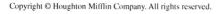

T I   E X P E R I M E N T 25

The Transmission Diffraction Grating: Measuring the Wavelengths of Light

Laboratory Report

Number of lines per millimeter on grating _____   Grating constant d _____ (mm)

DATA TABLE 1

Purpose: To determine the wavelength range of the visible spectrum.

Spectrum limit	Divided circle reading		2θ	θ	$\sin \theta$	Computed wavelength ( )
	Right	Left				
Violet end						
Red end						

Calculations
(show work)

Don't forget units

(continued)

/TI/ **DATA TABLE 2**

Purpose: To determine the wavelengths of spectral lines.

Mercury Lines		Divided circle reading		2θ	θ	$\sin \theta$	Computed λ ( )	Percent error
Color	Wavelength	Right	Left					
First-order spectrum								
Second-order spectrum								

Calculations
(show work)

EXPERIMENT 25   *Laboratory Report*

/TI/ QUESTIONS

1. If a grating with more lines per unit length were used, how would the observed angles or spread of the spectra be affected?

2. Was there any difference in the accuracy of the determination of the wavelengths of the mercury lines for the different-order spectra? If so, give an explanation.

3. Is it possible for the first-order spectrum to overlap the second-order spectrum? Explain, assuming a continuous spectrum.

4. Is there a theoretical limit to the order of the spectrum one would be able to observe with your diffraction grating? Justify your answer mathematically.

(continued)

5. (Reminder about rounding errors.) In Example 1, values were rounded to the proper number of significant figures in each step. Recall that Experiment 1 suggested that one or two insignificant (extra) figures usually be carried along and stated that if a calculator is used, rounding off may be done only on the final result of multiple calculations.

    If you applied these rules to the calculations in Example 1, would you get 589 nm? (Justify your answer.)

CI EXPERIMENT 25

Single-Slit and Double-Slit Diffraction

CI EQUIPMENT NEEDED

- Optics bench from the Basic Optics System (PASCO OS-8515)
- Diode laser (OS-8525)
- Linear translator (OS-8535)
- Aperture bracket (OS-8534)
- Rotary motion sensor (CI-6538)

- Light sensor (CI-6504A)
- Single-slit disk from Slit Accessory (OS-8523)
- Multiple-slit disk from Slit Accessory (OS-8523)

Note: The light sensor needs to be calibrated before use. Refer to the owner's manual for instructions on how to calibrate the sensor.

CI THEORY

A. Single-Slit Diffraction

Diffraction is the bending of a wave by means other than reflection or refraction. It occurs when a wave encounters an obstacle and bends around it, reaching places that would otherwise be shadowed. The amount of bending depends on the wavelength of the wave relative to the size of the obstacle. For waves of visible light that have wavelengths in the nanometer range (10^{-9} m), some obstacles that will produce diffraction are sharp edges, point objects, and thin slits.

Let's consider monochromatic light that passes through a single thin slit. The light "flares out" as it goes through, producing, on a screen a distance L away, what is called a single-slit diffraction pattern. A sketch of such a pattern is shown in ● CI Fig. 1. The diffraction pattern has a bright central region. Other, less intense regions, are symmetrically distributed around the central region. These bright regions, or bright fringes, are called maxima and are regions of constructive interference. The dark regions in between are called minima and are regions of destructive interference.

From an analysis of the geometry, it can be shown that the condition for dark fringes, or minima, is given by

$$w \sin\theta = m\lambda \qquad m = 1, 2, 3, \dots \qquad \textbf{(CI1)}$$
$$\text{(condition for dark fringes)}$$

where w is the width of the slit, λ is the wavelength of the light, and θ is the angle to a particular fringe minimum designated by $m = 1, 2, 3, \dots$ The m number is called the order number, and the fringes are referred to as first-order, second-order, third-order and so on.*

Note the geometry in ● CI Fig. 2, where $\tan\theta_1 = y_1/L$. Experimentally, $y \ll L$, and using the small-angle

---

*CI Eq. 1 is sometimes written $w \sin\theta = m\lambda \quad m = \pm 1, \pm 2, \pm 3, \dots$, where the plus and minus numbers are used to indicate dark fringes on opposite sides of the central maximum.

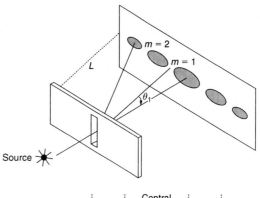

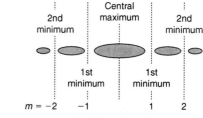

CI Figure 1   Single-slit diffraction. As the diagram shows, the minima are distributed symmetrically on both sides of the central maximum. The first minimum is designated $m = 1$ and occurs between the central maximum and the next bright fringe. After that, the other dark fringes are designated $m = 2, m = 3$, and so on. Traditionally, positive and negative numbers are used to distinguish one side from the other.

approximation $\tan\theta_1 \approx \sin\theta_1 \approx y_1/L$ for the first-order minimum, we can in general write CI Eq. 1 as

$$y_m \approx \frac{mL\lambda}{w} \qquad m = 1, 2, 3, \dots \qquad \textbf{(CI2)}$$
$$\text{(small angles only)}$$

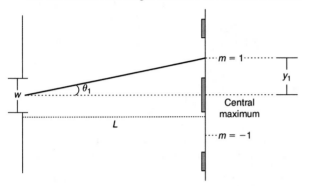

CI Figure 2 Geometry of the single-slit diffraction pattern. The first-order minimum, $m = 1$, is at a distance y_1 from the central maximum. In experimental conditions, L is much larger than w and y_1 ($L \gg y_1$).

where y_1 is the distance between the center of the central maximum and the center of the first-order minimum, and so on (see CI Fig. 2).

B. Double-Slit Interference

When light passes through two slits, the diffraction pattern is again bright-and-dark regions, but regions smaller than those seen with the single slit. ● CI Fig. 3 shows a diagram of the fringes of this interference pattern.

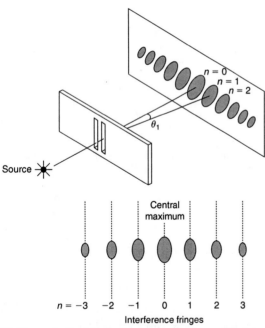

CI Figure 3 Double-slit interference pattern. The interference pattern from two slits produces a smaller and sharper set of bright and dark fringes than the diffraction pattern from a single slit.

From the geometry, it can be shown that the positions of the bright fringe maxima are given by

$$d \sin \theta = n\lambda \qquad n = 1, 2, 3, \dots \qquad \textbf{(CI3)}$$
(condition for bright fringes)

where d is the distance between the double slits, θ is the angular distance between the central maximum and another bright fringe of order n, and λ is the wavelength of the light.

Using a small-angle approximation as before, we find that CI Eq. 3 becomes

$$y_n = \frac{nL\lambda}{d} \qquad n = 0, 1, 2, 3, \dots \qquad \textbf{(CI4)}$$
(lateral distances to bright fringes, small angles only)

SETTING UP DATA STUDIO

1. Open Data Studio and choose "Create Experiment."
2. From the sensor list, choose a rotary motion sensor.
3. Double-click on the RMS icon. The sensor properties window will open. Make the following selections:

General:	Set the sample rate to 20 Hz, Fast
Measurement:	Select Position (m), and deselect all others.
RMS:	Divisions/Rotations: 1440
	Linear calibration: Rack

Click OK to exit the sensor properties window.

4. From the sensor list, choose a light sensor.
5. Double-click on the light sensor icon. The sensor properties window will open. Make the following selections:

General:	Set sample rate to 20 Hz, Fast
Measurement:	Select light intensity, and deselect all others.

Click OK to exit the sensor properties window.

6. Create a graph by dragging the "Light Intensity" icon from the data list and dropping it on top of the "Graph" icon of the displays list. A graph of intensity versus time will open in a window called Graph 1.
7. Now drag the "Position" icon from the data list, and drop it on top of the horizontal axis of the graph. The horizontal axis will change to measure position instead of time. ● CI Fig. 4 shows what the screen should look like after the setup is complete.

◉ EQUIPMENT SETUP

1. Mount the single-slit accessory to the optics bench. The slit disks are mounted on a ring that snaps into an empty lens holder. Rotate the ring in the lens holder so that the slits at the center of the ring are vertical in

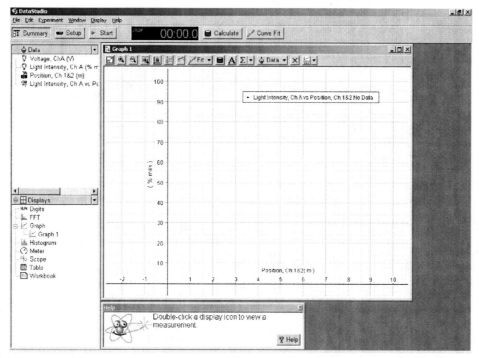

CI Figure 4 Data Studio setup. A light sensor, together with a rotary motion sensor, will be used to produce a plot of light intensity versus position for diffraction and interference patterns. (Data displayed using DataStudio Software. Reprinted courtesy of PASCO scientific.)

the holder. Then tighten the screw on the holder. (See ● CI Fig. 5.)

2. Align the laser beam with the slit.
   a. Mount the diode laser at one end of the bench. Put the slit holder a few centimeters away from the laser, with the disk side closer to the laser. Plug in the laser and turn it on.

b. Adjust the position of the beam from left-to-right and up-to-down until the beam is centered on the slit. The knobs to do this are on the back of the diode laser. (See ● CI Fig. 6.)

3. Prepare the rotary motion sensor and the light sensor.
   a. Mount the RMS in the rack of the linear translator. Then mount the linear translator to the end of the optics bench.

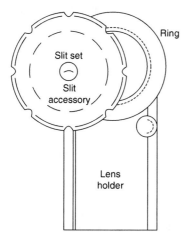

CI Figure 5 Slit accessory on lens holder. The slit accessory is mounted on a ring that snaps into the lens holder.

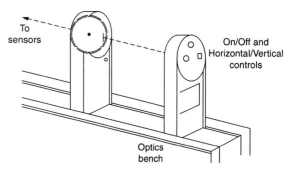

CI Figure 6 Diode laser and slits on track. The diode laser is placed in the track a few centimeters behind the lens holder with the slits.

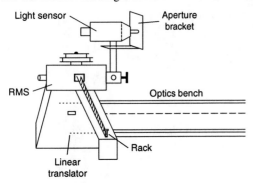

CI Figure 7 The RMS mounted in the rack with the light sensor. The RMS is mounted on the linear translator. The light sensor with the apertures is mounted on the RMS.

    b. The light sensor with the aperture bracket (set to slit 6) is mounted on the RMS rod clamp. (See ● CI Fig. 7.)

4. Plug the RMS and the light sensor into the interface, as shown in the setup window in the computer screen.

CI EXPERIMENTAL PROCEDURE

Start by making a note in the Laboratory Report of the wavelength (λ) of the laser light. It is printed on the back of the diode laser.

A. The Single-Slit Pattern

1. Select the 0.04-mm wide single slit from the disk. Make a note of the value in the Laboratory Report.
2. Place the laser on the side opposite the light sensor on the track. The slit disk should be a few centimeters in front of the laser. Record in the Laboratory Report the distance (L) between the slit and the sensor.
3. Set the light sensor aperture bracket to slit 6.
4. Turn the laser on and set the gain switch to ×10. If the light intensity goes off scale when you are measuring, turn it down to ×1.
5. The pattern should be visible on the aperture bracket of the light sensor. Move the light sensor to one side of the laser pattern.
6. Turn the classroom lights off.
7. Press the START button and *slowly* move the sensor across the pattern by rotating the large pulley of the RMS. Click the STOP button when you are finished.
8. Use the magnifier button (on the graph toolbar, a button that shows a magnifier lens) to enlarge the central maximum and the first maximum on each side.
9. Use the smart-tool (on the graph toolbar, a button labeled with *xy* axes) to measure the distance between the first minima on the two sides of the central maximum. That is, measure the distance between $m = -1$ and $m = 1$. Record the value in CI Data Table 1.

10. Determine the distance y_m from the center of the pattern to one of the $m = 1$ minima by dividing the previous distance by 2. Record the result in CI Data Table 1.
11. Repeat the measurements for the second-, the third-, and if possible the fourth-order minima. Use the magnifier button to enlarge the parts of the graph as needed.
12. Calculate $\sin \theta$ for each case, using the derived formula from the small-angle approximation. (See CI Data Table 1 for the formula.) Enter the values in both CI Data Table 1 and CI Data Table 2.
13. To check how well the observed pattern matches the theory, use the known wavelength of the light and the known width of the slit to calculate the theoretical value of $\sin \theta$ for each case. Compare the theory to the experiment by taking percent differences. Record all results in CI Data Table 1.
14. To demonstrate that the experimental data can also be used to find the wavelength of the light, use the data in CI Data Table 2 with CI Eq. 1 to calculate the wavelength of the light for each case; then find an average. Compare the average to the expected value by taking a percent error.
15. Cancel all zooms, and fix up the graph window so that all data collected can be seen. Print the graph and label each minimum, on both sides of the center, with the appropriate m value. Title this graph "Graph 1. Single-Slit Pattern, $w = 0.04$ mm." If no printer is available, make a careful sketch of the graph, paying attention to the location of the minima along the horizontal axis. Attach the graph to the Laboratory Report.

B. The Double-Slit Pattern

1. Change the slit accessory to a multiple-slit disk and realign the laser, if needed. Choose the double slit with slit separation 0.25 mm and slit width 0.04 mm.
2. Set the light sensor aperture bracket to slit 4.
3. The pattern should be visible on the aperture bracket of the light sensor. Move the light sensor to one side of the laser pattern.
4. Turn the classroom lights off.
5. Press the START button and *slowly* move the sensor across the pattern by rotating the large pulley of the RMS. Click the STOP button when you are finished.
6. Use the magnifier button to enlarge the central maximum and the first maximum on each side.
7. Use the smart-tool to measure the distance between the first maxima on the two sides of the central maximum. That is, measure the distance between $n = 1$ and $n = -1$. Record the value in CI Data Table 3.
8. Determine the distance y_n to one of the $n = 1$ fringes by dividing the previous distance by 2. Record the result in CI Data Table 3.
9. Repeat the measurements for the second-, third-, and if possible all the way to the sixth-order maxima. Use the

magnifier button to enlarge the parts of the graph as needed.

10. Calculate $\sin\theta$ for each case, using the derived formula from the small-angle approximation. Enter the values in both CI Data Table 3 and CI Data Table 4.

11. To check how well the observed pattern matches the theory, use the known wavelength of the light and the known separation of the slits to calculate the theoretical value of $\sin\theta$ for each case. Compare the theory to the experiment by taking percent differences. Record all results in CI Data Table 3.

12. To demonstrate that this experimental data can also be used to find the wavelength of the light, use the data in CI Data Table 4 to calculate the wavelength of the light for each case; then find an average. Compare the average to the expected value by taking a percent error.

13. Cancel all zooms, and fix up the graph window so that all data collected can be seen. Print the graph and label each maximum, on both sides of the center, with the appropriate n value. Title this graph "Graph 2.

Double-Slit Pattern, $d = 0.25$ mm, $w = 0.04$ mm." If no printer is available, make a careful sketch of the graph, paying attention to the location of the maxima along the horizontal axis. Attach the graph to the Laboratory Report.

C. Comparing Single-Slit Pattern to Double-Slit Pattern

1. Change the double-slit set to a set with slit separation 0.25 mm and slit width 0.08 mm.

2. Collect data as before and print the graph. Label on the graph the maxima with their appropriate n values. Title this graph "Graph 3. Double-Slit Pattern, $d = 0.25$ mm, $w = 0.08$ mm."

3. Repeat with the double-slit set of slit separation 0.50 mm and slit width 0.04 mm. This time, title the graph "Graph 4. Double-Slit Pattern, $d = 0.50$ mm, $w = 0.04$ mm."

Name _____ Section _____ Date _____

Lab Partner(s) _____

   C I   **E X P E R I M E N T 25**

Single-Slit and Double-Slit Diffraction

CI *Laboratory Report*

Wavelength of light λ _____

A. The Single-Slit Pattern

Slit width w _____

Distance between slit and pattern L _____

CI DATA TABLE 1

Purpose: To compare the experimental single-slit pattern with the pattern predicted by theory.

Order of minimum	Distance from $-m$ to m	Distance y_m (from center to m)	Calculated $\sin\theta \approx \dfrac{y_m}{L}$	Predicted $\sin\theta = \dfrac{m\lambda}{w}$	Percent difference
$m = 1$					
$m = 2$					
$m = 3$					
$m = 4$					

CI DATA TABLE 2

Purpose: To determine the wavelength of light using a single-slit diffraction pattern.

Order of minimum	Calculated $\sin\theta \approx \dfrac{y_m}{L}$	Calculated λ
$m = 1$		
$m = 2$		
$m = 3$		
$m = 4$		
Average =		
Percent error =		

Be sure to attach a copy of the graph to the report.

Don't forget units

(continued)

B. The Double-Slit Pattern

Slit width w _____

Separation of slits d _____

Distance between slit and pattern L _____

CI DATA TABLE 3

Purpose: To compare the experimental double-slit pattern with the pattern predicted by theory.

Order of maximum	Distance from $-n$ to n	Distance y_n (from center to n)	Calculated $\sin \theta \approx \dfrac{y_n}{L}$	Predicted $\sin \theta = \dfrac{n\lambda}{d}$	Percent difference
$n = 1$					
$n = 2$					
$n = 3$					
$n = 4$					
$n = 5$					
$n = 6$					

CI DATA TABLE 4

Purpose: To determine the wavelength of light using a double-slit interference pattern.

Order of maximum	Calculated $\sin \theta \approx \dfrac{y_m}{L}$	Calculated λ
$n = 1$		
$n = 2$		
$n = 3$		
$n = 4$		
Average =		
Percent error =		

Be sure to attach a copy of the graph to the report.

C. Comparing Single-Slit Pattern to Double-Slit Pattern

Attach the graphs to the report. Don't forget to label them appropriately so that it is easy to distinguish between them.

EXPERIMENT 25   *Laboratory Report*

CI QUESTIONS

1. Comparison between Graphs 1 and 2:
   (a) What parameters of the experiment were kept constant in producing Graphs 1 and 2? What parameters were changed?

   (b) Compare the locations of the first minima of diffraction ($m = 1$ and $m = -1$ on Graph 1) to the same positions along the x axis on Graph 2. Are the positions also minima in Graph 2?

   (c) In graph 2, how many interference fringes (bright) are in between the locations of $m = 1$ and $m = -1$ of the single-slit pattern?

2. Comparison between Graphs 2 and 3:
   (a) What parameters of the experiment were kept constant in producing Graphs 2 and 3? What parameters were changed?

   (b) Describe all things that look different between Graphs 2 and 3. What is the effect of changing the slit width?

(continued)

3. Comparison between Graphs 2 and 4:
   (a) What parameters of the experiment were kept constant in producing Graphs 2 and 4?
       What parameters were changed?

   (b) Describe all things that look different between Graphs 2 and 4. What is the effect of
       changing the separation between the slits?

EXPERIMENT 26

Detection of Nuclear Radiation: The Geiger Counter

[TI] *Advance Study Assignment*

Read the experiment and answer the following questions.

1. What is the principle of operation of the Geiger tube?

2. Define each of the following: (a) threshold voltage, (b) cumulative ionization, (c) plateau, and (d) dead time.

3. Are any radiations counted when the tube voltage is below the threshold voltage? Explain.

(continued)

4. Approximately how many volts above the threshold voltage is the normal operating voltage of the Geiger tube, and why is the operating voltage selected this way?

5. What is background radiation?

6. How does the count rate vary with distance from a point source? If the counter is moved twice the distance from the source, how is the count rate affected?

NUCLEAR SAFETY RULES

Radioactive sources will be used in the next few experiments. Some sources are solids and are encapsulated to prevent contact. However, liquid sources may also be used and transferred during an experiment. Some general safety precautions for the use of radioactive materials follow.

1. Radioactive materials should be used only by or under the supervision of a person properly informed about the nature of the material.
2. Care should be taken to avoid unnecessary handling or contact with the skin.
3. Mouth pipetting is strictly prohibited.
4. Eating, drinking, and smoking should not be permitted in any area where radioactive materials are being used.

5. Protective gloves or forceps should be used when the material is handled or transferred.
6. All persons working with radioactive material should thoroughly wash their hands afterward.
7. When not in use, radioactive materials should be stored in an appropriately labeled container and in a place of limited access.
8. Should an accident occur (particularly if it involves radioactive materials), it should be reported immediately to the laboratory instructor.
9. If you are pregnant, make your instructor aware of this, and do not go to the laboratory.

Detection of Nuclear Radiation: The Geiger Counter

INTRODUCTION AND OBJECTIVES

Nuclear radiations (alpha, beta, and gamma rays or particles) cannot be detected directly by our senses. Hence, we must use some observable detection method employing the interaction of nuclear decay particles with matter. There are several methods, but the most common is the **Geiger tube.***

In a Geiger tube, the particles from radioactive decay ionize gas molecules, giving rise to electrical pulses that can be amplified and counted. The total instrument is referred to as a Geiger counter.

In this experiment, we will investigate the characteristics of a Geiger tube and the inverse-square relationship for nuclear radiation.

After performing this experiment and analyzing the data, you should be able to:

1. Explain the principle of operation of the Geiger counter and its major disadvantage.
2. Describe how the count rate of a Geiger counter varies with its distance from a radioactive source.

---

* Sometimes referred to as a *Geiger-Müller tube* (or G-M tube). A prototype was developed in 1913 by the German physicist Hans Geiger (1882–1945), who worked in England on experiments that led to our present nuclear model of the atom. The tube was improved in 1928 in collaboration with the German physicist S. Müller.

EQUIPMENT NEEDED

- Geiger counter (rate meter or scaler type)
- Radioactive source [e.g., Cs-137 (beta-gamma)]
- Laboratory timer or stopwatch
- Calibrated mounting board or meterstick
- 2 sheets of Cartesian graph paper [or 1 sheet of Cartesian and (*optional*) 1 sheet of log (log-log) graph paper (3-cycle)]

THEORY

The three types of nuclear radiation—alpha, beta, and gamma—are all capable of ionizing a gas. The degree of ionization depends on the energy of the particles and the amount of radiation absorbed by the gas. The ionization of gas molecules by nuclear radiation is the principle of the Geiger tube.

A **Geiger tube** consists of a fine wire running axially through a metal cylinder filled with a gas, usually argon, at a pressure of about 0.1 atm (● Fig. 1). A potential difference or voltage is maintained between the central wire and the cylinder, the central wire being at a positive potential (+) with respect to the cylinder (−).

Energetic nuclear particles (ionizing radiation) passing through the cylinder and entering the tube ionize the gas molecules. The freed electrons are attracted toward the wire and the positive ions toward the cylinder. If the voltage between the wire and cylinder is great enough, the accelerated electrons acquire enough energy to ionize other gas molecules on their way to the positive wire. The electrons from the secondary ionizations produce additional ionizations. This process is called *cumulative ionization*.

As a result, an "avalanche" discharge sets in, and a current is produced in the resistor. This reduces the potential difference between wire and cylinder to the point where cumulative ionization does not occur. After the momentary

current pulse, which lasts on the order of microseconds, the potential difference between the wire and the cylinder resumes its original value.

A finite time is required for the discharge to be cleared from the tube. During this time, the voltage of the tube is less than that required to detect other radiation that might arrive. This recovery time is referred to as the *dead time* of the tube. If a large amount of radiation arrives at the tube, the counting rate (counts per minute, or cpm) as indicated on the counting equipment will be less than the true value.

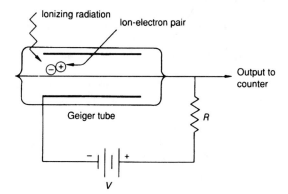

Figure 1 Geiger tube. A schematic diagram of the Geiger tube and circuit. See text for description.

Figure 2 Apparatus for radioactive experiments. The standard side-window Geiger tube probe on the mounting board is connected to a count-rate meter. A radioactive source is on the board in the foreground. Notice the radio-activity warning sign on the source.

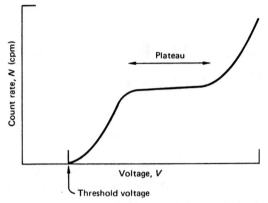

Figure 3 Count rate versus tube voltage. A typical graph showing how the count rate varies with Geiger tube voltage. Normal operation is the plateau region. See text for description.

There are two common types of Geiger tubes—a "normal" or side-window tube and an "end-window" tube. The side-window tube has a relatively thick wall that may not be penetrated by less penetrating radiation such as alpha particles (● Fig. 2). The end-window tube has a thin end window, usually of mica, and may be thin enough to be penetrated by very energetic alpha particles.

The brief change in the potential that occurs when a discharge takes place in the tube produces a voltage pulse that can be detected and counted by appropriate instrumentation. Common instruments used for counting are scalers and count-rate meters.

A *scaler* displays the cumulative number of counts on a lighted panel. By using a separate timer, the number of counts per minute (cpm) can be obtained. Some scalers have internal timers that stop the counting after a preset time interval.

A *rate meter* displays the average counting rate directly via a dial needle (Fig. 2). The needle reading fluctuates back and forth. This is due to the electronic averaging of the number of counts received during a short period of time. A scaler timer is usually preferred over a rate meter because of this effect.

A. Tube Voltage and Count Rate

When a Geiger tube is in the vicinity of a radiation source with particles of varying energy and there is no voltage on the tube, no counts are observed on the counter. (Counters usually have a loudspeaker circuit so that the counts may also be heard as audible "clicks.") If the tube voltage is slowly increased from zero, then at some applied voltage, counts will be observed. The lowest applied voltage that will produce a count in the instrument is called the *starting voltage* or **threshold voltage** (● Fig. 3).

As the tube voltage is increased above the threshold voltage, the number of counts per minute increases rapidly. In this region (about 50 V wide, beginning at about 600 to 700 V, depending on the tube), the count rate is almost linearly proportional to the voltage. This is because as the voltage increases, more of the less energetic particles are counted. Hence, in this region the tube discriminates between particles of different energy. At a given voltage, only particles above a certain energy are detected. The tube then acts as a proportional counter—the voltage being proportional to the energies of the incident particles.

Eventually, as the voltage is increased, the number of counts per minute becomes almost independent of the applied voltage (the level region in Fig. 3). This region (about 200 V wide) is called the **plateau** of the tube. A change in voltage has little effect on the number of counts detected. Normally, the Geiger tube is operated at a voltage in about the middle of the plateau. Fluctuations in the applied voltage from the power supply will then have little effect on the counting rate.

The tube voltage should never be raised to a value far above that of the end of the plateau. At such high voltages, a continuous discharge sets in, and if allowed to persist, this may destroy the tube.

B. Inverse-Square Relationship

In normal operation, the count rate depends on the number of particles per unit time entering the Geiger tube. Hence, the count rate depends on the distance of the tube from the source. For a point source emitting a total of N_o particles/min, the particles are emitted in all directions. The number of particles/min N' passing through a unit area of a sphere of radius r is

$$N' = \frac{N_o}{A} = \frac{N_o}{4\pi r^2} \quad \text{(counts/min/area)} \quad (1)$$

where $A = 4\pi r^2$ is the area of the sphere.

A Geiger tube with a window area A' at a distance r from a point source then intercepts or receives N counts/min, given by

$$N = N'A' = \frac{N_o A'}{4\pi r^2} \quad (2)$$

Although the effective area A' of the Geiger tube is usually not known, the equation shows that the count rate is inversely proportional to r^2 (inverse-square-law form):

$$N \propto \frac{1}{r^2} \quad (3)$$

Hence, for a point source, the count rate "falls off" as $1/r^2$ with the distance from the source.

EXPERIMENTAL PROCEDURE

Caution: Review the radiation safety procedures before performing this experiment.

1. Connect the Geiger tube probe to the counter by means of the coaxial cable. Before plugging the counter into an ac outlet, familiarize yourself with the controls, particularly the high-voltage control.

   Scaler: Set the high-voltage control to the minimum setting.

   Rate meter: Set the high-voltage control to the minimum setting. The off-on switch is commonly on the high-voltage control. A selector switch is labeled with volts and counts per minute multiplier positions ($\times 1$, $\times 10$, etc.). When the Geiger tube voltage is adjusted by means of the high-voltage control, the selector switch should *always* be set on "volts."

   The selector switch is then turned to the appropriate count multiplier range for counting. The meter display scale usually has dual calibrations in volts and counts per minute.

2. Plug in and turn on the counter. Place the radioactive source near the Geiger tube, with the source facing the probe opening as in Fig. 2. (A tube mount may be available for an end-window tube. The source is placed at the bottom of the tube mount in this case. *Note:* The end window is very fragile and can be punctured easily.)

Slowly increase the tube voltage by means of the high-voltage control until the first indication of counting is observed. Then increase the voltage to about 75 to 100 V above this value.

3. Set the counter to the counting mode, and adjust the distance of the source from the tube (or add aluminum sheets to the end-window tube mount) so that the count rate is several thousand (3000 to 5000) counts per minute. The Geiger tube is then operating normally, and the dead time will not cause serious error.

A. Tube Voltage and Count Rate

4. Lower the high-voltage control to the minimum setting. Then slowly raise the voltage until the first indication of counting is observed (rate meter selection on "volts"). Record this threshold voltage in Data Table 1.

5. Increase the voltage to 25 V above the threshold voltage and record the tube voltage. Measure and record the number of counts per minute at this voltage setting. (A rate meter is switched to a counting position. Because the meter needle fluctuates, it is best to watch the meter for 30 s and note the highest and lowest meter readings. The count rate is then taken as the mean or average of these readings.)

6. Continue to repeat procedure 5, increasing the voltage by 25 V each time. Record the voltage and the corresponding count rate for each voltage setting. You will notice that the count rate first increases rapidly with voltage. It then levels off, increasing only slightly with increases in voltage. This is the plateau region of the Geiger tube.

   Eventually, with a particular voltage step, a sharp increase in the count rate will be observed. *Do not increase the voltage above this value.* Quickly lower the tube voltage to the minimum setting after this reading to avoid damaging the tube.

7. Plot the count rate N (counts/min) versus voltage V on Cartesian graph paper. Include the threshold voltage. Draw a smooth curve that best fits the data.

B. Background Radiation

8. Remove the source several meters (across the room) from the Geiger tube, and apply the midplateau voltage to the tube as determined from the graph. (If using an end-window tube with a tube mount, remove the tube from the mount and lay the tube on the table.)

   You will observe an occasional count on the counter. This is due to background radiation arising from cosmic rays and radioactive elements in the environment (e.g., in building materials).

Let the counter run for a measured time, for example, 4 to 5 min, and determine the background count rate in counts per minute and record this value in part B of the Laboratory Report. If the background count rate is small compared to the source count, it may be considered negligible.

C. Inverse-Square Relationship

9. Bring the source toward the Geiger tube and locate the source at a distance from the tube where the counting rate begins to increase significantly over background. Record the distance r and the count rate N in Data Table 2. Record this r as the farthest distance.

10. Then bring the source relatively close to the tube and determine the distance from the source that gives a full-scale count rate. Record the count rate and distance (closest).

11. Divide the length between the two measured distances into eight intervals or steps. Measure and record the count rate and source distance from the tube for each step as the source is moved away from the tube.

12. The inverse-square relationship $N = A/r^2$ (where A is a constant) can be put into linear form by taking the logarithm of both sides:

$$\log N = \log(Ar^{-2}) = \log r^{-2} + \log A$$

or

$$\log N = -2 \log r + \log A \tag{4}$$

where log is the common logarithm (base 10). Note that Eq. 4 has the form of a straight line: $y = mx + b$. (See Experiment for general discussion.)

Take the logs of the r and N values in Data Table 2. On Cartesian graph paper, plot $\log N$ versus $\log r$, and draw a straight line that best fits the data. Determine the slope of the line and compare it to the theoretical value by finding the percent error.

(Optional) Your instructor may wish to introduce you to log-log graph paper. This special graph paper automatically takes the log values. See Appendix D for a discussion of graphing on log-log and semi-log graph papers. (There is optional use of the latter in Experiments and .)

Name _____ Section _____ Date _____

Lab Partner(s) _____

Detection of Nuclear Radiation: The Geiger Counter

$\boxed{\text{TI}}$ *Laboratory Report*

A. *Tube Voltage and Count Rate*

DATA TABLE 1

Purpose: To determine dependence of the count rate on tube voltage.

	Tube voltage	Count rate (cpm)
Threshold voltage		

B. *Background Radiation*

Number of counts _____

Counting time (min) _____

Counts/min _____

Don't forget units

(continued)

C. Inverse-Square Relationship

DATA TABLE 2

Purpose: To determine the count rate versus distance from source.

Source-to-counter distance r ( )	Count rate N (cpm)
Closest distance	
Farthest distance	

$\log r$	$\log N$

Calculations
(show work)

Slope of graph _____

Theoretical value _____

Percent error _____

EXPERIMENT 26

/TI/ QUESTIONS

1. What is the average percent increase in the count rate over the voltage range of the Geiger tube plateau? (Obtain from a graph of the data.)

2. If (a) dead time and (b) background radiation corrections were taken into account, how would each correction affect the graph of N versus V?

3. Give possible reasons why the experimental result of N versus r is not exactly an inverse-square relationship.

4. A count rate of 8000 cpm is recorded at a distance of 5.0 cm from a point source. What would be the observed count rate at a distance of 20 cm?

EXPERIMENT 26

EXPERIMENT 27

Radioactive Half-Life

/TI/ *Advance Study Assignment*

Read the experiment and answer the following questions.

1. What is the significance of the half-life of a radioactive isotope in terms of (a) the amount of sample or number of nuclei, and (b) the activity of the sample?

2. What is the *decay constant?* Is it the same for each decay process? What are the units of the decay constant?

3. How is the half-life related to the decay constant of a radioactive process?

4. What is meant by *milking a cow?* Give the technical terms for *milking* and *cow.*

(continued)

5. Ba-137m is a nuclear isomer of Ba-137. Explain what this means.

6. If a particular radioactive sample undergoes four half-lives, what fraction of the original material remains?

Radioactive Half-Life

INTRODUCTION AND OBJECTIVES

The decrease in the activity of a radioactive isotope is characterized by its half-life. This is the time required for one-half of the nuclei of a sample to decay.

Of course, the nuclei of a sample cannot be counted directly, but when one-half of the sample has decayed, the activity, or the rate of emission of nuclear radiation, has also decreased by one-half. Thus, as we monitor the sample with a Geiger counter, when the count rate (counts per minute, cpm) has decreased by one-half, one half-life has elapsed.

In this experiment, the half-life of a radioactive isotope will be determined.

After performing this experiment and analyzing the data, you should be able to:

1. Explain what is meant by the half-life of a radioactive isotope.
2. Distinguish between radioactive half-life and time constant.
3. Describe how the half-life of a short-lived radioactive isotope can be measured.

EQUIPMENT NEEDED

- Geiger counter (rate meter with clip mount or scaler type with tube mount)
- Cesium-137/Barium-137m Minigenerator with solution
- Laboratory timer or stopwatch

- Disposable planchet (small metal cuplike container to hold radioactive sample)
- 2 sheets of Cartesian graph paper [or (*optional*) 1 sheet of Cartesian and 1 sheet of semi-log graph paper (3-cycle)]

THEORY

The activity of a radioactive isotope is proportional to the quantity of isotope present, and the radioactive decay process is described by an exponential function:

$$N = N_0 e^{-\lambda t} = N_0 e^{-t/\tau} \tag{1}$$

where N is the number of nuclei present at time t, N_0 is the original number of nuclei present (at $t = 0$), λ is the decay constant of the process, and the time constant $\tau = 1/\lambda$. The variable N can also represent the activity (cpm) of an isotope sample.

The half-life $t_{1/2}$ is the time it takes for the number of nuclei present, or the activity, to decrease by one-half ($N = N_0/2$). Hence

$$\frac{N}{N_0} = \frac{1}{2} = e^{-t_{1/2}/\tau}$$

Because

$$e^{-0.693} = \frac{1}{2}$$

by comparison

$$0.693 = \frac{t_{1/2}}{\tau}$$

and

$$t_{1/2} = 0.693\,\tau = \frac{0.693}{\lambda} \tag{2}$$

Thus the half-life can be computed if the time constant or the decay constant is known.

Example 1   A radioactive sample has an activity of 4000 cpm. What is the observed activity after three half-lives?

Solution   After one half-life, the activity decreases by $\frac{1}{2}$, and after another half-life by another $\frac{1}{2}$, and so on. Hence, after three half-lives, the initial activity decreases by a factor of $\frac{1}{2} \times \frac{1}{2} \times \frac{1}{2} = \frac{1}{8}$. With $N_0 = 4000$ cpm,

$$N = \frac{1}{8} N_0 = \frac{1}{8}(4000) = 500 \text{ cpm}$$

Notice that, in general,

$$N = \frac{N_0}{2^n}$$

where n is the number of half-lives.

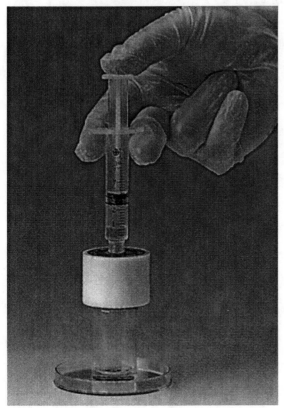

Figure 1 The Cesium-137/Barium-137m Minigenerator system. See text for description. (Reproduced with permission of Fisher Scientific.)

Theory of Minigenerator

The Cesium-137/Barium-137m Minigenerator* (● Fig. 1) is an eluting system, in which a short-lived daughter radioactive isotope is eluted (separated by washing) from a long-lived parent isotope. A small "generator" contains radioactive Cs-137, which has a half-life of 30 years. Cs-137 beta-decays into Ba-137m, which is an isomeric (excited) state of the stable nucleus Ba-137. The excited isomer Ba-137m has a relatively short half-life and gamma-decays into Ba-137.

The nuclear equation for the decay is

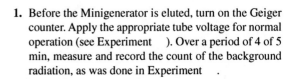

$$^{137}_{55}\text{Cs} \longrightarrow {}^{137}_{56}\text{Ba*} + {}^{0}_{-1}\text{e}$$

(cesium) (barium) (electron)
(excited)

$$\longrightarrow {}^{137}_{56}\text{Ba} + \gamma$$

(barium) (gamma ray)

where the asterisk (*) indicates an excited state. The original nucleus, ^{137}Cs, is commonly referred to as the *parent*

* Registered trademark, Union Carbide.

nucleus, and the resulting nucleus (^{137}Ba) is called the *daughter* nucleus.

The Ba-137m is washed or eluted from the generator by passing a hydrochloric acid–saline solution through the generator. Because of this process, the generator is commonly referred to as a "cow," and the Ba-137m is said to be "milked" from the cow. The generator "cow" may be milked many times, but as with an actual cow, a time interval must elapse between milkings.

Eluting removes the Ba-137m from the generator, and time is required for the "regeneration" of Ba-137m from the decay of Cs-137. Normally, the parent and daughter isotopes exist in equilibrium, with equal activities. After eluting (or milking the cow), it takes about 12 min for the Ba-137m to build up and again reach equilibrium with the Cs-137.

EXPERIMENTAL PROCEDURE

Caution: Review the radiation safety procedures at the beginning of Experiment .

1. Before the Minigenerator is eluted, turn on the Geiger counter. Apply the appropriate tube voltage for normal operation (see Experiment). Over a period of 4 of 5 min, measure and record the count of the background radiation, as was done in Experiment .

2. Mount the Geiger probe so that a planchet with the radioactive Ba-137m sample can be quickly and carefully placed below and near the probe opening at a fixed distance.

3. The counting procedure is as follows. When the sample is in place, the laboratory timer is started ($t = 0$) and allowed to run continuously. Simultaneously with the starting of the timer, the activity is measured on the Geiger counter for 15 s, and the count rate (cpm), together with the time elapsed on the timer, is recorded in the data table.

   Note: If using a rate meter, take the average of the high and low meter readings over the 15-s interval as the count rate. If a scaler is used (with or without an internal timer), the count rate in cpm must be computed. For example, suppose that 500 counts are observed for the 15-s ($\frac{1}{4}$-min) interval. The count rate is then 500 counts/$\frac{1}{4}$ min = $500 \times 4 = 2000$ cpm.

   Repeat the 15-s count of activity at the beginning of each minute of elapsed time for 10 to 12 min. A dry run of the counting procedure is helpful.

4. The instructor will "milk the cow" or supervise you in doing so. Only a few (2 to 3) drops of the eluate (milk) are needed.

   Caution: Care should be taken in handling the sample. The milking should be done over a sheet of

*paper that can be discarded in case of a spill, and **if you should come in contact with the sample, immediately wash your hands.***
 The instructor may wish to give you a sample for a trial run of the counting procedure.

5. When given the actual data sample, carry out the counting procedure as described above.

6. Correct for background radiation if necessary. Plot the sample activity (N) in cpm versus the elapsed time (t) in minutes on Cartesian graph paper, and note the shape of the curve.
 From the graph, make two determinations of the half-life by finding the time required for the sample activity to decay from its initial value to $\frac{1}{2}$ of the initial value, and from $\frac{1}{2}$ to $\frac{1}{4}$ of the initial value. Average and compare with the half-life for Ba-137m in Appendix Table A9 by computing the percent error. Also compute the decay constant (λ) from the average value of the half-life.

7. The decay constant may be found graphically by putting the exponential function, $N = N_0 e^{-\lambda t}$, into linear form by taking the natural logarithm (base e) of both sides:

$$\ln N = \ln(N_0 e^{-\lambda t}) = \ln(e^{-\lambda t}) + \ln N_0$$

or

$$\ln N = -\lambda t + \ln N_0 \qquad (3)$$

Note that Eq. 3 has the form of a straight line: $y = mx + b$. (See Experiment 1 for general discussion.)
 Find $\ln N$ for each value of N in the Data Table. (Make a column for these to the right of the table).
 Plot $\ln N$ versus t on Cartesian graph paper and draw a straight line that best fits the data. Determine the slope of the line and compare it to the value of the decay constant computed in the preceding procedure by finding the percent difference.

 (*Optional*) Your instructor may wish to introduce you to semi-log graph paper. This special graph paper automatically takes the log values of the variable plotted on the Y axis. See Appendix D for a discussion of graphing on semi-log (and log-log) graph paper.

Name _____ Section _____ Date _____

Lab Partner(s) _____

Radioactive Half-Life

TI/ *Laboratory Report*

Background count (cpm) ——————————

DATA TABLE

Purpose: To determine the half-life of Ba-137m.

t Elapsed time (min)	N Observed activity (cpm)	Corrected for background radiation (cpm)

Calculations
(show work)

Half-life Measurements for Ba-137m

Time from full to $\frac{1}{2}$ activity ——————————

Time from $\frac{1}{2}$ to $\frac{1}{4}$ activity ——————————

Average experimental half-life ——————————

Accepted half-life ——————————

Percent error ——————————

Decay constant from average half-life ——————————

Decay constant from graph ——————————

Percent difference ——————————

Don't forget units

(continued)

TI/ QUESTIONS

1. In the experiment, if the Ba-137m sample were placed closer to the Geiger tube, the measured activity would be greater (inverse-square relationship). Would this affect the result of the half-life? Explain how this would affect the N-versus-t graph on Cartesian graph paper (or on semi-log paper).

2. A cobalt-60 source has a measured activity of 12,000 cpm. After how long would the observed activity be 750 cpm? (The half-life of Co-60 is 5.27 y.)

3. An instructor buys a 10-μCi Cs-137 source for laboratory experiments. After 5 years, what is (a) the strength of the source in μCi; (b) the activity of the source in disintegrations per second? (1 Ci = 3.70×10^{10} disintegrations/s.) (c) What is the strength of the source in becquerels (Bq)? (The becquerel is the official SI unit; 1 Ci = 3.7×10^{10} Bq.)

4. Cesium-136 is also radioactive and decays into barium-136. Write the nuclear equation for this reaction.

Name _____ Section _____ Date _____

Lab Partner(s) _____

The Absorption
of Nuclear Radiation

/TI/ *Advance Study Assignment*

Read the experiment and answer the following questions.

1. On what parameters does the absorption of nuclear radiation depend?

2. Do the three types of nuclear radiation have definite ranges of penetration in materials? Explain. What is meant by *half-thickness?*

3. What is the mass absorption coefficient, and what are its units? Are there any advantages to using the mass absorption coefficient rather than the linear absorption coefficient? Explain.

(continued)

4. Explain how a source that has only one radioactive isotope can emit both beta and gamma radiation.

5. Why is a beta-gamma source that is shielded with a relatively thin sheet of aluminum effectively a gamma source?

The Absorption
of Nuclear Radiation

INTRODUCTION AND OBJECTIVES

The observed activity of a radioactive source of a given strength depends on several factors—for example, the distance of the counter from the source. For a point source, the observed activity varies inversely with the distance from the source (inverse-square relationship). This decrease is due to the geometrical spreading of the emitted nuclear radiation outward from the source.

If a Geiger probe is a fixed distance from a long-lived source, the observed activity is relatively constant. However, if a sheet of material is placed between the source and the counter, a decrease in the activity may be observed. That is, the nuclear radiation is absorbed by the material. The amount of absorption depends on the type and energy of the radiation and on the kind and density of the absorbing material.

The absorption or degree of penetration of nuclear radiation is an important consideration in applications such as medical radioisotope treatment and nuclear shielding (e.g., around a nuclear reactor). Also, in industrial manufacturing processes, the absorption of nuclear radiation is used to monitor and control automatically the thickness of metal and plastic sheets and films.

In this experiment, the absorption properties of various materials for different kinds of nuclear radiation will be investigated.

After performing this experiment and analyzing the data, you should be able to:

1. Describe the parameters on which the penetration of nuclear radiation in a material depends.
2. Explain the linear absorption coefficient, "half-thickness," and stopping range.
3. Explain the mass absorption coefficient.

EQUIPMENT NEEDED

- Geiger counter (rate meter or scaler type)
- Calibrated mounting board (or meterstick)
- Beta-gamma source (Cs-137 suggested)
- Set of cardboard, aluminum, and lead sheets (about 1 mm thick, 10 sheets of each)
- Laboratory timer or stopwatch
- Micrometer caliper
- 3 sheets of Cartesian graph paper [or (optional) 2 sheets of Cartesian graph paper and 1 sheet of semi-log graph paper (3-cycle)]

THEORY

The three types of nuclear radiation (alpha, beta, and gamma) are absorbed quite differently by different materials. The electrically charged alpha and beta particles interact with the material and produce ionizations along their paths. The greater the charge and the slower the particle, the greater the linear energy transfer (LET) and ionization along the path, and this determines the degree of penetration of the radiation. The absorption or degree of penetration of the radiation also depends on the density of the material.

Alpha particles are easily absorbed. A few centimeters of air and even a sheet of paper will almost completely absorb them. Hence, alpha particles do not generally penetrate the walls or window of an ordinary Geiger tube and so are not counted by this method.

Beta particles can travel a few meters in air or a few millimeters in aluminum before being completely absorbed. Beta radiation, then, does penetrate a Geiger tube.

Both alpha and beta particles of a given energy therefore have a definite *range* of penetration in a particular material.

● Figure 1 illustrates the radiation intensity (in cpm) versus absorber thickness for a relatively low-density absorber for radiation from a beta-gamma source. The "bend" in the curve indicates the range of the beta radiation. The penetration for thickness beyond this is due to gamma radiation.

Gamma rays, which consist of electromagnetic radiation of very short wavelength, are not readily absorbed. A significant number of high-energy gamma rays can penetrate 1 cm or more of a dense material such as lead. In a given material, a beam of gamma rays is absorbed exponentially. The intensity I (in cpm) of the beam after passing through a certain thickness x of a material is given by

$$I = I_0 e^{-\mu x} \qquad (1)$$

where I_0 is the original intensity (at $x = 0$) and the decay constant μ is called the **linear absorption coefficient.** The

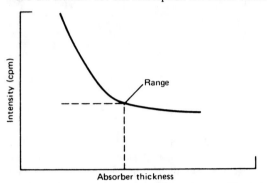

Figure 1 Radiation intensity versus absorber thickness. A typical graph of radiation intensity versus absorber thickness for beta-gamma radiation by a low-density absorber. The range is that of the beta radiation.

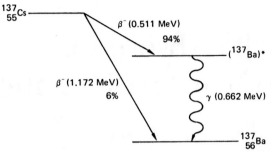

Figure 2 Decay scheme of Cs-137. Most of the cesium-137 (Cs-137) nuclei (94%) decay to an excited state of barium-137 (^{137}Ba*), which then gamma-decays to a stable state.

absorption coefficient is characteristic of the absorbing material (and the wavelength or energy of the gamma radiation). Notice that the unit of μ is inverse length (such as 1/cm or cm^{-1}).

The absorption of gamma radiation of a given wavelength or energy is related to the atomic number of a substance and, macroscopically, to the density ρ of the material. Thus it is convenient to define a **mass absorption coefficient** μ_m:

$$\mu_m = \frac{\mu}{\rho} \tag{2}$$

The mass absorption coefficient provides a "standardized" coefficient. Samples of a particular absorbing material may have different densities. Each sample would have a different linear absorption coefficient μ, but the mass absorption coefficient μ_m would have the same value for all the samples.

Notice from Eq. 2 that the units of μ_m are cm^2/g:

$$\mu_m = \frac{\mu\,(1/\text{cm})}{\rho\,(\text{g/cm}^3)} = \frac{\mu}{\rho}\,(\text{cm}^2/\text{g})$$

If μ_m is used in Eq. 1 in place of μ, then

$$I = I_0 e^{-\mu x} = I_0 e^{-(\mu/\rho)(x\rho)} = e^{-\mu_m x'} \tag{3}$$

and the absorber thickness $x' = x\rho$ is in g/cm^2. Absorber thicknesses are frequently expressed in these units.

A beta-gamma source will be used to study the absorption of nuclear radiations. The decay scheme of the suggested Cs-137 source is illustrated in ● Fig. 2.

The chief decay mode (94%) is beta decay to the excited (isomeric) state of Ba-137. This decays by gamma emission to the stable ground state of Ba-137. Only 6% of the Cs-137 beta-decays directly to ground-state Ba-137. Hence, for the most part, Cs-137 is a beta-gamma source of 0.511-MeV beta particles and 0.662-MeV gamma rays.

(The emitted beta particles actually have a spectrum of energies from 0 to 0.511 MeV.)

Since the gamma intensity decays exponentially, there is no definite penetrating or stopping range as there is in the case of beta radiation. Hence, it is convenient to speak in terms of a **half-thickness** $x_{1/2}$, the material thickness required to reduce the intensity by one-half (i.e., $I_{1/2} = I_0/2$ or $I_{1/2}/I_0 = \frac{1}{2}$). Then, by Eq. 1,

$$\frac{I_{1/2}}{I_0} = e^{-\mu x_{1/2}} = \frac{1}{2}$$

Taking the logarithm (base e) of both sides of the equation, we have

$$\ln\!\left(e^{-\mu x_{1/2}}\right) = \ln\frac{1}{2}$$

or

$$-\mu x_{1/2} = -\ln 2$$

and

$$x_{1/2} = \frac{\ln 2}{\mu} = \frac{0.693}{\mu} \tag{4}$$

Hence, knowing the absorption coefficient of a material, we can calculate the half-thickness.

EXPERIMENTAL PROCEDURE

Caution: Review the radiation safety procedures at the beginning of Experiment .

1. A radioactivity setup is shown in ● Fig. 3. First, measure the individual thickness of three different sheets of (a) cardboard, (b) aluminum, and (c) lead with the micrometer, and determine the average sheet thickness of each. Record in Data Table 1.

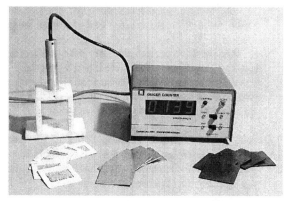

Figure 3 Geiger counter setup. (Reproduced with permission of Fisher Scientific.)

2. Set up the Geiger counter with the probe on the mounting board (see Fig. 3). If an end-window tube is used, lay the tube in the mounting board groove and tape it down to immobilize it (or tape it to a meterstick). Turn on the counter. Place the radioactive source near the probe, and adjust the tube voltage to the plateau operating voltage.

A. Absorption of Beta Radiation

3. Adjust the distance of the source from the probe so that the observed count rate is about 8000 cpm. (For a rate meter, the count rate is taken as the average of the high and low meter readings for 30-s time intervals.) Record the count rate I_0 in the cardboard column in Data Table 2.

   Place a sheet of cardboard between the source and the probe, and measure and record the count rate. (Allow a rate meter to come to equilibrium before taking a 30-s reading.)

4. Add cardboard sheets between the source and the probe one at a time, measuring and recording the count rate after the addition of each sheet. Continue until the count rate is relatively constant with the addition of four successive sheets.

5. Remove the cardboard sheets and repeat the procedure with aluminum sheets.

6. Without recording data, repeat the procedure with lead sheets and mentally note the degree of beta absorption or penetration in lead.

7. Plot the intensity I (in cpm) versus the number n of absorber sheets for both cardboard and aluminum on the same Cartesian graph. Dual-label the ordinate (Y)

axis so that the curve for each absorber occupies most of the graph paper.

   Determine the range of beta absorption for each absorber in sheet units from the graph, and record. Multiply each range (in sheet units) by the respective average sheet thickness to determine the range in length units.

B. Absorption of Gamma Radiation

8. Using the result of the range of beta absorption in aluminum from procedure 7, place in front of the probe the minimum number of sheets of aluminum that will completely absorb the beta radiation. Then move the source toward the probe until the intensity observed on the Geiger counter is 700 to 800 cpm.

   Record this intensity I_0 in Data Table 3. The observed intensity is then almost solely due to gamma radiation. Why?

9. Leaving the aluminum sheet(s) in place, insert lead sheets one at a time between the aluminum sheets and the source. Measure and record the count rate after each sheet is inserted. Be careful not to move the source. Insert a total of ten sheets of lead. After the sixth sheet, two sheets may be inserted at a time.

10. Remove all the sheets. Remove the source several meters (across the room) from the probe, and measure the background radiation intensity I_b over a 4- to 5-min interval. (See Experiment   for a description of the procedure, if necessary.)

11. Subtract the background count rate from each reading for the lead sheets to obtain the corrected intensities. To find the half-thickness, plot the corrected intensity (I_c) versus the number (n) of lead sheets on Cartesian graph paper and note the shape of the curve.

   From the graph, make a determination of the number of sheets ($n_{1/2}$) needed to reduce the intensity from its initial value to $\frac{1}{2}$ of the initial value. (Try to express this number to the nearest 0.05 of a sheet.) The half-thickness is $x_{1/2} = x_i n_{1/2}$, where x_i is the thickness of an individual sheet. From the half-thickness, compute the linear absorption coefficient μ from Eq. 4 and record this value in Data Table 3.

12. The absorption coefficient may be found graphically by putting the exponential Eq. 1 into linear form by taking the natural (base e) logarithm of both sides. But first note that in terms of the number of sheets (n), Eq. 1 has the form

$$I = I_0 e^{-\mu x} = I_0 e^{-\mu(n x_i)} = I_0 e^{-(\mu x_i)n} \qquad (5)$$

where x_i is the individual sheet thickness and the absorber thickness is $x = nx_i$.

Then, taking the natural log of both sides of Eq. 5 yields

$$\ln I = \ln I_0 e^{-(\mu x_i)n} = \ln e^{-(\mu x_i)n} + \ln I_0$$

or

$$\ln I = -(\mu x_i)n + \ln I_0 \qquad (6)$$

Note that Eq. 6 has the form of a straight line: $y = mx + b$. (See Experiment 1 for general discussion.)

Find $\ln I_c$ for each value of I_c in Data Table 3. (Make a column for these to the right of the table.)

Plot $\ln I_c$ versus n on Cartesian graph paper and draw a straight line that best fits the data. Determine the slope of the line and compute the linear absorption coefficient μ. (Note from Eq. 6 that the slope has a magnitude of μx_i.)

(*Optional*) Your instructor may wish you to use semi-log graph paper. This special graph paper automatically takes in values of the variable plotted on the y axis. See Appendix D for a discussion of graphing on semi-log (and log-log) graph paper.

13. Compute the mass absorption coefficient μ_m for lead ($\rho_{Pb} = 11.3$ g/cm^3). Compare the experimental values to the accepted value of $\mu_m = 0.10$ cm^2/g for gamma rays by computing the percent error for each.

EXPERIMENT 28

The Absorption
of Nuclear Radiation

/TI/ *Laboratory Report*

DATA TABLE 1

Purpose: To determine sheet thicknesses.

	Cardboard ( )	Aluminum ( )	Lead ( )
Average sheet thickness			

Calculations
(show work)

Don't forget units

(continued)

A. Absorption of Beta Radiation

DATA TABLE 2

Purpose: To determine the relation between intensity and thickness.

Cardboard		Aluminum	
Number of sheets n	Intensity I (cpm)	Number of sheets n	Intensity I (cpm)
0	(I_o)	0	(I_o)
Range of beta radiation (cm)		Range of beta radiation (cm)	

Calculations
(show work)

EXPERIMENT 28   *Laboratory Report*

B. Absorption of Gamma Radiation

DATA TABLE 3

Purpose: To determine the relationship of intensity and thickness.

Number of lead sheets n	Intensity I (cpm)	Corrected intensity $I_c = I - I_b$
0	(I_o)	

Calculations
(show work)

Background Radiation

Number of counts _____

Time interval _____

Intensity I_b (cpm) _____

(continued)

Absorption Coefficient Measurements for Gamma Rays

Number of sheets to reduce initial intensity to $\frac{1}{2}$ $n_{1/2}$ ————————————

Half-thickness $x_{1/2}$ ————————————

Linear absorption coefficient μ ————————————

Mass absorption coefficient μ_m ————————————

Percent error ————————————

Slope of graph x_i ————————————

Linear absorption coefficient μ ————————————

Mass absorption coefficient μ_m ————————————

Percent error ————————————

QUESTIONS

1. Was there a large difference in the percent errors of the experimental mass absorption coefficients? If so, why do you think this was the case?

2. Compute what percent of an incident beam of 0.662-MeV gamma rays is absorbed while passing through 2.5 mm of lead.

(continued)

3. Would the Cartesian graph of ln I_c versus n (or x) be a straight line if the gamma radiation contained gamma rays of two different energies? Explain.

4. The mass absorption coefficient of iron is 0.058 for 1.24-MeV gamma rays. What percentage (if any) of the beam of such gamma rays is transmitted through an iron plate 3 cm thick? ($\rho_{Fe} = 7.86$ g/cm^3.)

EXPERIMENT 28

EXPERIMENT 29

The Scientific Method:
The Simple Pendulum
/TI/ *Advance Study Assignment*

Read the experiment and answer the following questions.

1. Describe what is meant by the scientific method and how it is applied.

2. What are the physical parameters in the investigation of a simple pendulum?

3. A period is an interval of time. How is this applied to a pendulum?

4. What is the difference between an independent variable and a dependent variable? Give an example of each.

(continued)

5. How does the period of a pendulum vary, theoretically, with (a) length, (b) mass of bob, (c) angular displacement?

6. How will you experimentally check the theoretical predictions in the preceding question?

7. What is meant by a *small-angle approximation?*

8. How can the parabolic form $y = ax^2$ be plotted as a straight line on Cartesian graph paper?

The Scientific Method: The Simple Pendulum

INTRODUCTION AND OBJECTIVES

The laboratory is a place for the investigation of physical phenomena and principles. In the process, new discoveries are made, and technology is advanced. In some instances, while trying to invent things in the laboratory, scientists make various investigations at random. This might be called the *trial-and-error approach*. Edison's invention of the light bulb is an example. He kept trying until he found something that worked—a carbonized thread for a filament.

Today, the physics laboratory is used in general to apply what is called the **scientific method.** This principle states that no theory or model of nature is tenable unless the results it predicts are in accord with experiment.

Rather than applying the somewhat haphazard trial-and-error approach, scientists now try to predict physical phenomena theoretically and then test their theories against planned experiments in the laboratory. If the experimental results agree with the theoretical predictions, we consider the theory to be valid and believe that we have an accurate description of certain physical phenomena (until some other results demonstrate otherwise).

To illustrate the scientific method, in this experiment a theoretical expression or equation that describes the behavior of a simple pendulum will be given. You will test the validity of this relationship experimentally. In the process, you will learn what variables influence the period of a simple pendulum and how the physical relationship and experimental data can be used to find other useful information (for example, the value of the acceleration due to gravity).

After performing this experiment and analyzing the data, you should be able to do the following:

1. Apply the scientific method to theoretical predictions to check their validity.
2. Understand how physical parameters are varied so as to investigate theoretical predictions.
3. Appreciate the use of approximations to facilitate experimental investigations and analyses.

EQUIPMENT NEEDED

- Meterstick
- Laboratory timer or stopwatch
- Protractor
- String
- Three or more pendulum bobs of different masses
- Pendulum clamp (if available)
- 1 sheet of Cartesian graph paper

THEORY

A **pendulum** consists of a "bob" (a mass) attached to a string that is fastened such that the pendulum assembly can swing or oscillate in a plane (● Fig. 1). For a simple or ideal pendulum, all the mass is considered to be concentrated at a point at the center (of mass) of the bob.

Some of the physical properties or parameters of a simple pendulum are (1) the length L of the pendulum, (2) the mass m of the pendulum bob, (3) the angular displacement θ through which the pendulum swings, and (4) the period T of the pendulum, which is the time it takes for the pendulum to swing through one complete oscillation (for example, from A to B and back to A in Fig. 1).

From experience or preliminary investigation, we find that the period of a pendulum depends on its length (that is, the longer the pendulum, the greater its period). How do you think the other parameters (m and θ) affect the period?

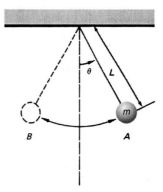

Figure 1 The simple pendulum. The physical parameters of a simple pendulum are the mass m of the bob, its length L, and the angle of swing θ. The period T of a pendulum is the time it takes for one completed oscillation—for example, the time it takes to swing from A to B and back to A.

From physical principles and advanced mathematics, the theoretical expression for the period of a simple pendulum oscillating in a plane is

$$T = 2\pi\sqrt{\frac{L}{g}}\left(1 + \frac{1}{4}\sin^2\frac{\theta}{2} + \frac{9}{64}\sin^4\frac{\theta}{2} + \cdots\right) \quad (1)$$

where g is the acceleration due to gravity and the terms in parentheses are part of an infinite series. In calculating T for a given angular distance θ, the more terms of the series that are evaluated, the greater the accuracy of the theoretical result.

For small angles ($\theta \lesssim 20°$), the θ terms in the series are small compared to unity (i.e., $<<1$), and in this case, to a good approximation,

$$T = 2\pi\sqrt{\frac{L}{g}} \quad (2)$$

(This is called a **first-order approximation.** If the second term in the series is retained, the approximation is to second order, and so on.)

Notice that even without an approximation (Eq 1), the period is theoretically independent of the mass of the pendulum bob. Also, within the limits of the small-angle approximation (Eq. 2), the period is independent of the displacement angle.

It is sometimes helpful to visualize a physical system as a "black box" with inputs and outputs.* The black box is the relationship between the input and output parameters. The term *parameter* refers to anything in the physical system that can be measured.

The input parameters are the physical variables that may *control* or *influence* the behavior of the output parameters (the physical quantities that are measured and that *describe* the resulting behavior of the system). The input parameters are often called **independent variables** because they can (and should) be varied independently of each other. The output parameters, on the other hand, may be called **dependent variables** because their values depend on the inputs. In any given system, some of the inputs may have little or no effect on the outputs.

In the case of a simple pendulum, a little insight suggests that a black box diagram might look like this:

Input

$m \quad \theta \quad L$
$\downarrow \quad \downarrow \quad \downarrow$

Simple Pendulum

$$T = 2\pi\sqrt{\frac{L}{g}}\left(1 + \frac{1}{4}\sin^2\frac{\theta}{2} + \cdots\right)$$

$\downarrow$

T

Output

* Suggested by Professor I. L. Fischer, Bergen Community College, New Jersey.

You may find that drawing black box diagrams will help you understand the physical systems investigated in later experiments.

EXPERIMENTAL PROCEDURE

1. Set up a simple pendulum arrangement. If a pendulum clamp is not available, the string may be tied around something such as a lab stand arm. Make sure that the string is secure and does not slip on the arm.

2. Experimentally investigate the small-angle approximation (Eq. 2) and the theoretical prediction (Eq. 1) that the period increases with larger angles. Do this by determining the pendulum period for the several angles listed in Data Table 1, keeping the length and mass of the pendulum constant.

   [Rather than timing only one oscillation, time several (e.g., four or five) and determine the average period. Timing is generally more accurate if you start the pendulum oscillating before the timing begins. Also, it is usually best to take the timing reference point as the lowest point of the swing.]

   Measure and record the pendulum length. The length should be measured to the center of the pendulum bob.

   Compute the percent error of the period for each angle θ, using Eq. 2 to calculate the accepted value. (In this case, do not use the absolute difference, so that each percent error will have a sign, + or −. (Further analysis will be done in the Questions section.) Proceed to the next step.

3. Experimentally investigate whether the period is independent of the mass of the pendulum bob. Using the three masses provided, determine the periods of a pendulum with each mass as the bob (keeping length L and the small angle of oscillation constant). Record your results in Data Table 2, and draw a conclusion from the data.

4. Experimentally investigate the relationship between the length and period of a pendulum. Using five different lengths (such as, 0.20, 0.40, 0.60, 0.80, and 1.00 m), determine the average period of a pendulum of each length (keeping mass and the small angle of oscillation constant). Record the data in Data Table 3.

5. Compute the theoretical period for each pendulum length (Eq. 2) and enter the results in Data Table 3 ($g = 9.80$ m/s^2 = 980 cm/s^2).

6. Compute the percent error between the experimental and the theoretical values of the period for each pendulum length, and record in Data Table 3. Draw conclusions about the validity or applicability of Eq. 2.

7. The object of the preceding experimental procedures was to determine the validity or applicability of Eq. 2— that is, whether the experimental results agree with the theoretical predictions as required by the scientific method. Once found acceptable, a theoretical expression can then be used to determine experimentally other quantities occurring in the expression. For example, Eq. 2 provides a means for experimentally determining g, the acceleration due to gravity, by measuring the pendulum parameters of length and period as was done previously.

Squaring both sides of Eq. 2, we have

$$T^2 = \frac{4\pi^2}{g} L \qquad (3)$$

or

$$L = \frac{g}{4\pi^2} T^2$$

Hence, the equation has the form $y = ax^2$, that of a parabola. This can be plotted as a straight line with the general form $y = mx'$ by letting $L = y$ and $x' = T^2$, or plotting T^2 on the X axis. The line will have a slope of $m = g/4\pi^2$.

8. Plot L versus T^2 from the experimental data in Data Table 3, determine the slope of the graph, and compute the experimental value of g. Record this in the Laboratory Report and compute the percent error of the result.

EXPERIMENT 29

The Scientific Method:
The Simple Pendulum

|TI| *Laboratory Report*

DATA TABLE 1

Purpose: To investigate the small-angle approximation.

Mass, m _____ Pendulum length, L _____

Angle θ	Period T ( )		Percent error
	Experimental	Theoretical	
5°			
10°			
20°			
30°			
45°			
60°			

Conclusion:

DATA TABLE 2

Purpose: To investigate period dependence on mass.

θ _____ L _____

m ( )	T ( )		Percent error
	Experimental	Theoretical	

Conclusion:

Don't forget units

(continued)

DATA TABLE 3

Purpose: To investigate period dependence on length.

θ _____ m _____

L ( )	T ( )		Percent error	T^2 ( )
	Experimental	Theoretical		

Conclusion:

Value of g from experimental _____
data (slope of graph)   (units)

Percent error   _____

/TI/ QUESTIONS

1. It was suggested that you measure the time for several periods and determine the average period, rather than timing only one period.
   (a) What are the advantages of this method?

   (b) How and why would the result be affected if a very large number of periods were timed?

2. In general, the results of procedure 2 may not have shown clear-cut evidence that the period increases as dramatically with the angle as Eq. 1 might suggest. To understand why, write Eq. 1 as

$$T = T_1\left(1 + \frac{1}{4}\sin^2\frac{\theta}{2} + \frac{9}{64}\sin^4\frac{\theta}{2}\right)$$

and compute T in terms of $\left(=2\pi\sqrt{\frac{L}{g}}\right)$ for angles of 5°, 20°, and 60°.

Comment on the theoretical predictions and experimental accuracy in relation to your results in Data Table 1.

3. Is air resistance or friction a systematic or a random source of error? Would it cause the period to be larger or smaller than the theoretical value? (*Hint:* Consider what would happen if the air resistance were much greater—for example, as though the pendulum were swinging in a liquid.)

4. Thomas Jefferson once suggested that the period of a simple pendulum be used to define the standard unit of length. What would be the length of a "2-second" pendulum (a pendulum with a period of 2 s)?

5. Suppose Jefferson's 2-second pendulum were operated on the Moon. What would its period be there?

EXPERIMENT 29

Name _____ Section _____ Date _____

Lab Partner(s) _____

EXPERIMENT 30

TI Rotational Motion and Moment of Inertia

CI The Conservation of Angular Momentum

TI *Advance Study Assignment*

Read the experiment and answer the following questions.

1. What are the rotational analogs of force, mass, and Newton's second law?

2. Define the term *moment of inertia.* How is torque related to moment of inertia?

3. Describe two methods by which the moment of inertia of a body may be experimentally determined.

(continued)

4. How is the frictional torque of a rotating system experimentally determined or compensated for?

5. Explain how tangential and angular velocities and accelerations are related to one another.

6. A solid cylindrical object has a mass of 2.0 kg, a diameter of 0.10 m, and a length of 0.18 m. What is the moment of inertia of the body about an axis along the axis of the cylinder? Show your work.

CI *Advance Study Assignment*

Read the experiment and answer the following questions.

1. What is required for an object to change its rate of rotation?

2. What is the rotational analog of Newton's second law?

TI Rotational Motion and Moment of Inertia

CI The Conservation of Angular Momentum

OVERVIEW

Experiment   is concerned with rotational motion, and the TI and CI procedures examine different aspects. These complementary experiments give a good overall insight into the fundamentals of rotational motion. The TI proce-

dure focuses on torque and the measurement of moments of inertia of various objects. The CI procedure is concerned with the conservation of angular momentum. This is examined experimentally by dropping a ring onto a rotating disk.

INTRODUCTION AND OBJECTIVES

In the case of linear motion, an unbalanced force **F** acting on an object gives it an acceleration **a,** and by Newton's second law,

$$\mathbf{F} = m\mathbf{a}$$

where m is the mass of the object. For *rotational motion,* an **unbalanced torque** τ acting on a body causes it to change its rate of rotation (i.e., gives it an angular acceleration α). The rotational analogue of Newton's second law is

$$\tau = I\alpha$$

where I is the moment of inertia of the body.

The **moment of inertia** of a body depends on its mass distribution and shape. For a symmetric, homogeneous object, the moment of inertia can be derived theoretically using calculus methods and is expressed in terms of the object's total mass and dimensions. By determining these quantities, we can calculate the moment of inertia.

However, the moment of inertia of an object can also be determined experimentally from the dynamical equation $\tau = I\alpha$ by measuring τ and α, or from energy considerations, because the rotational kinetic energy is $\frac{1}{2}I\omega^2$.

In the TI experiment, the theoretical and experimental values of the moments of inertia of some common, regularly shaped objects will be determined and compared. The experimental procedures will provide insight into the concepts of rotational motion and moment of inertia. Several types of apparatus are described. The experiment is rather

involved and will test your experimental skills. Read the TI Theory and TI Experimental Procedure sections well.

TI OBJECTIVES

After performing the TI experiment and analyzing the data, you should be able to do the following:

1. Explain the relationships involving torque, rotational motion, and rotational energy.
2. Tell what is meant by the moment of inertia of a rigid body and on what this quantity depends.
3. Describe the parallel axis theorem and how it can be practically applied.

CI OBJECTIVES

The CI procedure experimentally investigates the conservation of angular momentum. This is done by measuring first the angular momentum of a rotating disk and then the angular momentum of the system after a ring is dropped on the disk. With no external torque resulting from the dropping of the ring, the angular momentum should be conserved.

After performing the CI experiment, you should be able to do the following:

1. Describe the relationships among the quantities used to describe rotational motion.
2. Explain the conservation of angular momentum.

EQUIPMENT NEEDED

General

- Meterstick
- String
- Weight hanger and weights
- Laboratory timer or stopwatch
- Laboratory balance (kilogram range)
- Vernier calipers
- Large calipers
- Safety glasses

A. *Rotational Inertia Apparatus (TI Fig. 6) or Similar Apparatus*

- 1 sheet of Cartesian graph paper

B. *Rotating Support with Accessories (TI Fig. 4)*

- Cylindrical ring
- Disk
- Bar attachment (optional)
- Two cylinders
- Table clamp
- Support rod
- Two right-angle clamps
- One or two pulleys (depending on setup)

C. *Moment of Inertia Accessory (for centripetal force apparatus, see TI Fig. 7)*

- 2 sheets of Cartesian graph paper

THEORY

An unbalanced force **F** applied any distance from a body's axis of rotation produces an unbalanced torque **τ**, the magnitude of which is given by

$$\tau = rF \qquad \text{(TI 1)}$$

where r is the perpendicular distance from the axis of rotation to the force's line of action, a line running through the force vector (● TI Fig. 1). For a particle of mass m,

$$\tau = rF = r(ma) = mra$$

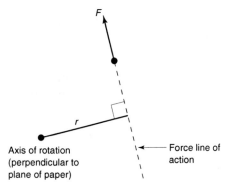

Axis of rotation
(perpendicular to
plane of paper)

Force line of
action

TI Figure 1 Torque. The magnitude of a torque is the product of the magnitude of the force F and the perpendicular (lever arm) r from the axis of rotation to the force's line of action ($\tau = rF$).

where a is the tangential acceleration. Using the relationship between the magnitudes of angular acceleration (α) and tangential acceleration, $a = r\alpha$, we obtain

$$\tau = mr(r\alpha) = (mr^2)\alpha = I\alpha$$

where the moment of inertia $I = mr^2$.

A rigid body is a system of particles, and summing this product for all the individual particles yields

$$\tau = (\Sigma_i m_i r_i^2)\alpha$$
$$= I\alpha \qquad \text{(TI 2)}$$

where the **moment of inertia** for a rigid body is $I = \Sigma_i m_i r_i^2$. The moments of inertia of several such bodies are given in ● TI Fig. 2.

When the torque is supplied by a constant-weight force as in ● TI Fig. 3, the magnitudes of the angular velocity and angular acceleration of the rotating body can be found from measurements of the distance y that the weight descends from rest and the time t that it takes the weight to travel this distance. The magnitude of the average velocity $\bar{v}$ is given by

$$\bar{v} = \frac{y}{t} \qquad \text{(TI 3)}$$

Using the kinematic equation for average velocity for motion with constant acceleration,

$$\bar{v} = \frac{v + v_0}{2}$$

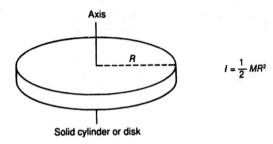

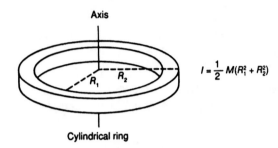

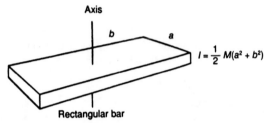

TI Figure 2 Moment of inertia. The moments of inertia for some uniform, regularly shaped objects.

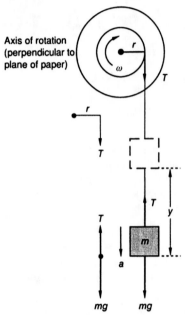

TI Figure 3 Applied torque. A torque being applied by a constant-weight force. See text for description.

we find that the magnitude of the final velocity v of the descending weight with $v_o = 0$ is given by

$$\bar{v} = \frac{v}{2}$$

or

$$v = 2\bar{v} = \frac{2y}{t} \qquad \textbf{(TI4)}$$

The angular velocity has a magnitude, then, of

$$\omega = \frac{v}{r} = \frac{2\bar{v}}{r} \qquad \textbf{(TI5)}$$

where r is the radius of the axle. Also,

$$v = at$$

or

$$a = \frac{v}{t}$$

and

$$\alpha = \frac{a}{r} \qquad \textbf{(TI6)}$$

From the forces acting on the descending mass (see TI Fig. 3), the acceleration is given by Newton's second law:

$$mg - T = ma$$

or

$$T = mg - ma = m(g - a) \qquad \textbf{(TI7)}$$

where T is the magnitude of the tension force in the string. The magnitude of the applied torque τ_a is then

$$\tau_a = rT \qquad \textbf{(TI8)}$$

Two torques are acting on the wheel and axle: the applied torque τ_a and an opposing frictional torque τ_f due to the friction on the bearings of the rotating axle. The frictional torque is assumed to be constant. The magnitude of the net torque, $\tau = I\alpha$, may then be written

$$\tau = \tau_a - \tau_f = I\alpha$$

so that

$$\tau_a = I\alpha + \tau_f \qquad \textbf{(TI9)}$$

DETERMINATION OF I

The moment of inertia I of a system can be found from TI Eq. 9 using experimentally determined values of τ_a,

α, and τ_f. The frictional torque can be determined or taken into account by either of the following means:

a. Determine the suspended mass m_0 needed to cause the body to rotate with a constant angular speed ω (i.e., $\alpha = 0$). Then $\tau_f = \tau_a = rT = r(m_0 g)$. In this case, the frictional torque is balanced or vectorially canceled by the applied torque due to m_0.

b. Since I and τ_f are constants, by determining the α's for two or more τ_a's, we can determine both I and τ_f from the set of equations (TI Eq. 9) for the different conditions, or we can determine them graphically.

In some instances, the rotating body may be symmetric, but a moment arm cannot be measured for the complete body (for example, the rotating support in ● TI Fig. 4). In this case, the moment of inertia can be found from energy considerations, which can also be used to find the moment of inertia of a body with a well-defined moment arm.

The potential energy lost by the descending mass in the setup in TI Fig 4 goes into the linear kinetic energy of the mass and the rotational kinetic energy of the rotating body (neglecting frictional losses). By the conservation of energy,

$$\text{Energy lost} = \text{energy gained}$$

$$mgy = \tfrac{1}{2}mv^2 + \tfrac{1}{2}I\omega^2 \qquad \textbf{(TI 10)}$$

where all the quantities except I are known or can be calculated from experimental data.

The energy lost to friction can be accounted for by adding the term $m_0 gy$ to the right side of TI Eq. 10, where m_0 is the mass determined in (a) above.

When a body is a composite of two or more regular bodies (such as a wheel and axle), the moment of inertia of the system about a coaxis along the component bodies' axes of symmetry is the sum of the moments of inertia of each of the component bodies (that is, $I = I_1 + I_2 + \cdots$).

When the system consists of a regularly shaped accessory body mounted on the rotating support with the axis of rotation through the center of mass (symmetry) of the accessory body, the total moment of inertia I of the system is $I = I_s + I_a$, where I_s and I_a are the moments of inertia of the support and accessory body, respectively. The moment of inertia of the accessory body is then $I_a = I - I_s$, where I and I_s are measured independently by one of the previously described methods.

PARALLEL-AXIS THEOREM

For a system of regular bodies, some of which do not have the axis of rotation through their centers of mass, the moment of inertia of the system can be calculated using the parallel-axis theorem. Such a system is that of the cylindrical masses placed on opposite arms of the metal cross

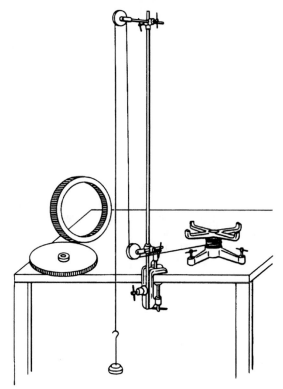

TI Figure 4 Moment of inertia apparatus. A rotating support and illustration of the experimental setup and accessories. (Photo courtesy of Sargent-Welch.)

of the rotating support or a bar attachment, as shown in ● TI Fig. 5.

The **parallel-axis theorem** states that the moment of inertia of a body about any axis O' parallel to an axis through the body's center of mass is equal to the moment of inertia about the axis through the center of mass plus the mass of the body times the square of the distance between

$$I_{o'} = I_{cm} + Md^2 \qquad \textbf{(TI 11)}$$

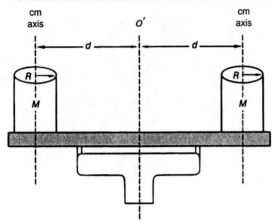

TI Figure 5 Parallel-axis theorem. The moment of inertia of a system of regularly shaped objects can be calculated using the parallel-axis theorem. See text for description.

Then the total moment of inertia I for the system shown in TI Fig. 5 (identical cylinders) about an axis through O' is

$$I = I_s + I_b + I_c + I_c$$
$$= I_s + I_b + 2I_c$$

where I_s is the moment of inertia of the rotating support, I_b is the moment of inertia of the bar, and I_c is the moment of inertia of one of the cylinders.

The axis of rotation is through the centers of mass of the support and the bar, so I_s and I_b can be determined by previously described methods. By the parallel axis theorem,

$$I_c = I_{cm} + Md^2$$

where I_{cm} for a cylinder is $I_{cm} = \frac{1}{2}MR^2$ (see TI Fig. 2). Although convenient, it is not necessary that the cylinders be placed equal distances from the axis of rotation (different d's are used if they are not at equal distances).

TI EXPERIMENTAL PROCEDURE

A. Rotational Inertia Apparatus (Wheel and Axle)

1. A rotational inertia apparatus is shown in ● TI Fig. 6. It is essentially a wheel and axle, a large heavy wheel or disk with a concentric axle hub that rotates freely on bearing mounts.*

2. Measure the diameter of the disk with the large calipers and the diameter of the axle hub with the vernier

---

* Some older inertia wheels have two concentric axles with radii of different sizes. In this case, record both radii and hang two masses from each in the following procedures.

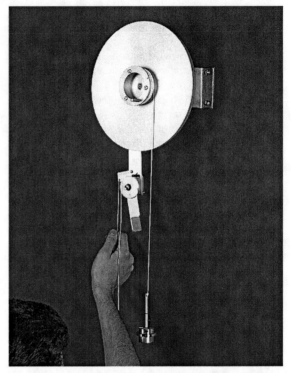

TI Figure 6 Rotational inertia and wheel apparatus. Wheel and axle used in the experimental setup. (Photo courtesy of Sargent-Welch.)

calipers, and record these diameters in TI Data Table 1. Also record the total mass of the wheel and axle, which is usually stamped on the wheel.

3. Fix one end of a length of string to the set pin in the axle, and wrap the string around the hub axle in a single layer. The string should be sufficiently long to allow weights attached to the free end of the string to descend to the floor.

    Making a loop in the free end of the string, attach just enough weight to the string so that the disk, when given a slight start, rotates with a uniform angular speed as the weight descends. (The ordinary 50-g weight hanger will probably be too large. Put smaller weights in the string loop.)

    When the disk rotates with a uniform angular speed, the weight descends with uniform linear speed, and this may be easier to judge than the disk's rotational speed.

    Record the mass of the descending weight m_o in TI Data Table 1. (This will be used in a direct calculation of τ_f.)

4. Remove the weights from the string loop and attach a 0.050-kg (50-g) weight hanger. Measure the distance

of the bottom of the weight hanger from the floor with the string wrapped around the smaller axle, and record it as the distance of descent y in TI Data Table 1.

Releasing the weight hanger, measure the time it takes to reach the floor, and record. [Newer models of the apparatus have a platform on which the mass initially rests. The experimental run is initiated by tripping the platform (see TI Fig. 6). Also, some rotational inertia apparatuses may be equipped with spark timers. Consult your instructor for operating instructions.] Repeat the timing procedure twice more for the same distance of descent.

5. Repeat the timing procedures (three trials each) for three more *total* masses of 0.075 kg, 0.10 kg, and 0.125 kg (or different appropriate masses).

6. Compute the average time of descent for each case, and, using TI Eqs. 3 through 8, calculate the magnitudes of the quantities called for in the Laboratory Report.

7. Notice that TI Eq. 9, $\tau_a = I\alpha + \tau_f$, is of the form of a straight line,

$$y = mx + b$$

where m is the slope of the line and b its y intercept. Plot τ_a versus α for the data in TI Data Table 1, and draw the straight line that best fits the data. Determine the slope I and the y intercept τ_f. Calculate the theoretical value of I using the disk equation from TI Fig. 2, and compare the experimental value determined from the slope of the graph to it by calculating the percent difference.

Use m_o and r to make a direct calculation of τ_f. $\tau_f = m_o gr$. Compare with this the value obtained from the y intercept of the graph by calculating the percent difference.

8. *Moment of inertia from conservation of energy.* Using TI Eq. 10, find the moment of inertia of the wheel and axle using the data from the final descending weight case trial. (*Note:* The mass used on the left side of this equation should be the mass of the descending weight *minus* the mass of the descending weight in the constant angular speed case of procedure 3. Why?)

Compare the moment of inertia determined from TI Eq. 10 with the computed theoretical value by calculating the percent difference.

B. Rotating Support and Accessory Bodies

9. A rotating support and some accessories are shown in TI Fig. 4. The rotating support consists of a light metal cross mounted on ball bearings so as to rotate in a horizontal plane about a vertical axis.

10. With the vernier calipers, measure the diameter of the drum of the rotating support, and record this diameter in TI Data Table 2. (The string unwinds from the drum.) With the appropriate instruments, measure the linear dimensions of the accessory bodies needed to calculate the theoretical moments of inertia. (Refer to TI Fig. 2.) Also, determine the masses of the accessory bodies on a laboratory balance, and record these masses.*

11. Use a pulley arrangement similar to the one in TI Fig. 4 so that the descent distance of the driving weight is at least 1 m. Attach one end of a length of string to the rotating support drum, and wrap the string around the drum in a single layer. The length of the string should be sufficient to allow weights attached to the free end of the string to descend to the floor.

Making a loop in the free end of the string, attach just enough weight (mass m_o) to the string so that, when given a slight start, the cross of the rotating support rotates with a uniform speed as the weight descends. (The ordinary 50-g weight hanger will probably be too large. If so, put some smaller weights in the string loop.)

The weight should also descend with a uniform speed, which may be easier to judge. Make certain that the string unwinding from the drum is parallel to the table top.

Record the mass of the descending weight m_o in TI Data Table 2.

12. Remove the weights from the string loop, attach a 50-g weight hanger to the loop, and place the removed weights on the weight hanger. Record the total mass (m_1) in the data table. Measure the distance y from the bottom of the weight hanger to the floor with the string wrapped around the drum, and record in the data table.

After releasing the weight hanger, measure the time it takes to reach the floor, and record. Repeat the timing procedure twice more for the same distance of descent, and record.

13. Using the distance of descent and the average time of descent of the three trials, calculate the magnitude of the final velocity v of the descending weight and the final angular speed ω of the rotating support from the measured data, using TI Eqs. 4 and 5.

Using TI Eq. 10, calculate the moment of inertia I_s of the rotating support. [*Note:* The mass m in this equation does *not* include the mass on the weight hanger used to determine the constant speed in procedure 11 (that is, $m = m_1 - m_o$). Why?]

14. Place one of the regularly shaped accessory bodies (for example, the cylindrical ring) on the metal cross of the

---

* The values of the masses may be given to you by your instructor if the masses of the accessories are too large for regular laboratory balances.

rotating support so that the axis of rotation is through the body's center of mass. Add just enough weight to the weight hanger so that the system rotates (or the weight hanger descends) with a constant speed when given a slight start. Record the total mass m_0 of the descending weight.

15. Add additional mass (0.10 to 0.20 kg) to the weight hanger so that it descends slowly to the floor when released from rest. Record the total mass m_1 of the descending weight. Measure the time t it takes for the weight hanger to descend a distance y to the floor for three trials, and record these data.

16. Repeat procedures 14 and 15 for the other accessory bodies (disk and bar) if available.

17. Compute the average time of descent, and, using TI Eqs. 4 and 5, find the magnitude of the final velocity of the descending weight and the final angular speed ω of the rotating system for each case.

    Calculate the moments of inertia I for the combined accessory-support systems from the experimental data, using TI Eq. 10. Determine the moments of inertia of the individual accessory objects ($I_a = I - I_s$), using the previously determined value of I_s.

18. Using the measured masses and appropriate length measurements of the accessory objects, compute the theoretical values of their moments of inertia (see TI Fig. 2), and compare with the experimentally determined values by calculating the percent differences.

19. *Parallel axis theorem.* Set up the system as shown in TI Fig. 5, using the bar attachment if available. Measure the distance d from the axis of rotation to the parallel axes through the centers of masses of the cylinders, and record. Repeat procedures 14, 15, and 17 for this system, where, in procedure 17, $I_a = I_c$ and

$$I = I_s + I_b + 2I_c$$

20. Applying the parallel axis theorem (TI Eq. 11), calculate the theoretical moment of inertia of the cylinders about the axis of rotation. Compare this result with the experimentally determined value of I by calculating the percent difference.

C. Moment of Inertia Accessory (for Centripetal Force Apparatus)

21. The moment of inertia accessory for a centripetal force apparatus is shown in ● TI Fig. 7. Using the

TI Figure 7   Moment of inertia apparatus.   A hand-operated centripetal force apparatus with accessory may be used to measure the moment of inertia of rotating masses. Descending weights supply the torque. (Photo courtesy of Sargent-Welch.)

vernier calipers, measure the diameter of the vertical shaft, and record the radius r in TI Data Table 3.

Set up the apparatus with 100-g slotted weights about 12 cm from the vertical shaft on the horizontal rod. Secure the weights with the wing nuts, making certain that they are equidistant from the shaft. Record the mass of a single weight ($M = 100$ g) in the first column. Measure the distance R from the *center* of the vertical shaft to the *center* of one of the slotted weights, and record.

22. Secure one end of a length of string to the screw in the vertical shaft and the other end to a weight hanger. The string should be long enough to allow the weight hanger to reach the floor. Twisting the shaft, wind the string evenly around it, so that the weight hanger is near the base of the apparatus at a convenient distance from the floor. Measure and record the distance y of the bottom of the weight hanger from the floor.

23. Add mass to the weight hanger so that it accelerates at a moderate rate as it descends (e.g., add sufficient mass to make the total $m = 150$ g, including the mass of the weight hanger). Record this mass in the data table.

Rewind the string around the shaft so that the bottom of the weight hanger is again at the previous height (y) from the floor. Releasing the weight hanger, measure the time t that it takes to descend to the floor, and record. Repeat the timing procedure twice for the same distance of descent in order to get an average.

24. Remove the two 100-g weights from the horizontal arm, leaving the wing nuts on the arm. (The wing nuts should be loosened only slightly and turned back so that they are at their previous positions.) Repeat the timing procedures with no weights (other than wing nuts) on the horizontal arm. Use columns for the data, cross out the spaces not needed.

25. Using TI Eqs. 4, 5, and 10, compute the moments of inertia of the rotating system with the masses on the arm (I) and without them (I_o).

The moment of inertia I of the rotating system with the masses can be considered the sum of two parts: $I = I_1 + I_o$, where I_1 is the moment of inertia of the two-mass system. Subtract the results to obtain the moment of inertia I_1 of the two-mass system alone ($I_1 = I - I_o$).

26. Repeat procedures 22 and 23 for various values of R with constant M. ($M = 100$ g is convenient, since you already have one data point.) Assume the changes in I_o to be negligible.

27. Repeat procedures 22 and 23 for various values of M with the same constant R. Assume the changes in I_o to be negligible.

28. Using the data from the preceding procedures, plot (a) I_1 versus R, with constant M, and (b) I_1 versus M, with constant R.

29. When we consider the slotted weights to be point masses (that is, concentrated at their centers of mass), the moment of inertia of the two-mass system is given theoretically by $I_1 = 2MR^2$. Using this relationship, compute the moment of inertia I_1 for each combination of M and R, and record in TI Data Table 3.

Compare the computed theoretical values with the experimental values by plotting the computed values on the respective graphs. Comment on any noticeable similarities or differences. (Note the general forms of the graphs, and keep in mind that the theoretical I is an approximation for the extended masses and that the experimental values are subject to experimental error.)

EXPERIMENT 30

Rotational Motion
and Moment of Inertia

/TI/ *Laboratory Report*

A. *Rotational Inertia Apparatus (Wheel and Axle)*

/TI/ **DATA TABLE 1**

Diameter of axle _____

Radius, r _____

Total mass of
wheel and axle _____

Suspended mass
for constant
velocity, m_o _____

Thickness of
wheel (needed
for some
apparatus) _____

Length of axle
(beyond wheel) _____

Diameter of
wheel _____

Suspended mass ( )				
Distance of descent y ( )				
Time of descent t ( ) 1.				
2.				
3.				
Average				
Magnitudes of: Final velocity $v = 2y/t$ ( )				
Final angular velocity $\omega = v/r$ ( )				
Acceleration $a = v/t = 2y/t^2$ ( )				
Angular acceleration $\alpha = a/r$ ( )				
Tension $T = m(g - a)$ ( )				
Applied torque, $\tau_a = rT$ ( )				

(Space for calculations and results on the following pages)   *(continued)*

429

Calculations
(Show calculations for all cases, attaching additional sheet if necessary.)

1. Magnitudes of final velocity v and angular velocity ω

2. Magnitudes of acceleration a and angular acceleration α

3. Magnitudes of tension in string T and applied torque τ_a

4. Theoretical moment of inertia I (TI Fig. 2) and frictional torque τ_f

5. Moment of inertia I (energy method)

EXPERIMENT 30   *Laboratory Report*

Results

Using the torque method:

	Experimental value from plotted data	Computed value	Percent difference
Moment of inertia I	_____	_____	_____
Frictional torque τ_f	_____	_____	_____

Using the energy method:

Moment of inertia I _____

Percent difference (between experimental and theoretical values)   _____

B. Rotating Support and Accessory Bodies

$\boxed{\text{T1}}$ **DATA TABLE 2**

Diameter of support drum _____   Radius r _____

Accessory object:

1. Cylindrical ring   Mass _____   Inner radius R_1 _____

                                                 Outer radius R_2 _____

2. Disk   Mass _____   Radius R _____

3. Bar   Mass _____   Width _____

                                                 Length _____

4. Solid cylinder   Mass _____   Diameter _____   Radius R _____

    (Both cylinders should be approximately the same.)

Procedure 19: Distance from axis of rotation to parallel axis through center of mass of cylinders d _____

Don't forget units

(continued)

	(a) Rotating support	(b) Support with cylindrical ring	(c) Support with disk	(d) Support with bar	(e) Support with (bar and) cylinders
m_0 ( )					
m_1 ( )					
$m = m_1 - m_0$					
Distance of descent y ( )					
Time of descent t ( ) 1.					
2.					
3.					
Average					
Magnitudes: Final velocity v ( )					
Magnitudes: Final angular velocity ω ( )					
Calculated I of system ( )	(I_s)				
Moment of inertia of accessory I_a $(I_a = I - I_s)$	✕				$(I_a = I_c)$
Theoretical I_a	✕				
Percent difference	✕				

(Space for calculations on following page)

EXPERIMENT 30

Calculations
(Show calculations for all cases, attaching additional sheet if necessary.)

1. Magnitudes of final velocity v and final angular velocity ω

2. I of the system

3. Theoretical I_a

(continued)

C. Moment of Inertia Accessory

/TI/ DATA TABLE 3

Diameter of shaft _____   Radius r _____

			Variable R			Variable M			
Mass M ( )			Constant M _____ ( )						
Distance R ( )						Constant R _____ ( )			
Height y ( )									
Mass m ( )									
Time of descent t ( ) 1.									
2.									
3.									
Average									
Magnitudes: Velocity v ( )									
Magnitudes: Angular velocity ω ( )									
I ( )									
I_o ( )									
$I_1 = I - I_o$ ( )									
$I_1 = 2MR^2$ ( )									

Comments on graphs:

EXPERIMENT 30   *Laboratory Report*

Calculations
(Show calculations for all cases, attaching additional sheet if necessary.)

1. Magnitudes of final velocity v and final angular velocity ω

2. I of the system

3. I_1 of the two-mass system

4. I_0 of the system without masses

5. I_1 theoretical

(continued)

/T1/ QUESTIONS

A. Rotational Inertia Apparatus (Wheel and Axle)

1. Discuss the probable sources of error in the experiment. Which are systematic, and which are random? Which are most important?

2. The energy lost by the descending mass m_o is $m_o g y$. Express this energy loss mathematically in terms of the frictional torque τ_f and the angle termed, θ.

3. In a graph of τ_a versus α, what would be the implications if (a) the line went through the origin, and (b) the y intercept were negative?

B. Rotating Support and Accessory Bodies

4. Answer Question 1 above.

5. Answer Question 2 above.

6. If the mass for the disk is M_d and the cylinder has a mass M_c, find the expression for the moment of inertia of the system shown in ● TI Fig. 8: (a) about the axis O', and (b) about the axis O?

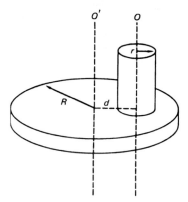

TI Figure 8

C. Moment of Inertia Accessory

7. Answer Question 1 above.

(continued)

8. The moment of inertia of a two-mass system was approximated by
   $I_1 = 2MR^2$. Justify this approximation.

9. Are the shapes of the curves on the graphs for the experimental values consistent with what
   would be expected theoretically? (Explain what the theoretical shapes of the curves would
   be.) For each nonlinear curve, explain how you might obtain a straight-line graph by
   plotting some function of the dependent variable other than itself (for example, ln m
   instead of m).

EQUIPMENT NEEDED

- Rotary motion sensor (PASCO CI-6538)
- Disk and ring: Mini-rotational accessory (PASCO CI-6691)

- Base and support rod
- Balance
- Calipers

THEORY

An object can change its rate of rotation when acted on by a net external torque. This is the rotational equivalent to Newton's second law: A rotational acceleration α is produced on an object with a moment of inertia I by a net external torque, which is the sum of all torques produced by all external forces acting on the object:

$$\tau_{\text{net}} = \Sigma\tau = I\alpha \qquad \textbf{(CI 1)}$$

(Note the similarity to $F = ma$.)

The rotational acceleration is the rate of change of the angular velocity of the object:

$$\tau_{\text{net}} = I\frac{d\omega}{dt} = \frac{d(I\omega)}{dt}$$

where the moment of inertia I is a measure of rotational inertia.

Note that in the absence of external forces, the external torque is zero, in which case

$$\frac{d(I\omega)}{dt} = 0$$

and

$$I\omega = \text{constant} \qquad \textbf{(CI 2)}$$

The quantity $I\omega$ that is conserved in the absence of external forces is called the angular momentum of a system, L. For a system made of n objects, all rotating with the same angular speed, the magnitude of the total angular momentum is

$$L_{\text{TOTAL}} = I_{\text{system}}\omega = (I_1 + I_2 + \cdots + I_n)\omega$$

In this experiment, a ring of rotational inertia, $I_{\text{ring}} = \frac{1}{2}M(R_{\text{inner}}^2 + R_{\text{outer}}^2)$, will be dropped on a rotating disk of rotational inertia $I_{\text{disk}} = \frac{1}{2}MR^2$. See ● CI Fig. 1.

Before the ring falls on the disk, the only thing rotating is the disk, so the magnitude of the initial total angular momentum of the system is given by

$$L_i = L_{\text{disk}} = I_{\text{disk}}\omega_i \qquad \textbf{(CI 3)}$$

After the ring falls on the disk, both rotate at the same final angular velocity, so the final total angular momentum of the system has a magnitude of

$$L_f = L_{\text{disk}} + L_{\text{ring}} = (I_{\text{disk}} + I_{\text{ring}})\omega_f \qquad \textbf{(CI 4)}$$

A rotary motion sensor will keep track of the angular speed of the system before and after the ring drops on the disk. Knowing the rotational inertias of the disk and the ring, we can determine the total angular momentum before and after and compare those values.

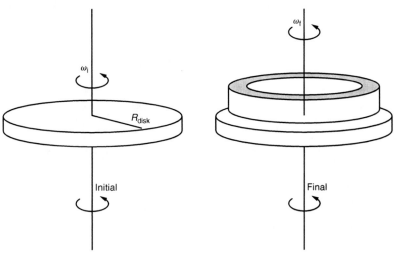

CI Figure 1   Rotating disk and rotating disk with ring. A ring will fall centered on an initially rotating disk. The initial and final angular speeds will be measured using sensors.

SETTING UP DATA STUDIO

1. Open Data Studio and choose to "Create Experiment".
2. From the sensor list, choose a rotary motion sensor. Connect the sensor to the interface, as shown in the computer screen.
3. Double-click on the RMS icon. The sensor properties window will open. Make the following selections:

   General: 10 Hz, Fast

   Measurement: Angular velocity (rad/s) (Deselect all others.)

   RMS: Divisions/Rotations: 360 Linear Calibration: Rack

   Click OK to exit the sensor properties window.
4. Create a graph by dragging the "Angular Velocity" icon in the data list and dropping it on top of the "Graph" icon of the displays list. A graph of angular velocity versus time will open in a window called Graph 1.

CI EXPERIMENTAL PROCEDURE

1. Use calipers to measure the inner and outer diameters of the ring and the diameter of the disk. Determine the corresponding radii, and enter the values, in meters, in CI Data Table 1.
2. Use the scale to measure the masses of the disk and the ring. Report the results, in kilograms, in CI Data Table 1.
3. Calculate the rotational inertia of the disk, the rotational inertia of the ring, and the total rotational inertia of disk and ring together. Enter the results in CI Data Table 1.
4. Mount the disk on the pulley of the rotary motion sensor, as shown in ● CI Fig. 2. The center screw must be secured tightly. Make sure the rotating disk will not collide with the support rod.

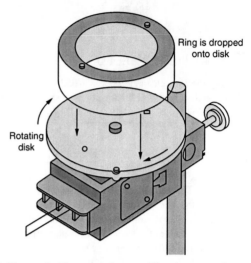

CI Figure 3 The experiment. The ring must drop vertically and fall centered on the spinning disk. For best results, drop it from not too far above the disk, as shown here, and practice several times first. A bit of tape on the ring will help it stick to the disk.

5. Practice first! Give the disk a spin, and drop the ring from directly above the disk, as shown in ● CI Fig. 3. The disk must drop vertically and must land centered on the ring. A bit of adhesive tape on the ring may help prevent it from slipping off the center when it falls.
6. To collect the data, give the disk a spin, and then press the START button. Drop the ring on the disk, and press the STOP button.
7. Use the smart-tool on the graph toolbar (it is a button labeled with *xy* axes) to determine the angular speed just before and just after the ring landed on the disk. Enter these values in CI Data Table 2.
8. Repeat the procedure three more times.
9. Using the known values for the rotational inertias of the disk and the ring, calculate the total angular momentum of the system before and after the collision for each trial. Enter the results in CI Data Table 2.
10. Compare the initial angular momentum and the final angular momentum by computing the percent difference.

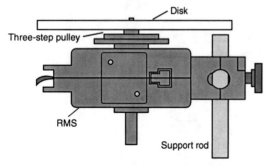

CI Figure 2 Experimental setup. The disk is mounted on the center screw of the rotary motion sensor pulley.

EXPERIMENT 30

The Conservation of Angular Momentum

CI *Laboratory Report*

CI DATA TABLE 1

Purpose: To determine the rotational inertias of the ring and the disk.

		Radius (m)	Mass, M (kg)	Rotational inertia, I
Ring	Inner:			
	Outer:			
Disk				
			Total rotational inertia, I	

CI DATA TABLE 2

Purpose: To verify the conservation of angular momentum.

	Angular speed		Total angular momentum		Percent difference
	Initial	Final	Initial	Final	
Trial 1					
Trial 2					
Trial 3					
Trial 4					

Don't forget units

(continued)

CI QUESTIONS

1. Discuss how well the results of the experiment match the theory. Was angular momentum conserved?

2. What is causing the difference between the initial and final values of the angular momentum?

3. What are some sources of experimental error?

4. How would the rotational inertia of the ring be affected if it did not land exactly centered on the disk? Discuss how this would affect the results of the experiment.

EXPERIMENT 31

Elasticity: Young's Modulus

TI/ *Advance Study Assignment*

Read the experiment and answer the following questions.

1. What is meant by *elasticity?*

2. Define the terms *stress, strain,* and *Young's modulus.*

3. Show that Hooke's law has the form of the general equation for a straight line. What would be the slope and *y*-intercept of this straight-line relationship?

(continued)

4. What is the elastic limit of a material? Does Hooke's law hold for a material stressed beyond its elastic limit?

5. In the experiment, why is it permissible not to include the initial 1-kg mass and weight hanger in the loads for the corresponding elongations?

Elasticity: Young's Modulus

INTRODUCTION AND OBJECTIVES

All bodies are deformed to some extent when acted on by forces. If, after deformation, a body returns to its original dimensions, it is said to be **elastic.** Elasticity is a material property and is characterized by an **elastic modulus,** defined as the ratio of the stress to the strain. **Stress** is related to the deforming force, and **strain** is the relative changes in the dimensions of the body when it is subjected to a stress.

You may not have realized that materials such as steel are elastic. However, this is the case, as will be demonstrated. The deformation and recovery of a steel wire

will be observed, and by experimentally determining the longitudinal strain for an applied tensile stress, you will compute the elastic modulus, called **Young's modulus.**

After performing this experiment and analyzing the data, you should be able to:

1. Clearly distinguish between stress and strain by citing experimental examples.
2. Explain how Young's modulus and Hooke's law are related and how they are limited in describing elasticity.
3. Appreciate that all solids, even metal wires, are elastic to some extent.

EQUIPMENT NEEDED

- Young's modulus apparatus
- Optical lever attachment (mirror on tripod base and telescope and scale; a desk lamp may be needed to illuminate the scale if laboratory lighting is poor) *or*

- Micrometer screw attachment
- Slotted weights (ten 1-kg masses)
- Meterstick and 2-m stick (if available)
- Micrometer calipers
- Safety goggles

THEORY

Elasticity is characterized by elastic moduli: Young's modulus (linear), shear modulus (two-dimensional), and bulk modulus (three-dimensional). Each modulus describes a different type of elastic deformation. Only Young's modulus will be considered in this experiment.

In general, an elastic modulus is defined as

$$\text{Elastic modulus} = \frac{\text{stress}}{\text{strain}} \qquad (1)$$

For the linear case, the stress is equal to the magnitude of the applied longitudinal force F divided by the object's cross-sectional area A normal to the force (● Fig. 1). This is commonly called a **tensile stress** (or **normal stress**):

$$\text{Tensile stress} = \frac{F}{A} \qquad (2)$$

The forces also may be opposite for a **compressive stress.**

A **strain** is the effect of relative changes in the dimensions or shape of a body subjected to a stress. For the linear case, this involves a change in length, and the tensile strain

is defined as the ratio of the change in length ΔL to the original length L_0 (Fig. 1):

$$\text{Tensile strain} = \frac{\Delta L}{L_0} = \frac{L - L_0}{L_0} \qquad (3)$$

Longitudinal strains may be either tensile (elongating) or compressive, depending on the applied force.

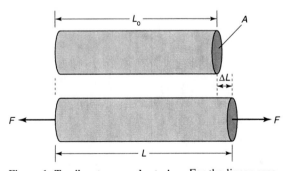

Figure 1 Tensile stress and strain. For the linear case, the stress is equal to the magnitude of the applied longitudinal force F divided by the cross-sectional area A normal to the force (F/A). The strain is equal to the ratio of $\Delta L/L_0$.

The elastic modulus for the linear case is called **Young's modulus** Y, and

$$Y = \frac{\text{tensile stress}}{\text{tensile strain}} = \frac{F/A}{\Delta L/L_o} \qquad (4)$$

[It should be noted that it is the increment, or change, in stress $\Delta(F/A)$ that produces the strain ΔL.]

Since the strain is a ratio of lengths, it is dimensionless, and the dimensions of Young's modulus are N/m^2 or lb/in^2. Provided that the elastic limit of the material is not exceeded, the ratio of the stress to the strain—Young's modulus—is often found to be constant and characteristic of a given material. Such materials obey **Hooke's law,** which, for a spring force, is commonly expressed as $F = -k\Delta x$, where k is a constant and the minus sign indicates that the spring force is opposite to the displacement. Equation 4 may be rewritten

$$F = \left(\frac{YA}{L_o}\right)\Delta L$$

A typical plot of stress versus strain is shown in ● Fig. 2. In the Hooke's law region, there is a straight-line relationship with a slope equal to Young's modulus. If the stress is increased beyond the elastic limit (point A), the sample does not return to its original length when the stress is removed but retains a permanent strain, or is permanently deformed. If enough stress is applied, the sample will fracture or break (point B).

For a wire of radius r with an applied longitudinal weight force mg, Young's modulus has the form

$$Y = \frac{F/A}{\Delta L/L_o} = \frac{mg/\pi r^2}{\Delta L/L_o} \qquad (5)$$

where ΔL is the elongation of the wire from its original length L_o.

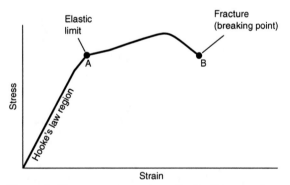

Figure 2 Stress versus strain. In the linear region, where Hooke's law is applicable, the slope of the line is typically equal to Young's modulus.

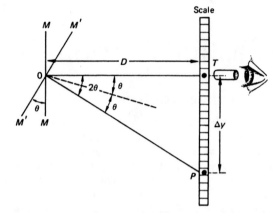

Figure 3 Optical lever. An illustration of the optical lever method for determining elongation. See text for description.

OPTICAL LEVER

(For experimental apparatus with elongation measurements made by the optical lever method)

In the optical lever method, small elongations are measured by means of the deflection of a mirror. If the plane of a mirror is originally vertical (MM), as illustrated in ● Fig. 3, then when you look through a telescope, the cross-hairs of the telescope will be on a mark T on the scale. However, if the mirror becomes tilted through an angle θ (because of a change in the length of the wire), then a reflected beam is turned through an angle of 2θ.

The cross-hairs of the telescope will then be on a mark on the scale at point P. That is, looking through the telescope, you would see point P. From Fig. 3, it can be seen that the double angle of deflection 2θ can be written in terms of the deflection distance Δy on the scale and the distance D from the mirror to the scale:

$$\tan 2\theta = \frac{\Delta y}{D} \qquad (6)$$

The tilting of the mirror results from the lowering of one end of the tripod base of the mirror mount (see ● Figs. 4 and 5). As seen in Fig. 4, the elongation ΔL is related to the deflection angle by

$$\Delta L = d \sin \theta \qquad (7)$$

where d is the apex length of the tripod triangle. For a small angle, as is the case here, $\frac{1}{2} \tan 2\theta \simeq \tan \theta \simeq \sin \theta$. From Eq. 6, the elongation becomes

$$\Delta L = \frac{d\Delta y}{2D} \qquad (8)$$

and the small elongation can be determined from easily measured experimental parameters.

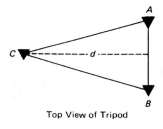

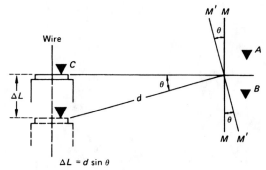

$\Delta L = d \sin \theta$

Figure 4 Optical lever and elongation. The tilting of the optical lever mirror results from the lowering of one side of the tripod base of the mirror mount due to the elongation of the wire. (See Fig. 5a.)

EXPERIMENTAL PROCEDURE

1. A common type of Young's modulus apparatus is shown in ● Fig. 6. The wire is supported by a chuck in a clamped yoke at the upper end of the apparatus support rods. A lower chuck that tightly grips the wire is positioned in the hole of an adjustable platform that is clamped to the support rods. If an optical lever attachment is used, the mount rests in the groove on the platform and on the face of the chuck (Fig. 5a).

   If a micrometer screw attachment is used (Fig. 5b), a lever with a spirit level on one end and a micrometer screw on the other is pivoted on the chuck. The micrometer screw is referenced to the adjustable platform. A weight hanger is suspended from the lower end of the wire, with the bottom of the hanger about 2 or 3 cm above the heavy tripod base of the apparatus.

   A direct-reading dial micrometer is also available (Fig.5c).

2. Hang a 1-kg mass on the weight hanger, and level the tripod base of the apparatus by means of the leveling screws. The mass should be centered between the support rods when the base is level. Measure the length L_0 of the wire from the bottom of the upper chuck to the top of the lower chuck, and record in the appropriate data table.

   [*Note:* The length of wire between the gripping chuck holders, not the total length of the wire, is taken as L_0. Also, the measurement attachment zero reading, y_0 below, is taken with an initial 1-kg mass on the weight hanger. These weights are *not* included as part of the individual load increments since the individual load increments (1 kg each) produce the corresponding ΔL's.]

3. (a) *Optical lever attachment* (if used).
   (i) Place the optical lever mount on a piece of paper and press lightly so that the pointed feet make indentations on the paper. Mark the indentations with a pencil. Then draw a line between the two front points (*AB* in Fig. 4), and draw a line perpendicularly from this line to point *C*. With a vernier caliper, measure the length of the latter line (*d* in Fig. 4), and record in Data Table 1.
   (ii) Carefully place the optical lever on the platform with its two front feet in the platform groove and its rear foot on the chuck (see Fig. 5). Rotate the mirror to the vertical position. Set the scale and telescope at least 1 m away, with the telescope at about the same height as the mirror. Adjust the telescope so that a mark on the scale as indicated by the telescope cross-hairs is easily seen. Record the zero reading (y_0).
   (iii) Carefully measure the distance between the mirror and the scale with a meterstick (distance *D* in Fig. 3), and record. It may be necessary to illuminate the scale with a lamp.

   (b) *Micrometer screw attachment* (if used). Pivot the micrometer attachment lever on the lower chuck, and adjust the platform and the micrometer screw so that it has a midscale reading when the screw is balanced against the platform with the spirit level in a level position. Record this zero reading (L_0') of the micrometer in Data Table 2.

      (*Note:* Always move the screw in the same direction just before each reading (for example, toward the higher numbers) so as to avoid screw thread "backlash" or hysteresis. If a direct-reading dial micrometer attachment is used, see Fig. 6c for arrangement.)

   (c) *Direct-reading dial micrometer attachment* (if used). The zero reading (L_0') may be read directly from the micrometer and recorded in Data Table 2.

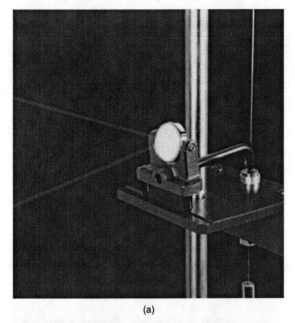

(a)

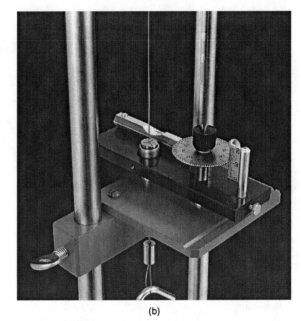

(b)

(c)

Figure 5 Elongation measurement. (a) Optical lever attachment. (b) Micrometer screw attachment. (c) Direct-reading dial micrometer attachment. (Courtesy of Sargent-Welch.)

4. Carefully add another 1-kg mass to the load, and record, in the appropriate data table, the scale reading seen through the telescope or the micrometer reading when again brought into level balance. The reading should not be taken immediately after the load is increased. A short time is required for the wire to stretch to equilibrium; this is called **elastic lag.**

 Caution: *If the wire is rusted, kinked, or in some way weakened, it could break and possibly damage*

the eyes. It is advisable to wear safety goggles when adding weights.

5. Add 1-kg masses, one at a time, recording in the appropriate data table the scale reading or the micrometer reading after adding each 1-kg mass until there is a total of 10 kg on the hanger. (*Note:* These loads are appropriate for a steel wire. If another kind of wire is used, ask your instructor for the proper maximum load.)

Figure 6 Young's modulus apparatus. A great deal of weight is needed to elongate the wire appreciably. (Courtesy of Sargent-Welch.)

At some point in this procedure after a 5-kg load is on the wire, measure the diameter of the wire with the micrometer calipers, and record in the appropriate data table. With a substantial load, error resulting from kinks in the wire is avoided.

6. Remove the weights one at a time and record the scale reading or balanced micrometer reading after each weight is removed (allowing for the elastic lag). Stop when only one 1-kg mass is left on the weight hanger. At this point, measure L_o again and record.

7. Compute the elongations ΔL_i for each load from
   (a) *Optical lever method:* the average of the scale readings for increasing and decreasing loads using Eq. 8, where $\Delta y_i = \bar{y}_i - \bar{y}_o$ and $\bar{y}_i$ and $\bar{y}_o$ are the averages of two scale readings (increasing and decreasing). That is, $\bar{y}_o$ is subtracted from each of the other average scale readings.
   (b) *Micrometer method:* the average micrometer readings for increasing and decreasing loads, $\Delta L_i = \bar{L}_i - \bar{L}'_o$, where $\bar{L}_i$ and $\bar{L}'_o$ are the averages of two micrometer readings (increasing and decreasing). That is, $\bar{L}'_o$ is subtracted from each of the other average lengths.
   Notice in the data table that the initial 1-kg mass and the mass of the weight hanger are not to be included as part of the individual loads.

8. Compute the stress F/A and the strain $\Delta L_i/L_o$ for each load.
   Use an average L_o if the initial and final measurements are different. Plot a graph of stress versus strain. Draw a straight line that best fits the data points, and determine the slope of the line. Compare this result with the value of Young's modulus of the wire material given in Appendix A, Table A2 by calculating the percent error.

Name _____ Section _____ Date _____

Lab Partner(s) _____

Elasticity: Young's Modulus

TI/ *Laboratory Report*

d _____

D _____

Wire diameter _____

L_o (initial) _____

L_o (final) _____

DATA TABLE 1 (*Optical Lever Method*)

Purpose: To determine Young's modulus using the optical lever method.

Load ( )	Increased load scale reading ( )		Decreased load scale reading ( )		Average scale reading	Computed elongation ΔL_i ( )		Strain $\Delta L_i / L_o$	Stress $F/A = mg/\pi r^2$ ( )
*	y_o		y_o			✕		✕	✕
1	y_1		y_1			ΔL_1			
2	y_2		y_2			ΔL_2			
3	y_3		y_3			ΔL_3			
4	y_4		y_4			ΔL_4			
5	y_5		y_5			ΔL_5			
6	y_6		y_6			ΔL_6			
7	y_7		y_7			ΔL_7			
8	y_8		y_8			ΔL_8			
9	y_9		y_9			ΔL_9			

* The initial 1-kg mass and weight hanger are not included as part of the loads.

Calculations
(continue work on following page)

Slope of graph _____

Accepted Young's modulus _____

Percent error _____

Don't forget units

(continued)

Calculations

Wire diameter _____

L_o (initial) _____

DATA TABLE 2 (*Micrometer Method*)

Purpose: To determine Young's modulus using the micrometer method.

L_o (final) _____

Load ( )	Increased load micr. reading ( )		Decreased load micr. reading ( )		Average reading	Elongation $\Delta L_i = \bar{L}_i - \bar{L}'_o$ ( )		Strain $\Delta L_i/L_o$	Stress $F/A = mg/\pi r^2$ ( )
*	L'_o		L'_o						
1	L_1		L_1			ΔL_1			
2	L_2		L_2			ΔL_2			
3	L_3		L_3			ΔL_3			
4	L_4		L_4			ΔL_4			
5	L_5		L_5			ΔL_5			
6	L_6		L_6			ΔL_6			
7	L_7		L_7			ΔL_7			
8	L_8		L_8			ΔL_8			
9	L_9		L_9			ΔL_9			

* The initial 1-kg mass and weight hanger are not included as part of the loads.

EXPERIMENT 31 · *Laboratory Report*

Calculations
(show work)

Slope of graph _____

Accepted Young's modulus _____

Percent error _____

/TI/ QUESTIONS

1. (a) Why is it proper to take L_o as the length of wire between the chucks rather than the total length of the wire?

   (b) Why aren't the initial 1-kg mass and weight hanger included as part of the load?

(continued)

2. Suppose that the stress-versus-strain graph for another wire had a steeper slope than that of the wire used in this experiment. What would this mean physically?

3. Figure 2 shows the general stress-versus-strain curve for an elongating stress on a wire or block of material. What would the curve look like for a compressive stress on a block?

4. (a) Suppose that the length of the wire between the chucks L_o were twice the length used in the experiment (same wire). Would a particular load produce the same elongation as found in the experiment? If not, what would it be?

   (b) Suppose that the experiment were repeated (same L_o) with a wire of the same material, but with a diameter twice that of the wire used in the experiment. Would a particular load produce the same corresponding elongation as found in the experiment? If not, what would it be?

EXPERIMENT 31   *Laboratory Report*

5. What should happen to the diameter of the wire as the load on the wire is increased? Why?

6. *Optional (optical lever method)* Prove explicitly that when a mirror is rotated (tilted) through an angle θ, a reflected beam is deflected by an angle 2θ (see Fig. 3). *Hint:* Draw reflected rays for the same incident ray for each orientation.

7. *(Optional)* In the interest of speed and simplicity, "raw data" are often recorded directly on a graph. In this experiment the "increasing load" data might have been recorded on a graph where x = load = total mass (2 kg to 11 kg) and y = actual micrometer reading (in mm). Draw such a graph and use it to compute Y. Comment on the speed and validity of this method.

EXPERIMENT 31

Air Column Resonance:
The Speed of Sound in Air

 *Advance Study Assignment*

Read the experiment and answer the following questions.

1. What causes resonance in a pipe, and how is a resonance condition detected in the experimental setup?

2. Why are only odd harmonics observed for the resonance tube apparatus?

3. How does the speed of sound vary with temperature?

4. Should a tuning fork be set into oscillation by striking it with or on a hard object? Explain.

(continued)

5. For a resonance tube apparatus with a total tube length of 1 m, how many resonance positions would be observed when the water level was lowered through the total length of the tube for a tuning fork with a frequency of (a) 500 Hz, (b) 1000 Hz? (Show your calculations.)

Air Column Resonance: The Speed of Sound in Air

INTRODUCTION AND OBJECTIVES

Systems generally have one or more natural vibrating frequencies. When a system is driven at a natural frequency, there is a maximum energy transfer and the vibrational amplitude increases to a maximum. For example, when you push a person on a swing, if you apply the pushes at the proper frequency, there will be maximum energy transfer and the person will swing higher (greater amplitude). When a system is driven at a natural frequency, we say that the system is in **resonance** (with the driving source) and refer to the particular frequency at which this occurs as a **resonance frequency.**

A swing or pendulum has only one natural frequency. However, a stretched string and an air column can have several natural frequencies, depending on the number of wavelength segments that can be "fitted" into the system length. From the relationship among the frequency f, the wavelength λ, and the wave speed v, which is $\lambda f = v$, it can be seen that if the frequency and wavelength are known, the wave speed can be determined. And if the wavelength and wave speed are known, the frequency can be determined.

As an application of resonance, in this experiment the speed of sound in air will be determined by driving an air column in resonance. Also, knowing the speed of sound, we will find the unknown frequency of a tuning fork from resonance conditions.

After performing this experiment and analyzing the data, you should be able to:

1. Explain why a closed organ pipe has particular resonant frequencies.
2. Tell how the speed of sound in air is affected by temperature and what effect this has on air column resonance.
3. Describe how a resonance tube can be used to measure the speed of sound in air or to determine the unknown frequency of a tuning fork.

EQUIPMENT NEEDED

- Resonance tube apparatus
- Three tuning forks (500- to 1000-Hz range), with stamped frequency of one fork covered or masked so as to serve as an unknown. It is helpful to have known-frequency tuning forks with frequencies different from those being used by neighboring groups or to have nearby groups alternate their use of frequencies so that the resonances coming from a particular tube can be distinguished.
- Rubber mallet or block
- Meterstick (if there is no measurement scale on the resonance tube)
- Thermometer (to determine air temperature—one per class is sufficient)
- Vernier calipers
- Rubber bands (3 or 4)

THEORY

Air columns in pipes or tubes of fixed lengths have particular resonant frequencies. For example, in a closed organ pipe (closed at one end*) of length L, when the air column is driven at particular frequencies, it vibrates in resonance. The interference of the waves traveling down the tube and the reflected waves traveling up the tube produces (longitudinal) standing waves, which must have a node at the closed end of the tube and an antinode at the open end

* A closed tube or organ pipe is open at one end. An open pipe is open at both ends.

● Fig. 1. This corresponds to transverse standing waves in a metal strip or rod fixed at one end.

Keep in mind that the waves in the air column are longitudinal—particle displacement is parallel to the direction of wave propagation. The maximum *longitudinal* displacements of the air molecules are plotted (transversely) in the drawings and show the close relationship between standing waves in a column of air and standing waves in a string. (Review standing waves in a stretched string in your textbook.)

The resonance frequencies of a pipe or tube (air column) depend on its length L. As illustrated in Fig. 1, only a certain number of wavelengths, or "loops," can be

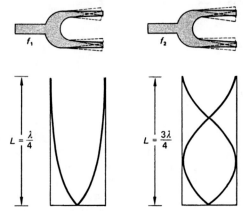

Figure 1   Standing waves.   Illustrations of longitudinal standing waves of different frequencies in tubes of fixed length. There must be a node at the closed end and an antinode at the open end of the tube.

"fitted" into the tube length with the node-antinode requirements (only the two lowest-frequency waves or allowed wavelength loops are shown in the figure).

Since each loop corresponds to one half wavelength, resonance occurs when the length of the tube is nearly equal to an odd number of quarter wavelengths* (that is, $L = \lambda/4$, $L = 3\lambda/4$, $L = 5\lambda/4$, etc.) or in general $L = n\lambda/4$, where $n = 1, 3, 5, \ldots$, and $\lambda = 4L/n$. Incorporating the frequency f and speed v_s through the general relationship $\lambda f = v$, or $f = v/\lambda$, we have

$$f_n = \frac{nv}{4L} \qquad n = 1, 3, 5, \ldots \qquad (1)$$

Hence, an air column (tube) of length L has particular resonance frequencies and will be in resonance with the corresponding odd-harmonic driving frequencies.

As in the case of standing waves in a string, the lowest natural or resonant frequency f_1 is called the fundamental frequency or first harmonic. Successively higher frequencies are higher harmonics and are sometimes called overtones (f_2, the second harmonic, is the first overtone, etc.).

As can be seen from Eq. 1, the three experimental parameters involved in the resonance condition of an air column are f, v, and L. To study resonance in this experiment, we will vary the length L of an air column for a given driving frequency, instead of varying f for a fixed L as in the case of the closed organ pipe described above. (The wave speed in air is relatively constant in either case.) The length of an air column will be varied by raising and lowering the water level in a tube (● Fig. 2).

As the length of the air column is increased, more wavelength segments will fit into the tube, consistent with the node-antinode requirements at the ends (● Fig. 3).

* The antinode does not occur exactly at the open end of the tube but at a slight distance above it, a distance that depends on the diameter of the tube.

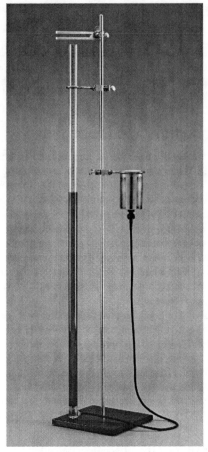

Figure 2   A resonance tube apparatus.   The "length" of the tube is varied by raising and lowering the reservoir can to adjust the water level in the tube.

The difference in the tube (air column) lengths when successive antinodes are at the open end of the tube and resonance occurs is equal to a half wavelength; for example,

$$\Delta L = L_2 - L_1 = \frac{3\lambda}{4} - \frac{\lambda}{4} = \frac{\lambda}{2}$$

(for the case illustrated in Fig. 3), and

$$\Delta L = L_3 - L_2 = \frac{5\lambda}{4} - \frac{3\lambda}{4} = \frac{\lambda}{2}$$

for the next resonance.

When an antinode is at the open end of the tube, a loud resonance tone is heard. Hence, the tube lengths for antinodes to be at the open end of the tube can be determined by lowering the water level in the tube and "listening" for successive resonances. No end correction is needed for the antinode occurring slightly above the end of the tube in this case, since the differences in tube lengths for successive antinodes are equal to $\lambda/2$.

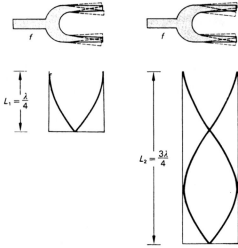

Figure 3   Standing waves.   Illustrations of longitudinal standing waves of the same frequency in tubes of different lengths.

If the frequency f of the driving tuning fork is known and the wavelength is determined by measuring the difference in tube length between successive antinodes, $\Delta L = \lambda/2$, or $\lambda = 2\Delta L$, the speed of sound in air v_s can be determined from

$$v_s = \lambda f \qquad (2)$$

Once v_s is determined (for a given ambient temperature), the unknown frequency of a tuning fork can be computed from Eq. 1 by using the experimentally determined resonance wavelength in an air column for the unknown tuning fork frequency.

The speed of sound in air is temperature-dependent and is given to a good approximation over the normal temperature range by

$$v_s = (331.5 + 0.6T_c) \text{ m/s} \qquad (3)$$

where T_c is the air temperature in degrees Celsius. The equation shows that the speed of sound at 0°C is 331.5 m/s and increases by 0.6 m/s for each degree of temperature increase over a normal temperature range. For example, if the air temperature is $T_c = 20$°C (room temperature, 68°F), then

$$v_s = [331.5 + (0.6)(20)] \text{ m/s} = 343.5 \text{ m/s}$$

EXPERIMENTAL PROCEDURE

1. A resonance tube apparatus is shown in Fig. 2. Measure the inside diameter (I.D.) of the tube with a vernier caliper, and record in the data table, along with the room temperature.

   Place several rubber bands around the tube at various heights. (You will use the rubber bands later to

mark resonance positions.) Then raise the water level to near the top of the tube by raising the reservoir can. This can be done by depressing the can clamp and sliding it on the support rod. However, it may be more convenient to remove the reservoir can from the support rod so that it can be raised and lowered freely.

With the water level near the top of the tube, there should be little water in the reservoir can. If this is not the case, remove some water from the can to prevent overflow and spilling when the can becomes filled on lowering. Practice lowering and raising the water level in the tube to get the "feel" of the apparatus.

2. With the water level in the tube near the top, take one of the tuning forks of known frequency (it is usually stamped on the fork) and set it into oscillation by striking it with a rubber mallet or on a rubber block, whichever is available. *Never strike the tuning fork on a hard object* (e.g., a table). This may damage the fork and cause a change in its characteristic frequency.*

   Hold the vibrating fork slightly above the top of the tube so that the sound is directed down the tube. (A fork has directional sound-propagation characteristics. Experiment with a vibrating fork and your ear to determine these directional characteristics.) *If the tube is glass, be careful not to let the vibrating fork strike the top of the tube.* This may chip the tube. Some resonance tubes are made of plastic.

3. With the vibrating tuning fork above the tube, *quickly* lower the reservoir can. The water level in the tube will fall slowly, and successive resonances will be heard as the level passes through the resonance-length positions. It may be necessary to strike the tuning fork several times to keep it vibrating sufficiently.

   As the water level passes through successive resonance positions, mark the approximate position of each by sliding a rubber band down to that position. Repeat, moving the water both up and down and adjusting the rubber band heights as needed, until you are satisfied that the rubber bands are at the heights for the water levels where resonances are obtained.

   It is convenient to have one lab partner raise and lower the can and determine the resonance positions while another lab partner handles the tuning fork.

4. Determine the length from the top of the tube for the first resonance condition, and record this length in the data table. Repeat this procedure for the other observed resonance positions. (The number of observed lengths may be less than provided for in the data table.)

---

* It may be helpful to touch the vibrating tuning fork briefly about one-third of the way up from the base of the fork to dampen out the tinny sounds caused by the fork's overtones (resonant frequencies $> f_1$).

5. Repeat procedures 3 and 4 for the other tuning fork of known frequency.

6. (a) Compute the average wavelength for each fork from the average of the differences in the tube lengths between successive antinodes.
   (b) Using the known frequency of each fork, compute the speed of sound for each case.
   (c) Compare the average of these two experimental values with the value of the speed of sound given by Eq. 3 by computing the percent error.*

* If there is a large percent error and the average experimental value is a multiple of the accepted value, it may be that you have observed the resonances of the tuning forks' overtone frequencies. Consult your instructor in this case.

7. Repeat procedures 3 and 4 for the tuning fork of unknown frequency.

8. Compute the frequency of the tuning fork using the average experimental value of the wavelength and the speed of sound as given by Eq. 3. Your instructor may supply you with the known frequency of the tuning fork. If so, compare your experimentally determined value with the known frequency by computing the percent error.

EXPERIMENT 32

Air Column Resonance:
The Speed of Sound in Air
/TI/ *Laboratory Report*

DATA TABLE

Purpose: To determine the speed of sound and the
unknown frequency of a tuning fork.

Temperature _____

Inside diameter of tube, I.D. _____

	Tuning fork 1		Tuning fork 2		Tuning fork 3	
	Frequency f _____		Frequency f _____		Unknown frequency	
	Position of resonance ( )	ΔL ( )	Position of resonance	ΔL ( )	Position of resonance	ΔL ( )
L_1						
L_2						
L_3						
L_4						
Average ΔL						
Average λ						

Calculations
(show work)

1. Speed of sound v_s

v_s (from fork 1 data) _____

v_s (from fork 2 data) _____

Average v_s _____

v_s (from Eq. 3) _____

Percent error _____

Don't forget units

(continued)

2. Frequency of tuning fork

Experimental value of f _____

Known value _____

Percent error _____

/TI/ QUESTIONS

1. Suppose that the laboratory temperature were 5°C higher than the temperature at which you performed the experiment. Explain what effect(s) this would have on the experimental results.

2. Say the water in the tube was at the level for the tube length L_1 of the first resonance of the first tuning fork.
   (a) Could another tuning fork with a frequency lower than that of the first tuning fork produce a resonance?

   (b) Could another tuning fork of some higher frequency produce a resonance?

   (c) If your answer is yes in either case, what would be the frequency of the other fork?

3. What are the resonant frequencies for an open organ pipe of fixed length L? (Show your sketches and calculations.)

4. (*Optional*) The uppermost antinode (Fig. 1) is slightly above the open end of the tube by an amount E (called the *end correction*).
   (a) Compute E from the data for each tuning fork.

   (b) Compare the computed E's with the theoretical values given by the equation $E = 0.4 \times$ I.D., where I.D. is the inner diameter of the tube.

EXPERIMENT 32

466

Name _____ Section _____ Date _____

Lab Partner(s) _____

EXPERIMENT 33

Latent Heats: Heats of Fusion and Vaporization of Water

◢ Latent Heat of Fusion of Water

[TI] *Advance Study Assignment*

Read the experiment and answer the following questions.

1. What is latent heat?

2. The heat of vaporization of water is almost seven times its heat of fusion. What does this imply?

3. Why is the water used in the experimental procedures for the heat of fusion and heat of vaporization initially heated above room temperature and cooled below room temperature?

(continued)

467

4. Why are the pieces of ice in the heat-of-fusion procedure dried and handled with a paper towel? Explain the effect on the experimental result if this were not done.

5. What is the purpose of the water trap in the steam line in the heat-of-vaporization procedure? Explain the effect on the experimental result if it were not used.

6. Explain how latent heat is computed from the experimental data.

◾ *Advance Study Assignment*

Read the experiment and answer the following questions.

1. What is the definition of the latent heat of fusion of water?

2. Describe the mixture that will be made in this experiment by listing all the ingredients.

EXPERIMENT 33   *Advance Study Assignment*

3. What ingredients of the mixture will lose heat?

4. What ingredients of the mixture will gain heat?

EXPERIMENT 33

/TI/ Latent Heats: Heats of Fusion and Vaporization of Water

CI Latent Heat of Fusion of Water

OVERVIEW

Experiment examines the latent heats of water. The TI procedure considers both the latent heat of fusion and the latent heat of vaporization. The CI procedure considers only the latent heat of fusion. Calorimeter methods are used in both procedures, and the temperature in the CI procedure is measured by a temperature sensor.

(An addendum at the end of this experiment gives a TI procedure for the calibration of a thermometer. It is a short but interesting procedure that enhances an understanding of calibration.)

INTRODUCTION AND OBJECTIVES

When heat is added to a substance, its temperature normally rises. However, when a substance undergoes a change of phase (for example, solid to liquid or liquid to gas), the heat energy goes into doing work against the intermolecular forces and is not reflected in a change in the temperature of the substance. This heat energy is called the **heat of fusion** and the **heat of vaporization** for the phase changes that occur at the melting- (or freezing-) point temperature and boiling- (or condensation-) point temperature, respectively.

The energy involved in a phase change is commonly referred to as **latent heat,** because the heat energy is seemingly hidden or concealed in that it is not evidenced by a temperature change. For the inverse processes, when a vapor or gas condenses or a liquid freezes, by the conservation of energy the (latent) heat of vaporization is given up or the (latent) heat of fusion must be extracted.

In this experiment, the heats of fusion and vaporization of water will be determined through the calorimetry method of mixtures.

/TI/ OBJECTIVES

After performing this experiment and analyzing the data, you should be able to:

1. Clearly distinguish between latent heat and specific heat. (Consult your text book if necessary.)
2. Explain the role of latent heat in a phase change.
3. Describe how latent heat can be determined experimentally.

CI OBJECTIVES

After performing this experiment and analyzing the data, you should be able to:

1. Understand and use calorimetry methods.
2. Experimentally determine the latent heat of fusion of water.

<u>TI</u> EQUIPMENT NEEDED

- Calorimeter
- Steam generator and stand
- Hot plate or Bunsen burner and striker
- Thermometer (0 to 110°C)
- Rubber hose

- Water trap
- Ice
- Paper towels
- Laboratory balance
- Beaker
- Safety glasses

<u>TI</u> THEORY

A general graph of temperature versus heat energy for a given mass m of water is shown in ● TI Fig. 1. The sloping phase lines follow the relationship

$$\Delta Q = mc\,\Delta T \qquad \textbf{(TI 1)}$$

The slopes of the lines are mc, where c is the specific heat of the particular phase:

$c_{water} = 1.0$ kcal/kg-C°,   or   1.0 cal/g-C°

$c_{ice} \approx c_{steam} = 0.5$ kcal/kg-C°,   or   0.5 cal/g-C°.

Actually,

$$c_{ice} = 0.50 \text{ kcal/kg-C}°$$
$$c_{steam} = 0.48 \text{ kcal/kg-C}°.$$

At the freezing and boiling points, the addition of certain amounts of heat has no effect on the temperature, as indicated by the vertical lines. At these points, the heat energy goes into the work of effecting the phase changes and not into increasing the molecular activity, which would be reflected as a temperature increase.

The **latent heat** L is defined as the amount of heat required to change the phase of a unit mass of a substance (without a change of temperature) and has the common units of kcal/kg (or cal/g) and J/kg in the SI system. Hence the amount of heat absorbed or given up by a quantity of a substance when it undergoes a change of phase is

$$\Delta Q = mL_i \qquad \textbf{(TI 2)}$$

where m is the mass of the substance and L_i is the latent heat for the particular phase change.

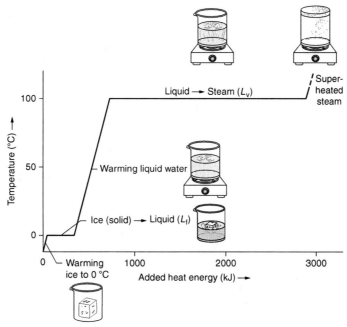

TI Figure 1   Graph of temperature versus heat for water.   As heat is added to the various phases of water, the temperature increases. However, during a phase change (horizontal lines), the temperature remains constant.

The latent heats are, of course, characteristic of the substance. For example, the **(latent) heat of fusion** L_f of water is 80 kcal/kg (or 80 cal/g) or 3.33×10^5 J/kg, and its **(latent) heat of vaporization** L_v is 540 kcal/kg (or 540 cal/g) or 2.26×10^6 J/kg. This means that 80 kcal is required to melt 1 kg of ice at 0°C (or 80 kcal must be removed from 1 kg of water at 0°C to freeze it). Also, 540 kcal is required to convert 1 kg of water at 100°C to steam (or 540 kcal is released when 1 kg of steam at 100°C condenses to water). Other substances have different latent heats.

In this experiment, the heats of fusion and vaporization of water will be measured by a calorimetry procedure called the methods of mixtures (see Experiment 15). If several substances at various temperatures are brought together, the hotter substances lose heat and the colder substances gain heat until all the substances reach a common equilibrium temperature.

This is generally true even if one or more of the substances undergoes a phase change in the process. If the system is insulated and no heat is lost to the surroundings, then by the conservation of energy, the heat lost by the hot substances will equal the heat gained by the colder substances.

A. Heat of Fusion

If a quantity of ice of mass m_i at 0°C is added to a sufficient quantity of warm water in a calorimeter cup with a stirrer at an initial temperature T_h, the ice melts and the system comes to equilibrium at some intermediate or final temperature T_f. (The calorimeter insulates the system against heat loss. See TI Fig. 15.1.) Then, by the conservation of energy,

$$\begin{matrix} \text{Heat of fusion} \\ \text{to melt the ice} \end{matrix} + \begin{matrix} \text{heat gained by the} \\ \text{ice water} \end{matrix}$$

$$= \begin{matrix} \text{heat lost by the} \\ \text{warm water} \end{matrix} + \begin{matrix} \text{heat lost by the} \\ \text{calorimeter cup and stirrer} \end{matrix}$$

or

$$\Delta Q_\text{fusion} + \Delta Q_\text{ice water} = \Delta Q_{\substack{\text{warm} \\ \text{water}}} + \Delta Q_{\substack{\text{calorimeter cup} \\ \text{and stirrer}}}$$

and

$$m_i L_f + m_i c_w(T_f - 0)$$
$$= m_w c_w(T_h - T_f) + m_{cs} c_{cs}(T_h - T_f)$$
$$= (m_w c_w + m_{cs} c_{cs})(T_h - T_f) \tag{TI 3}$$

where the m's and c's are the masses and specific heats, respectively, of the various components as indicated by the subscripts i = ice, w = water, cs = calorimeter-stirrer. Hence, TI Eq. 3 may be used to determine the heat of fusion L_f of water if all the other quantities are known.

B. Heat of Vaporization

If a quantity of steam of mass m_s at 100°C is added to cool water in a calorimeter cup with a stirrer at an initial temperature T_c, the steam condenses and the system comes to equilibrium at some intermediate or final temperature T_f. Assuming no heat loss to the surroundings, by the conservation of energy,

$$\begin{matrix} \text{Heat of vaporization lost} \\ \text{by steam in condensing} \end{matrix} + \begin{matrix} \text{heat lost by hot} \\ \text{condensed water} \end{matrix}$$

$$= \begin{matrix} \text{heat gained by the} \\ \text{cool water} \end{matrix} + \begin{matrix} \text{heat gained by the} \\ \text{calorimeter cup and stirrer} \end{matrix}$$

or

$$\Delta Q_\text{vaporization} + \Delta Q_{\substack{\text{hot} \\ \text{water}}} = \Delta Q_{\substack{\text{cool} \\ \text{water}}} + \Delta Q_{\substack{\text{calorimeter cup} \\ \text{and stirrer}}}$$

and

$$m_s L_v + m_s c_w(100 - T_f)$$
$$= m_w c_w(T_f - T_c) + m_{cs} c_{cs}(T_f - T_c)$$
$$= (m_w c_w + m_{cs} c_{cs})(T_f - T_c) \tag{TI 4}$$

where the m's and c's are the masses and specific heats, respectively, of the various components, as indicated by the subscripts. Hence, CI Eq. 4 may be used to determine the heat of vaporization L_v of water if all the other quantities are known.

/TI/ EXPERIMENTAL PROCEDURE

A. Heat of Fusion

1. Heat some water in the beaker (enough to fill the inner calorimeter cup about half full) to about 10 to 15°C above room temperature. While the water is heating, determine the mass of the inner calorimeter cup (without ring) and stirrer (total mass m_{cs}) on a laboratory balance. Also, note and record the type of metal and specific heat of the cup and stirrer, usually stamped on the cup.*

2. Fill the inner calorimeter cup about half full of the warm water and weigh it (with the stirrer) to determine the mass of the water m_w.

   Place the calorimeter cup with the water and stirrer in the calorimeter jacket and put on the lid with a thermometer extending into the water. Stir the water gently and record its temperature T_h.

3. Select several small pieces of ice about the size of the end of your thumb and dry them with a paper towel. (It is important that the ice be dry.) Without touching the ice with your bare fingers (use the paper towel), carefully add the pieces of ice to the calorimeter cup one at a time without splashing. (It is good procedure to stir

---

* If the cup and stirrer are not of the same material, they must be treated separately, and the mass term in TI Eqs. 3 and 4 becomes

$$(m_w c_w + m_c c_c + m_s c_s).$$

the water and check its temperature again just before adding the ice. If that temperature has changed, take the later reading as T_h.)

4. Gently stir the water-ice mixture while adding the ice. Add enough ice so that the temperature of the mixture is about 10 to 15°C below room temperature after the ice has melted. Add the ice more slowly toward the end so that you can better control the final temperature. Continue to stir gently, and read and record the equilibrium temperature T_f when the ice has melted completely.

   Then weigh and record the mass of the inner calorimeter cup with its contents (water and stirrer) so as to determine the mass of the melted ice water or mass of the ice m_i.

5. Compute the heat of fusion L_f using TI Eq. 3, and compare it with the accepted value, $L_f = 80$ kcal/kg = 80 cal/g = 3.33×10^5 J/kg, by finding the percent error.

B. Heat of Vaporization

6. Set up the steam generator as shown in ● TI Fig. 2 with the boiler about two-thirds full of water, and begin heating the boiler. A water trap is used in the steam line (which should be as short as practically pos-

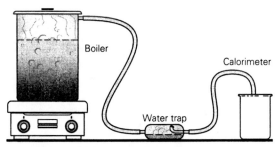

TI Figure 2 Steam generator arrangement.   For the heat-of-vaporization measurement, steam from boiling water is transferred to the calorimeter through a water trap. The trap prevents hot water condensed in the tube from entering the calorimeter, which would introduce error.

sible) to prevent hot water condensed in the tube from entering the calorimeter.

It is important that the hot water condensed in the tube be prevented from entering the calorimeter—note the hose arrangement in TI Fig. 2.

7. Fill the inner calorimeter cup about two-thirds full of cool water at about 15°C below room temperature. Either the cool water from the heat-of-fusion experiment or fresh tap water may be used with some ice added to obtain the desired temperature. (Make sure that all the ice has melted, however.)

   Weigh and record the mass of the calorimeter cup with the water and stirrer. Then place the cup in the calorimeter jacket and replace the lid. Gently stir the water, and read and record the equilibrium temperature T_c.

8. With the water in the steam generator boiling gently and steam flowing freely from the steam tube, as evidenced by water vapor coming from the tube (remember that steam itself is invisible), introduce steam into the calorimeter water and stir.

   Caution: Be careful not to get a steam burn when placing the steam line into, or removing it from, the cup.

   It is good procedure to stir the water and check its temperature again just before introducing the steam. If that temperature has changed, take the later reading as T_c.

   When the temperature of the water is about 15°C above room temperature, turn off the heat source and carefully remove the steam line from the calorimeter. Stir gently, and read and record the equilibrium temperature T_f.

   Finally, reweigh the inner calorimeter cup and its contents so as to determine the mass of the condensed steam m_s.

9. Compute the heat of vaporization L_v using TI Eq. 4 and compare it with the accepted value $L_v = 540$ kcal/kg = 540 cal/g = 2.26×10^6 J/kg by finding the percent error.

Name _____ Section _____ Date _____

Lab Partner(s) _____

EXPERIMENT 33

Latent Heats: Heats of Fusion and Vaporization of Water

/TI/ *Laboratory Report*

Mass of calorimeter
cup and stirrer m_{cs} _____

A. Heat of Fusion

Mass of calorimeter
plus water _____

Mass of water m_w _____

Mass of calorimeter,
water, and melted ice _____

Mass of ice m_i _____

Calculations
(show work)

Type of metal and specific
heat of calorimeter cup
and stirrer c_{cs} _____

Initial temperature of
water T_h _____

Final equilibrium
temperature T_f _____

Experimental L_f _____

Accepted L_f _____

Percent error _____

Don't forget units

(continued)

B. Heat of Vaporization

Mass of calorimeter
plus water _____

Mass of water m_w _____

Mass of calorimeter,
water, and condensed
steam _____

Mass of steam m_s _____

Initial temperature of
water T_c _____

Final equilibrium
temperature T_f _____

Calculations
(show work)

Experimental L_v _____

Accepted L_v _____

Percent error _____

EXPERIMENT 33   *Laboratory Report*

/TI/ QUESTIONS

1. Discuss the most important sources of error in the experimental procedures. If the temperature of the ice added to the calorimeter in the heat-of-fusion experiment were less than 0°C, how would this affect the results? Why should you use small pieces of ice rather than large pieces?

2. Explain why burns caused by steam at 100°C are more serious than those caused by an equal mass of boiling water at the same temperature.

3. How is the latent heat of water used in the cooling mechanism of our bodies?

4. What is the heat of sublimation? [*Hint:* Dry ice (solid CO_2) sublimes.]

(continued)

5. A piece of ice with a mass of 30 g at 0°C is added to 100 mL of water at 20°C. Assuming that no heat is lost to the surroundings, what is the situation when thermal equilibrium is reached? (Ignore the calorimeter or container.)

- 1 Temperature sensor (PASCO CI-6505A)
- Calorimeter

- Hot water ($\sim$ 50° to 60°C)
- Ice
- Paper towels

CI THEORY

In this experiment, we will measure the latent heat of fusion of water by finding how many calories of heat are needed to melt 1 g of ice completely. A temperature sensor will be used to monitor the temperature changes of a mixture of warm water and ice. The water (in a calorimeter cup) will lose heat when mixed with the ice. This heat will be used in melting the ice and in warming up the resulting water-ice mixture. By the conservation of energy, the heat lost by one part of the system must be equal to the heat gained by other parts of the system, if the system is isolated.

In this case, the amount of heat transferred will be determined by measuring the heat lost by the water and the calorimeter cup, as follows:

$$Q = Q_{\substack{\text{lost by} \\ \text{water}}} + Q_{\substack{\text{lost by} \\ \text{cup}}} \qquad \text{(CI 1)}$$
$$= m_{\text{water}} c_{\text{water}} \Delta T_{\text{water}} + m_{\text{cup}} c_{\text{cup}} \Delta T_{\text{cup}}$$

Assuming that the water and the calorimeter cup both start at the same initial temperature, and assuming that they both reach the same final, equilibrium temperature as the rest of the mixture, $\Delta T_{\text{water}} = \Delta T_{\text{cup}}$ and

$$Q = (m_{\text{water}} c_{\text{water}} + m_{\text{cup}} c_{\text{cup}}) \Delta T \qquad \text{(CI 2)}$$
$$= (m_{\text{water}} c_{\text{water}} + m_{\text{cup}} c_{\text{cup}})(T_{\text{initial}} - T_{\text{final}})$$

To determine how much of this heat goes into melting the ice, let's examine the other side of the heat transfer. When the ice melts and the resulting water-ice mixture warms up, the heat that is transferred can be expressed as

$$Q = Q_{\substack{\text{to melt} \\ \text{the ice}}} + Q_{\substack{\text{gained by} \\ \text{ice water}}}$$
$$= Q_{\substack{\text{to melt} \\ \text{the ice}}} + m_{\text{ice}} c_{\text{water}} \Delta T_{\text{ice}} \qquad \text{(CI 3)}$$
$$= Q_{\substack{\text{to melt} \\ \text{the ice}}} + m_{\text{ice}} c_{\text{water}} (T_{\text{final}} - 0°\text{C})$$

Note that the initial temperature of the ice is 0°C but that the final temperature of the resulting water-ice mixture, which has the same mass, m_{ice}, as the original ice, is the same final equilibrium temperature of the water and cup.

By conservation of energy, the heat calculated with CI Eq. 2 is the same as that calculated with CI Eq. 3.

Combining both expressions, we can find the heat needed to melt the ice completely:

$$Q_{\substack{\text{to melt} \\ \text{the ice}}} = Q - m_{\text{ice}} c_{\text{water}} T_{\text{final}} \qquad \text{(CI 4)}$$

For a particular mass of ice, m_{ice}, the heat used to melt each gram of the sample is easily calculated. This quantity is called the latent heat of fusion of water:

$$L_{\text{fusion}} = \frac{Q_{\substack{\text{to melt} \\ \text{the ice}}}}{m_{\text{ice}}} \qquad \text{(CI 5)}$$

SETTING UP DATA STUDIO

1. Open Data Studio and choose "Create Experiment."
2. From the sensor list, choose a temperature sensor. Connect the temperature sensor to channel A of the interface, as shown in the computer screen.
3. Double-click on the "Temperature Sensor" icon. The sensor properties window will open. Set the sample rate to slow, and set the timer to 5 s between samples. Click OK.
4. Drag the "Temperature, ChA" icon from the data list, and drop it on top of the "Digits" icon in the displays list. A digits display will open in a window called Digits 1.
5. Drag the "Temperature, ChA" icon from the data list, and drop it on top of the "Graph" icon in the displays list. A graph of temperature versus time will open in a window called Graph 1.
6. There should be a digits display and a graph on the screen. Adjust their sizes and locations so that they can be seen simultaneously. ● CI Fig. 1 shows what the screen should look like after the setup is complete.

CI EXPERIMENTAL PROCEDURE

1. Measure the mass of the calorimeter cup, and record it in the laboratory report.*
2. Make a note, in the report, of the specific heat of the calorimeter cup, including units. It is usually printed on the cup by the manufacturer.
3. Set aside about 50 g of ice.† The ice has to be in pieces small enough to fit through the calorimeter lid. Wrap

---

* If the calorimeter comes with a stirrer, remove the stirrer. The temperature sensor will be the stirrer in this experiment.
† Anything from 40 g to 70 g will work fine; 50 g is just a suggestion.

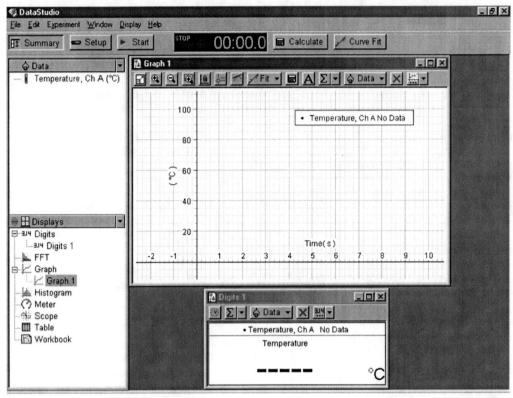

CI Figure 1   Data studio setup.   The temperature of the water and ice mixture will be displayed simultaneously on a graph and on a digits display. A new data point will appear every 5 s. (Data displayed using DataStudio Software. Reprinted courtesy of PASCO scientific.)

the ice with paper towels to keep it dry until you are ready to use it. (Alternatively, keep it in the freezer or in a cooler until just before you use it.)

4. Put about 200 g of warm water (50 to 70° C) in the cup. Measure the mass of the cup with the water, and determine the mass of the water. Record it in CI Data Table 1.

5. Place the cup with the water in the calorimeter and put the lid on. Insert the temperature sensor into the cup through the hole in the lid, and gently stir the water for a few seconds with the sensor. This is to allow both the water and the cup to reach the same temperature.

6. While gently stirring the water, press the START button. (Keep stirring the water throughout the experiment.) The temperature readings will appear on the digits display and on the graph. Remember that a new data point for the graph will appear every 5 s.

7. Press the Scale-to-Fit button. This is the leftmost button on the graph toolbar. This will help scale the graph as data are collected.

8. If the temperature is observed to increase slightly, wait until the readings level off before going to the next step.

9. Add a few pieces of the ice to the water and continue stirring until it melts. Continue adding the ice and stirring until all the ice has melted.

10. Continue stirring for a few seconds after all the ice has melted, and observe the graph. The temperature plot should level out as the mixture reaches the equilibrium temperature. Press STOP.

11. Go to the digits display. On the toolbar there is a drop menu labeled with the Greek letter sigma (Σ). Choose "Maximum" from the menu. The display will show what was the maximum temperature measured during the experiment. Record this temperature as the initial temperature of the water and the cup.

12. Repeat, this time choosing "Minimum." Record the minimum temperature as the final temperature of the mixture. It should correspond to the temperature at which the plot leveled out at the end.

13. Carefully take the water cup out of the calorimeter and measure its mass again, with all of its contents. Record this value in CI Data Table 1.

14. Use the measured mass of the cup, the mass of the water, and the mass of the cup with water and with melted ice to determine the mass of ice, m_{ice}. Record it in CI Data Table 1.

15. Use CI Eq. 2 to determine the amount of heat lost by the water and the cup during the experiment.

16. Use CI Eq. 4 to calculate how much of the heat was used to melt the ice.
17. Use CI Eq. 5 to calculate the heat used per gram of water, L_f.
18. Repeat the experiment, this time using a little less water (or a little more water) and a little less ice (or a little more ice). Also vary the initial temperature of the water. Record all results and calculations as Trial 2 in the Laboratory Report.
19. Average the results for the latent heat of fusion from Trial 1 and Trial 2. Compare the average to the accepted value of the heat of fusion, 79.7 cal/g, by taking the percent error.

Name _____ Section _____ Date _____
Lab Partner(s) _____

EXPERIMENT 33

Latent Heat of Fusion of Water

CI *Laboratory Report*

CI DATA TABLE 1

Purpose: To determine the latent heat of fusion of water.

$$c_{cup} = \underline{\hspace{1cm}} \qquad c_{water} = \frac{1.0 \text{ cal}}{g \cdot C°}$$

	Trial 1	Trial 2
m_{cup}		
$m_{cup + water}$		
m_{water}		
$m_{cup + water + icewater}$		
m_{ice}		
$T_{initial}$		
T_{final}		
Heat lost by the water and the cup, Q		
Heat used to melt the ice, $Q_{to\ melt}$		
Latent heat of fusion of water, L_f		

Average L_f _____ Percent error _____

Don't forget units

(continued)

CI QUESTIONS

1. Was energy conserved in this experiment as theory predicts? Discuss your experimental results and the sources of uncertainty.

2. Why is it important that all the ice that is added to the water melt? How would the results of the experiment be affected if some ice remained in the mixture?

3. Why is it important that the ice be dry?

Name _____ Section _____ Date _____

Lab Partner(s) _____

/TI/ Calibration of a Thermometer

INTRODUCTION

With ice and a means to boil water available, you can quickly calibrate a thermometer, which is quite interesting and instructive.

EQUIPMENT NEEDED

- Uncalibrated thermometer (glass)
- Beaker (500 mL)
- Tripod stand and wire gauze pad (if Bunsen burner is used)

- Felt-tip pen or wax pencil
- Ruler
- Classroom Celsius thermometer

THEORY

A thermometer may be calibrated by using two fixed-point temperatures and then making a graduated scale between these points.* For example, the ice point and steam point of water are convenient fixed points (0°C or 32°F and 100°C or 212°F, respectively).

If the length between these two points is divided into 100 equal intervals, or degrees, the result corresponds with the Celsius temperature scale. Similarly, if the length between the fixed points is divided into 180 intervals, or degrees, the result corresponds to the Fahrenheit temperature scale.

By analogy, the meter standard was originally referenced to two marks on a metal bar. The length between the marks may be divided into 100 intervals, or centimeters. It could be divided into 180 intervals, but that would not produce a decimal scale.

---

* On the SI Kelvin (absolute) scale, the interval is known as the kelvin and is defined as 1/273.16 of the temperature of the triple point of water (the point where water coexists in three phases, 0.01°C and 4.58 mm Hg). Here the temperature scale fixed points are the triple point and absolute zero.

EXPERIMENTAL PROCEDURE

1. Start heating a *small* amount of water in the beaker. The water depth in the beaker should be several centimeters, enough to cover the thermometer bulb. (More water may be used, but it is unnecessary and will take longer to boil.)*

   While the water is heating, draw an outline of the thermometer on a sheet of paper. The temperature scale will be marked off on the paper, and the outline will allow the thermometer to be replaced on the paper in the proper position. (The scales on many thermometers are engraved not on the glass tube of the thermometer but on an attached frame.)

2. Place the uncalibrated thermometer into the heating water, and heat until the water boils vigorously. With

---

* The steam generator from the main experiment may also be used. Remove the cover, make sure there are several centimeters of water left in the bottom, and heat.

the pen or pencil supplied, mark the height of the thermometer's liquid column on the stem. Turn off the burner. Remove the thermometer and place it on the outline, making a mark on the paper at the mark of the fixed steam point on the thermometer.

3. Pour the hot water from the beaker into a sink.

   Caution: Use paper towels or tongs to transport the hot beaker, and pour carefully so as not to burn yourself. Recall that steam has a large latent heat.

   Add a small amount of cold water to the beaker (1–2 cm in depth) and enough ice so that the beaker is one-third to one-half full.

4. Insert the uncalibrated thermometer into the water-ice mixture and stir. The liquid column of the thermometer should reach a low stationary point after a few minutes. Make a mark on the thermometer stem at this lowest point. (Leave the thermometer in the ice water for another minute or so to make sure the column goes

487

no lower. Adjust the mark accordingly if it does.) In thermal equilibrium, ice and water coexist at the ice point.

5. Remove the thermometer, place it on the outline, and mark the ice point on the paper. Lay the thermometer aside and, with a ruler, divide the length between the two fixed points into 10 equal intervals (deka-degree or 10-degree intervals). Using the Celsius values for the fixed points, label these marks on one side of the thermometer outline.

    Then divide the interval between the 20- and 30-degree marks into 10 equal intervals, or degrees. (The same could be done for the other 10-degree intervals, giving a centigrade, or 100-degree, scale, but this is unnecessary for the purpose of the experiment.)

6. Place the uncalibrated thermometer (which should now be at room temperature) on the outline and mark and read the temperature. Then, read the room temperature from the classroom thermometer and mark and record the value on the paper scale. Compare the temperature reading on your scale with the classroom (accepted) thermometer reading by finding the percent error. Show this calculation on the outline paper.

7. On the other side of the thermometer outline, make a graduated Fahrenheit scale with a fine scale in the vicinity of the room temperature reading. Note the value of your thermometer's room temperature reading on the Fahrenheit scale and record on the paper.

    Compare the experimental Fahrenheit reading with the experimental Celsius reading by computing T_F from $T_F = (9/5)T_C + 32$ using the experimental value of T_C in the equation. How close did you come?

8. Attach the outline paper to the Laboratory Report for the latent heat portion of the experiment.

Name _____ Section _____ Date _____

Lab Partner(s) _____

EXPERIMENT 34

The Voltmeter and Ammeter

/TI/ *Advance Study Assignment*

Read the experiment and answer the following questions.

1. What is the function of the galvanometer in a voltmeter or ammeter?

2. What is the purpose of the "multiplier" resistor in a voltmeter? Is the voltmeter a high- or a low-resistance instrument?

3. How is the voltmeter connected in the circuit? Explain why.

(continued)

4. What is the purpose of the "shunt" resistance in an ammeter? Is the ammeter a high- or a low-resistance instrument?

5. How is the ammeter connected in a circuit? Explain why.

6. What would happen if the ammeter were connected in parallel with a circuit element?

The Voltmeter and Ammeter*

INTRODUCTION AND OBJECTIVES

Voltmeters and ammeters are widely used instruments for detecting and measuring voltages and currents. Electrical quantities cannot be seen directly by looking at a wire. Like a weather vane, these instruments are used to help us observe that which we cannot see directly.

This experiment is intended to provide a behind-the-scenes look at the construction of these meters and insight into their proper use. The design of a meter and the calculation of the appropriate series or parallel resistance not only illustrates the inner workings of the voltmeter and ammeter but also provides experience analyzing a circuit using Ohm's law.

An understanding of these instruments will be useful in future experiments to avoid the damage resulting from misuse and to recognize the conditions under which the meters introduce errors as their internal resistance becomes a part of the circuit.

In this experiment you will measure the characteristics of a galvanometer and use resistances to convert it into several voltmeters and a three-range ammeter. After performing the experiment, you should be able to:

1. Explain the principle of the galvanometer and how voltmeters and ammeters are made by combining galvanometers with resistors.
2. Understand why voltmeters are high-resistance instruments and ammeters are low-resistance instruments.
3. Describe how ammeters and voltmeters are correctly connected in circuits and tell what would happen if they were incorrectly connected.

EQUIPMENT NEEDED

- Galvanometer[†]
- Decade resistance box (1 to 999999 Ω)
- Voltmeter ($\approx$ 10 V, triple-range preferred)
- Millivoltmeter ($\approx$ 100 mV)
- Power supply or battery ($\approx$ 1.5 V and 6 V)
- Wheatstone-bridge slide wire

- Triple-range milliammeter
- Single-pole switch
- Miscellaneous wires, including at least 6 clip leads

[†] A left-zero galvanometer like those used in meters is preferred. The galvanometer from an open bench-type voltmeter can be used by clipping directly to the galvanometer terminals on the back, or a microammeter can be used as a galvanometer.

THEORY

The **galvanometer** is the chameleon of scientific instruments. With the addition of a resistor and a new face, it can become a voltmeter or ammeter of whatever scale you want. The galvanometer is the basic indicating component of all deflection-type meters. The galvanometer is an electromagnetic device capable of detecting very small currents. (How small depends on how much you want to pay.)

The basic design of a moving-coil galvanometer is shown in ● Fig. 1 (sometimes called a *D'Arsonval galvanometer*, after the French physicist who invented it around 1882). It consists of a coil of wire mounted on bearings between two poles of a permanent magnet. When a current passes through the coil, it experiences a torque and rotates the pointer along the scale until balanced by the torque supplied by two small springs. The deflection is proportional to the current in the coil.

Some galvanometers have a scale marked with zero at the center. These are used in circuits like the potentiometer and Wheatstone bridge to detect a balance point. The galvanometer used in most voltmeters and ammeters has the zero at the left.

The galvanometers used for meter construction are characterized by their coil resistance r and the coil current required to give a full-scale deflection of the pointer I_c. Once these are known, the resistances necessary to convert the galvanometer to the desired voltmeter or ammeter can be calculated.

A. The dc Voltmeter

The goal here is to design a circuit that can be connected across a resistance R that will allow a small current that is proportional to the voltage difference to flow through the galvanometer, and that will divert as little current as possible from the resistance R. The circuit in ● Fig. 2a accomplishes both of these goals. The series resistor R_m is called a *multiplier* resistor because it multiplies the voltage range of the galvanometer. The combination of the

* This experiment was written and designed by Professor Fred B. Otto, Maine Maritime Academy, Castine, Maine.

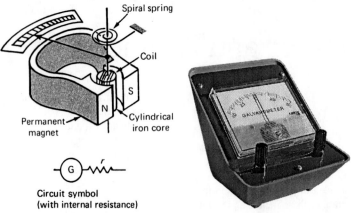

Figure 1 Galvanometer. The drawing shows the basic design of a moving-coil (D'Arsonval) galvanometer. An actual galvanometer is shown in the photo. (Photo courtesy of Sargent-Welch.)

multiplier resistance and the galvanometer is the voltmeter. These are usually combined in a single box.

The resistance of the voltmeter can be calculated using Ohm's law. If V_{max} is to be the maximum scale reading on the meter, the total resistance of the meter equals V_{max}/I_c. Since the galvanometer resistance, r, represents part of the total resistance of the voltmeter, it must be subtracted to give the value of the multiplier resistance R_m to be added in series to make a voltmeter.

$$R_m = \frac{V_{max}}{I_c} - r \qquad (1)$$

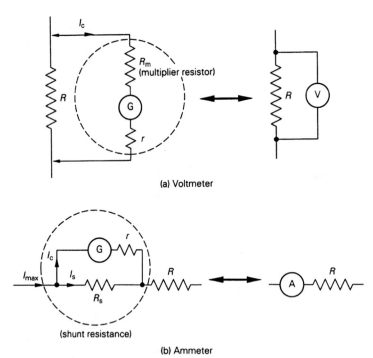

Figure 2 Circuit diagram for voltmeter and ammeter. (a) A voltmeter has a large "multiplier" resistor R_m in series with the galvanometer. (b) An ammeter has a "shunt" resistor R_s in parallel with the galvanometer.

Example 1 Suppose that we wish to calculate the multiplier resistor to convert a galvanometer with a full-scale current (I_c) of 2.0×10^{-3} A and a coil resistance of 100 Ω into a 3-V voltmeter—that is, a voltmeter with a $V_{max} = 3.0$ V.

Solution

$$R_m = \frac{V_{max}}{I_c} - r$$

$$= \frac{(3.0 \text{ V})}{(2.0 \times 10^{-3} \text{ A})} - 100 \text{ Ω} = 1400 \text{ Ω}$$

The deflection is proportional to the voltage, and a scale is provided showing the value of V_{max} at the right end. The scale is subdivided and graduated so that the voltage can be read directly from the scale.

*A voltmeter is **always** connected "across," or in parallel with, a circuit component to measure the potential difference, or voltage, drop across the component.*

If a voltmeter were connected in series with a circuit component, its high resistance would reduce the current in the circuit and the voltage drop across that component.

B. The dc Ammeter

The challenge here is to design a circuit that will divert a small amount of current through the galvanometer that is proportional to the much larger current flowing through the resistor R. This diversion is accomplished by placing a small "shunt" resistance R_s in series with R as shown in Fig. 2b.

The small resistance connected in parallel with the galvanometer is called a *shunt* because it directs or "shunts" most of the current away from the galvanometer. As the current flows through the shunt resistance, a small voltage is developed that diverts a small amount of the current through the galvanometer, causing it to have a controlled deflection proportional to the current through the circuit. The galvanometer and shunt resistance combination is the ammeter.

As with the voltmeter, the value of the shunt resistance required to give a full-scale deflection for a desired maximum current can be calculated using Ohm's law. With the pointer reading full scale, the current through the galvanometer is I_c and the voltage drop across the galvanometer is $I_c r$. Because the galvanometer and the shunt resistor are in parallel, there must be the same voltage drop across the shunt, so

$$I_c r = I_s R_s \qquad (2)$$

The current I_s through the shunt must be much larger than the current through the galvanometer coil; therefore the shunt resistance, R_s, must be much smaller than r of the coil. With the pointer deflecting to maximum, the current

flowing in the resistor R is that of the maximum scale reading of the ammeter. Since $I_s = I_{max} - I_c$, Eq. 2 can be solved for R_s to give

$$R_s = \frac{I_c r}{I_{max} - I_c} \qquad (3)$$

Example 2 Suppose that we wish to calculate the shunt resistor to convert a galvanometer with a full-scale current I_c of 2.0×10^{-3} A and a coil resistance of 100 Ω to a 3-A ammeter—that is, an ammeter with an $I_{max} = 3.0$ A.

Solution

$$R_s = \frac{I_c r}{I_{max} - I_c} = \frac{(2.0 \times 10^{-3} \text{ A})(100 \text{ Ω})}{3.000 \text{ A} - 0.0020 \text{ A}}$$

$$= \frac{0.20 \text{ V}}{2.998 \text{ A}} = 0.067 \text{ Ω}$$

(where any number of significant figures is assumed for I_{max}).

A scale is designed for the ammeter with the value of the maximum current marked at the right-hand end. The scale is subdivided so that we can read the current directly.

The low shunt resistance in parallel with the galvanometer makes the resistance of the ammeter small enough that, in calculating circuit currents, we can usually consider it negligible relative to R.

*An ammeter is **always** connected in line (in series) with a circuit component to measure the current flowing through that component.*

Since an ammeter is a low-resistance instrument, if an ammeter were connected in parallel with a circuit component, the meter would carry a large current and could burn out. Some meters have fuses to save the meter. Some others have diodes across the galvanometer that protect it from voltages above about 0.5 V.

EXPERIMENTAL PROCEDURE

1. Determine the full-scale current of the galvanometer by setting up the circuit as shown in ● Fig. 3. Set the resistance box to maximum. If you are using the galvanometer from a voltmeter, use clip leads to connect directly to the terminals on the back. Do not connect the meter to measure V_2 yet.

2. After the instructor has checked the circuit, close the switch and reduce the resistance box until the pointer of the meter is on the last mark. The full-scale current I_c is now flowing through both the resistance box and

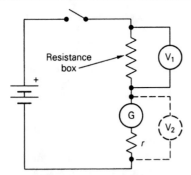

Figure 3 **The circuit for measuring the full-scale current and the resistance of a galvanometer.** See text for description.

the galvanometer. Record the resistance box setting and the voltmeter reading V_1. Open the switch after the reading is taken. Use Ohm's law to calculate the current I_c:

$$I_c = \frac{V_1}{R_{box}} \qquad (4)$$

and record it in the Laboratory Report.

3. Connect the millivoltmeter or a low-range digital voltmeter to measure the voltage V_2. Millivoltmeters tend to have somewhat low resistance and may draw some current away from the galvanometer. If the galvanometer reading falls below the maximum, adjust the resistance box until the galvanometer needle indicates that the full maximum coil current is again flowing through the galvanometer.

Record the voltage across the galvanometer V_2. Open the switch after the reading is taken. Use Ohm's law to calculate the resistance of the galvanometer coil:

$$r = \frac{V_2}{I_c} \qquad (5)$$

and record it in the Laboratory Report.

A. The dc Voltmeter

4. Use Eq. 1 to calculate the multiplier resistance for a 1.5-V voltmeter (a voltmeter with a maximum reading of 1.5 V).

5. Set the resistance box as a multiplier resistor by setting it to the calculated value, and use this "home-grown" meter to measure the voltage of a dry cell. Compare this reading to the value read by the voltmeter provided and find the percent difference.

6. Design and build a 6-V and a 15-V voltmeter. Test each by using it to measure the voltage of the 6-V sup-

ply, and compare your measurement to that of the provided voltmeter.

7. Devise and draw a circuit for building a triple-scale voltmeter where the scale is changed by connecting to different terminals. If possible, look inside a triple-scale voltmeter and compare its circuit to yours.

B. The dc Ammeter

8. Use Eq. 3 to calculate the shunt resistance necessary to make a 300-mA (0.3-A) ammeter from your galvanometer. This can be recorded in Data Table 3 as the first scale of the three-scale meter you will be making later.

9. Small resistances like those used for ammeter shunts are hard to obtain from resistance boxes. The slide wire that is used for the Wheatstone bridge or the potentiometer experiments can be used for this purpose. We will measure the resistance of the 1-meter-long wire and then calculate the length required to provide the required shunt resistance.

To measure the resistance of the wire, connect the voltmeter and ammeter as shown in ● Fig. 4a. Initially set the resistance box to 20 Ω. Use the 500-mA (0.5-A) scale of the ammeter.

10. After the instructor has checked your circuit, close the switch and carefully reduce the resistance until the ammeter reads about 500 mA (0.5 A). Record the current and voltage under slide-wire data. Open the switch and calculate the resistance of the wire using Ohm's law.

11. Compute the length of wire that will have a resistance equal to the required shunt resistance.

$$\text{Length} = \frac{\text{shunt } R}{R \text{ of one meter}} \qquad (6)$$

12. Near the "zero" end of the wire, construct a shunt by attaching two clip leads as shown in Fig. 4b. Make the distance between the leads equal to the length computed above. Attach the other ends of these leads to your galvanometer. Your ammeter is complete.

13. Test your ammeter. Set the resistance box to 100 Ω. As you reduce the resistance, compare the readings of your meter with that of the ammeter at about 100, 200, and 300 mA, and record the results in Data Table 2. **Note:** Whenever a connection is made, there is an unpredictable amount of contact resistance. Contact resistance can give large errors when dealing with small resistances. The better high-current shunts are four-terminal resistors, providing separate terminals for the current-carrying wires and the wires to the galvanometer.

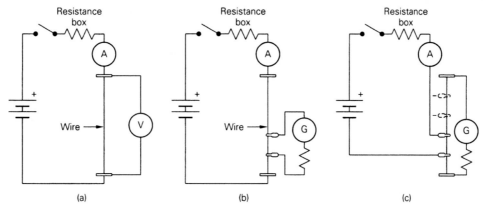

Figure 4   Experiment circuits.   (a) Circuit to measure the resistance of a wire. (b) Single-range ammeter. (c) Multirange ammeter. See text for description.

This way, the extra voltage drop at the current-carrying connections is not included in the shunt voltage drop seen by the galvanometer. Our slide-wire shunt has provided us with these separate connections.

14. Making a three-range ammeter is a little more complex than making a three-range voltmeter. The first step toward expanding your ammeter into a three-range ammeter will be to reverse the connections as shown in Fig. 4c.

    Connect the galvanometer to the ends of the wire, and connect the circuit so that the current flows in and out through the clip leads. It may be less obvious, but the contact resistance at the current connections is still not a part of the shunt resistance seen by the galvanometer.

15. Test your ammeter again against the laboratory meter. It will probably read a little lower this time. The extra wire of the slide wire in effect increases the r of the galvanometer. Adjust the length of the shunt by sliding the clip up the wire until both meters read the same. Record this as the experimental length in Data Table 3 and compute the percent difference.

16. To add a second scale to your meter, first calculate the shunt resistance and wire length for a 150-mA meter. Leave the first two clip leads in place. Attach a third lead at the calculated distance from the zero-end lead.

    You now have a two-scale meter with the zero-end lead acting as the common terminal. Test your meter against the laboratory meter and adjust the length until the meters agree. Record the computed and experimental lengths.

17. Repeat the process, and add a third scale with a full-scale reading of 60 mA.

18. If possible, look inside a laboratory triple-scale ammeter and compare its circuit with yours. Draw the circuit for the triple-scale ammeter.

EXPERIMENT 34

The Voltmeter and Ammeter

/TI/ *Laboratory Report*

Galvanometer characteristics:

Resistance box setting _____

V_1 _____

Full-scale coil current I_c _____

Full-scale voltage drop V_2 _____

Galvanometer resistance r _____

A. The dc Voltmeter

DATA TABLE 1

Scale	R_m ( )	Your meter ( )	Lab voltmeter ( )	Percent difference

Circuit diagram for a triple-scale voltmeter:

Don't forget units

(continued)

B. The dc Ammeter

Slide-wire data:

Current _____   Voltage _____

Computed resistance for one meter _____

DATA TABLE 2

Your meter ( )	Lab voltmeter ( )	Percent difference

DATA TABLE 3

Scale	Shunt resistance ( )	Computed length ( )	Experimental length ( )	Percent difference

Circuit diagram for a triple-scale ammeter:

EXPERIMENT 34   *Laboratory Report*

 QUESTIONS

1. Indicate how the dial for the triple-scale voltmeter with 1.5-V, 6-V, and 15-V scales might be labeled.

2. What would be the lowest voltage scale that could be made using your galvanometer?

3. The sensitivity of a voltmeter is commonly expressed in *ohms per volt* (Ω/V), which is the total resistance of the meter ($r + R_m$) divided by the full-scale reading. Calculate the *ohms per volt* sensitivity for each of the three voltage scales that you designed for your galvanometer. Are they the same? Are they equal to $1/I_c$?

(continued)

4. Indicate how the dial for the triple-scale ammeter you designed might be labeled.

5. Ammeter shunts are rated in terms of the voltage drop when the rated full-scale current is flowing. For example, it might be 200 A, 50 mV. What would be the millivolt rating on shunts designed for use with your galvanometer?

6. **Challenge question:** If ρ is the linear resistivity of the slide wire used for the ammeter shunts, the shunt resistance in Eq. 3 is $\rho\ell$, and the resistance of the rest of the wire, $\rho(1 - \ell)$, is added to the galvanometer resistance r. Make these changes to Eq. 3, and solve for ℓ to obtain an equation that compensates for the extra wire length when the lengths for shunts are calculated.

Name _____ Section _____ Date _____

Lab Partner(s) _____

EXPERIMENT 35

Introduction to
the Oscilloscope
/TI/ *Advance Study Assignment*

Read the experiment and answer the following questions.

1. What is a CRT?

2. What is the purpose of the deflection plates in the oscilloscope CRT?

3. How is a graph of voltage versus time obtained on the oscilloscope screen?

4. Distinguish between peak voltage and rms voltage.

(continued)

5. What is the condition for a stationary trace of an ac voltage input signal?

6. What are Lissajous figures, and how are they produced? (*See the exercise below.*)

ADVANCE STUDY EXERCISE

To help you better understand the answer to Question 6, here is an exercise to show you how the x-y position of the oscilloscope beam varies with time and gives different patterns or figures.

In each of the three parts of the exercise, graphs of x and y versus time are given below and beside an x-y graph. For a particular time, find the respective x and y coordinates on the time graphs, and plot this point (x, y) on the x-y graph. For example, in (A) at $t = 0$, $(x, y) = (0 \text{ cm}, 0 \text{ cm})$. Select 8 to 10 increasing times over each complete cycle, and plot the corresponding points. Number each point sequentially so you can see how they progress with time.

Finally, connect the points with a smooth curve to see what kind of figure or pattern is produced. The oscilloscope beam can trace out similar patterns in very short times. Continuous, cyclic tracings give rise to continuous patterns on the fluorescent screen of the oscilloscope.

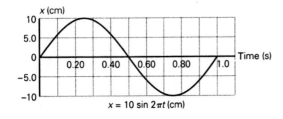

$x = 10 \sin 2\pi t$ (cm)

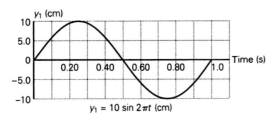

$y_1 = 10 \sin 2\pi t$ (cm)

y_1 versus x

Figure A

(continued)

EXPERIMENT 35

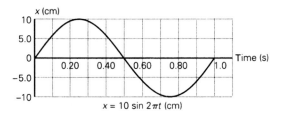

$x = 10 \sin 2\pi t$ (cm)

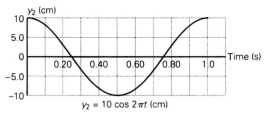

$y_2 = 10 \cos 2\pi t$ (cm)

Figure B

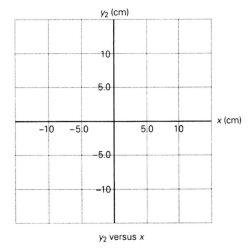

y_2 versus x

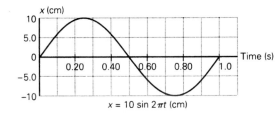

$x = 10 \sin 2\pi t$ (cm)

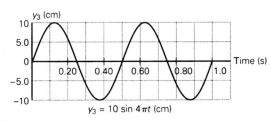

$y_3 = 10 \sin 4\pi t$ (cm)

Figure C

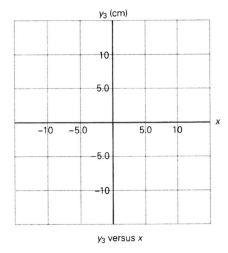

y_3 versus x

EXPERIMENT 35

Introduction to the Oscilloscope

INTRODUCTION AND OBJECTIVES

The cathode-ray oscilloscope is one of the most versatile laboratory instruments for studying ac circuits. Having long been quite common in the physics laboratory, the oscilloscope is finding increasing uses and applications in biology and medicine.

In its most basic application, the oscilloscope is used to display a graph of voltage versus time on its screen. The signal may be the voltage across a component in an electrical circuit or that generated by a nerve impulse or a heartbeat.

In this experiment, an introduction to the basic principles of the oscilloscope is presented, and you will operate an oscilloscope so as to become familiar with its controls and characteristics.

After performing this experiment and analyzing the data, you should be able to:

1. Describe the basic principle of the cathode-ray tube of an oscilloscope.
2. Explain how time and voltage measurements may be made with an oscilloscope.
3. Explain how Lissajous figures are formed.

EQUIPMENT NEEDED

- Cathode-ray oscilloscope (student model)
- Operating instruction manual for the particular oscilloscope used
- Audio-signal function generator (sine and square waves)
- (60-Hz sine wave source, or second generator if oscilloscope does not have internal line input)
- Connecting wires

THEORY

The basic component of an oscilloscope (or "scope") is a **cathode-ray tube** (CRT) or electron-beam tube (● Fig. 1). The name *cathode-ray tube* comes from early experiments with gas discharge tubes in which "rays" coming from the cathode or negative electrode were observed.

The beam of electrons (cathode rays) in a CRT is formed by an "electron gun," in which electrons thermally emitted from a cathode filament are accelerated through a potential difference of several thousand volts and focused into a beam. The electron beam strikes a fluorescent screen coated with a phosphor that emits visible light, and a spot of light is seen on the screen.

The CRT is also equipped with sets of vertical and horizontal deflection plates. If no voltage signals are applied to the deflection plates, the beam is undeflected and strikes the center of the screen. However, if a voltage signal is applied to the horizontal deflection plates, the electron beam experiences a force and is deflected horizontally.

A constant dc voltage deflects the beam spot on the screen a fixed distance. An ac voltage, on the other hand, deflects the beam back and forth, since the polarity is continually changing. If the frequency of the ac voltage signal is high enough, the beam spot traces out an observable continuous horizontal line. This is due to the relatively slow decay of the brightness of the fluorescent screen after each excitation and the persistence of vision in the human eye.

Similarly, a voltage signal applied to the vertical deflection plates causes the beam to move vertically. In either case, the magnitude of the deflection of the beam spot from the center of the screen is proportional to the magnitude of the voltage applied to the deflection plates. The cathode-ray oscilloscope is an extremely fast *X-Y* plotter that is capable of plotting an input signal versus either time or another signal.

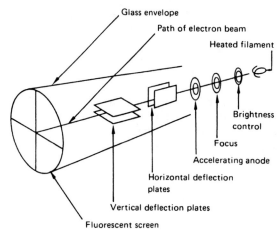

Figure 1   Cathode-ray tube.   The basic components of an oscilloscope cathode-ray tube (CRT). See text for description.

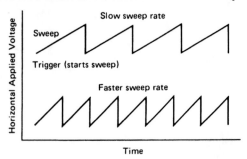

Figure 2 Voltage versus time.   Illustration of sawtooth voltage functions that supply different oscilloscope sweep rates.

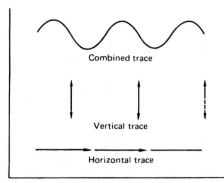

Figure 3 Sinusoidal waveform.   The combination of a horizontal sweep signal and a vertical sinusoidal signal produces a sine-wave trace on an oscilloscope screen.

In ac voltage applications, it is usually desired to display the voltage on the screen as a function of time (i.e., a graph of voltage versus time). The signal to be studied is applied to the vertical deflection plates. The oscilloscope has an internal amplifier or gain to amplify weak vertical input signals. A horizontal linear time axis is obtained if the beam spot moves horizontally (left to right) with a constant speed.

The time axis is generated by applying a deflecting voltage to the horizontal plates, which increases linearly with time (e.g., a voltage signal with a "sawtooth" wave form, as illustrated in ● Fig. 2). As the voltage increases uniformly, the spot sweeps uniformly (with constant speed) across the screen from left to right. When the voltage suddenly drops, the beam flies back to its initial position and begins another horizontal sweep with the "triggering" of the next sawtooth.

With a slow sweep rate, the spot can be observed moving periodically from left to right. However, with a fast sweep rate, a continuous trace is observed, because the eye cannot follow the motion. The oscilloscope is equipped with an internal variable sawtooth generator to supply such horizontal signals with no externally supplied signal on the horizontal plates.

When a sinusoidal voltage having the form $V = V_0 \sin 2\pi ft$ is applied to the vertical plates, the beam moves up and down, as well as side to side. If the sweep is triggered to start each sweep at the same point on the cycle of the signal applied to the vertical plates, the combined motions of the beam spot will trace out a graph of the applied voltage versus time, as shown in ● Fig. 3.

A. Time and Frequency Measurements

The sweep rate is given in time per division. (These are major divisions of about 1 cm, not one of the small divisions.) To make time measurements on the signal, count the number of divisions between two points and multiply that number by the sweep time or rate (time per division, which is typically of the order of milliseconds/div (ms/div). For example, if there are 4 divisions between the points of interest and the sweep time is 2 ms/div, then

$$t = (4 \text{ div}) \times (2 \text{ ms/div}) = 8 \text{ ms}$$

and 4 divisions correspond to 8 ms on the horizontal time base. Care must be taken to see that the sweep controls are in a calibrated position.

To measure frequency (f), first measure the time for one full wave (one wavelength), which is the period T. Then use the relationship $f = 1/T$ to calculate the frequency. Another method of measuring frequency will be discussed in Section C on Lissajous figures.

B. Voltage Measurements

In addition to frequency measurements, the oscilloscope can be used as a voltmeter to read the peak-to-peak voltages of ac signals as well as dc voltages. The voltage readings are read directly from a plastic screen or grid attached to the face of the oscilloscope tube. Recall that the vertical deflection of the beam is proportional to the voltage of the signal applied to the vertical deflection plates.

The peak-to-peak height of the trace of an ac input signal is then measured, and the peak-to-peak voltage is computed from the calibration. Dividing by 2 gives the peak voltage V_0, and if it is a sine wave multiplying by $1/\sqrt{2} = 0.707$ gives the root-mean-square (rms) value of the voltage.*

There is a large variety of oscilloscopes, and the controls of each model and make cannot be discussed here. All oscilloscopes have some operating controls in common; however, some have controls that others lack, depending on the purpose and sophistication of the scope.

Also, controls and connectors are not always found in the same location or in the same form on all oscilloscopes.

---

* For alternating current, Ohm's law is written $V_{rms} = I_{rms}R$, where $V_{rms} = V_0/\sqrt{2} = 0.7070 V_0$, and $I_{rms} = 0.7071_0$; here V_0 and I_0 are the peak voltage and peak current, respectively. It is customary to measure and specify rms values for ac quantities.

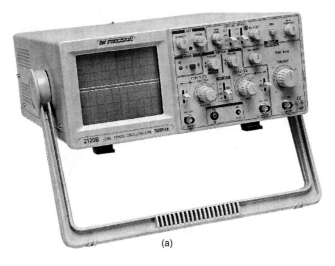

(a)

(b)

Figure 4   Oscilloscopes.   (a) A typical student model oscilloscope. (Photo courtesy of B + K Precision, Dynamics Corporation.) (b) A sine-wave trace on an oscilloscope from a function generator input.

An oscilloscope that is commonly found in an introductory physics lab is shown in ● Fig. 4a.

C. Lissajous Figures

The frequency of a sinusoidal voltage also can be determined by comparing it with another calibrated sinusoidal signal. The simplest oscilloscope pattern results when the two sinusoidal signals have the same frequency and are in phase—for example,

$$x = A \sin 2\pi f t$$

and

$$y = A \sin 2\pi f t \qquad (1)$$

A diagonal line appears on the screen (● Fig. 5). The pattern observed when the frequency of the *y* signal is three times that of the *x* signal is also shown in Fig. 5. Such patterns are called **Lissajous figures.** As before, a stationary pattern results when the frequency of the *X*-axis signal and the frequency of the *Y*-axis signal are integral or a rational number.

Suppose that two sinusoidal signals have the same frequency but different phases—for example,

$$x = A \sin 2\pi f t$$

and

$$y = A \sin(2\pi f t - \delta) \qquad (2)$$

where δ is the *phase angle* or *phase difference*. The phase angle of the *x* signal is equal to zero. Then an ellipse will be traced out on the screen (● Fig. 6). If the phase difference is 90°, the pattern is a circle. (Why?)

EXPERIMENTAL PROCEDURE

Note: Because different brands and models of oscilloscopes may have somewhat different control names and settings, the described procedures may need modification.

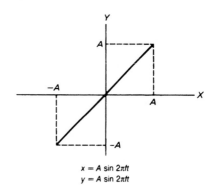

$x = A \sin 2\pi f t$
$y = A \sin 2\pi f t$

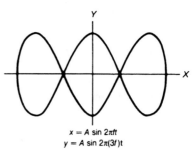

$x = A \sin 2\pi f t$
$y = A \sin 2\pi(3f)t$

Figure 5   Lissajous figures.   Lissajous figures for different *x* and *y* signals. See text for description.

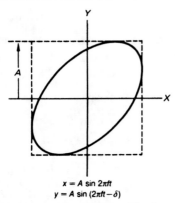

$$x = A \sin 2\pi ft$$
$$y = A \sin (2\pi ft - \delta)$$

Figure 6 Lissajous figure. The Lissajous figure for signals that have the same frequency but different phases is an ellipse.

Even so, the general operation of all oscilloscopes is the same. Consult the operating instructions manual for your particular instrument if there is a question.

Keep in mind that this is an *introduction* to the oscilloscope. You may encounter a variety of oscilloscopes in your career, some very sophisticated, but being familiar with the theory and functions of the basic controls will aid you greatly in their operation.

1. Locate and familiarize yourself with the following general controls and connections on your oscilloscope. The operating instructions manual of your instrument should be consulted for specific and detailed explanations of the operating controls and for initial control settings.

DISPLAY CONTROLS
(a) POWER SWITCH (OFF-ON)—turns the oscilloscope on and off. This switch is often on the same shaft as one of the following controls.
(b) INTENSITY—adjusts the brightness of the spot (by adjusting the rate of electron emission from the cathode).
(c) FOCUS—adjusts the sharpness of the spot.
(d) ASTIG (astigmatism)—adjusts the vertical and horizontal focus of the spot to the same position. If a well-defined trace cannot be obtained with focus control, it may be necessary to adjust the astigmatism control.
(e) SCALE LIGHT (*optional*)—adjusts a lamp that lights the screen scale.

POSITION CONTROLS
(a) HORIZONTAL POSITION (H-POS)—adjusts the horizontal position of the spot or trace on the screen.
(b) VERTICAL POSITION (V-POS)—adjusts the vertical position of the spot or trace on the screen.
(c) HORIZONTAL GAIN—provides continuous adjustment of the gain of the horizontal amplifier (i.e., amplifies or "magnifies" the trace).
(d) VERTICAL GAIN—provides continuous adjustment of the gain of the vertical amplifier (i.e., amplifies or "magnifies" the trace).
(e) VOLTS/DIV or VERTICAL ATTENUATOR—provides step adjustment of vertical sensitivity in calibrations of volts per grid division or V/cm.
(f) VARIABLE VOLTS/DIV (VERTICAL GAIN)—provides fine control of vertical sensitivity. In CAL (calibrated) position, the VERTICAL ATTENUATOR is calibrated. May also serve as vertical gain in some operations. (On uncalibrated scopes, the VERTICAL GAIN provides fine control of vertical sensitivity between steps of VERTICAL ATTENUATOR settings, usually in steps of 10.)

SWEEP TIME AND TRIGGERING MODE
(a) SWEEP TIME/DIV or SWEEP (HOR) SELECTOR—selects calibrated horizontal sweep rates in terms of time per grid division.
(b) VARIABLE TIME/DIV (*X* or HORIZONTAL GAIN or SWEEP VERNIER)—provides fine sweep time adjustment. In CAL (calibrated) position, the sweep time is calibrated in time/div steps. (On uncalibrated scopes, horizontal gain adjustment is provided by a HORIZONTAL GAIN control.)
(c) TRIG MODE—three-position switch that selects triggering mode.
 AUTO —triggers sweep operation when trigger signal is present and automatically generates sweep in absence of trigger signal.
 NORM—normal triggered sweep operation. No trace unless proper trigger signal is applied.
 X-Y —vertical input signal produces vertical (*Y*-axis) deflection; EXT input signal produces horizontal (*X*-axis) deflection.
(d) SYNC (HOR) SELECTOR or SOURCE—selects the triggering source (the source that determines *when* the spot is triggered or starts sweeping across the screen).
 INT —wave form being observed is used as a sync trigger.
 LINE—sweep is triggered by line voltage or in step with (60-Hz) line frequency.
 EXT —sweep is triggered by signal applied to EXT jack.

EXTERNAL CONNECTIONS

(a) VERTICAL INPUT—applies an external signal to the vertical amplifier. The lower terminal is usually grounded (GND) to the instrument case. Sometimes there are two terminals above ground for balanced inputs.

The lower terminal must be grounded for grounded-side input and is usually ganged to the ground terminal by a connector. See the operating instructions manual for the instrument.

(b) HORIZONTAL INPUT and/or EXT TRIG—applies an external signal to the horizontal amplifier. May also be input terminal for external trigger signal.

2. Consult the instrument's operating instructions manual for initial starting procedure and control settings. Make the appropriate control adjustments. Turn on oscilloscope. A trace should appear on the screen.

Adjust the INTENSITY and FOCUS. *Never advance the intensity control to the point where an excessively bright spot or trace appears on the screen. A bright spot can burn the screen and decrease its useful life.* Adjust the HORIZONTAL POSITION and VERTICAL POSITION controls so that the spot is in the center of the screen. If you have trouble obtaining a spot, ask the instructor for assistance.

3. With a low SWEEP TIME/DIV setting, adjust the VARIABLE TIME control and note the effect.

A. Time and Frequency Measurements

4. Set the TRIG MODE to NORM, the SOURCE SWITCH to INT, and the SWEEP TIME variable control to CAL. Connect the function generator sine-wave output to the VERTICAL (Y) INPUT of the scope. If one terminal of the function generator is grounded, be sure that this is the terminal connected to the grounded terminal of the oscilloscope. Turn on the function generator and set the generator frequency at 90 Hz. Adjust the function generator amplitude control so that the sine-wave pattern is almost full-scale on the screen. (Switch the TRIG MODE momentarily to *X-Y.* What do you observe and why?)

5. Adjust the SWEEP TIME/DIV control so that a wave pattern with two peaks appears on the screen. Read the number of divisions for one full sine-wave cycle, and record this number in the Laboratory Report.

6. (a) Compute the time period (T) of one cycle of the wave pattern using the calibrated SWEEP TIME/DIV setting, and record.

(b) Compute the sine-wave frequency ($f = 1/T$) and compare with the function generator setting by computing the percent difference.

7. Repeat the preceding procedures for a function generator output of 300 Hz.

8. With the generator output still at 300 Hz, adjust the SWEEP TIME/DIV control in various steps, and note the relationship of the number of wave cycles to the sweep time/div. (Can you explain? See Question 1 at the end of the experiment.)

9. Set the SYNC (HOR) SELECTOR to LINE. Adjust the function generator frequency to 60-Hz so a stationary pattern appears on the screen. This matches the generator frequency to the relatively stable 60-Hz line frequency, which is more accurate than the calibration markings on the function generator. Compare the generator frequency setting to the line frequency by finding the percent error.

10. Adjust the calibrated SWEEP TIME/DIV control until one or more full sine-wave cycles appear on the screen. Then compute the frequency of the sine wave appearing on the scope as before.

B. Voltage Measurement

11. (a) With the full sine-wave cycle(s) on the screen (procedure 10), record the VOLTS/DIV control setting (with variable control at CAL).

(b) Read the number of peak-to-peak divisions for the height of the wave pattern on the screen.

(c) Compute the rms voltage of the wave.

C. Lissajous Figures

12. Set TRIG MODE to *X-Y* and apply a 60-Hz sine wave to the HORIZONTAL or *X* INPUT. A second function generator may be used.* (Voltages of 25 V or 50 V are recommended.)

With either the VERTICAL or the HORIZONTAL INPUT set at 60 Hz, adjust the other input frequency to 30 Hz, making a fine adjustment of this generator frequency to obtain a stationary pattern. Adjust the generator and/or the *X* and *Y* input controls until the pattern is a good size, taking up between half and full height, and the width of the screen. Observe the pattern on the screen.

13. Adjust the generator frequency to 60 Hz and obtain a stationary wave pattern. Note that by slight adjustment you can vary the pattern between a straight line and a circle. (Can you explain why?)

---

* With two function generators, it is important to have a common ground for both generators and the oscilloscope.

14. Continue to increase the generator frequency and observe the various stationary and moving patterns. Record in the Laboratory Report.

D. General

15. You should now be getting a feeling for the oscilloscope controls and operations. Adjust the various control knobs so as to understand their functions better. Also, connect the square-wave output of the function generator to the VERTICAL INPUT and investigate this wave form.

Name _____ Section _____ Date _____

Lab Partner(s) _____

Introduction to the Oscilloscope

/TI/ *Laboratory Report*

A. *Time and Frequency Measurements*

Generator frequency 90 Hz

No. of divisions _____

Time sweep/div _____

Calculations
(show work)

Period _____

Frequency _____

Percent difference _____

Generator frequency 300 Hz

No. of divisions _____

Time sweep/div _____

Calculations
(show work)

Period _____

Frequency _____

Percent difference _____

Don't forget units

(continued)

Source—Line (60 Hz)

Generator frequency setting _____

Percent error _____

No. of divisions _____

Time sweep/div _____

Period _____

Frequency _____

Calculations
(show work)

B. *Voltage Measurement*

Calculations
(show work)

Volts/div _____

No. of divisions _____
(peak to peak)

rms voltage _____

EXPERIMENT 35   *Laboratory Report*

C. *Lissajous Figures*

X-input frequency	*Y*-input frequency	Sketches of patterns (attach separate sheet)
_____	_____	
_____	_____	
_____	_____	
_____	_____	
_____	_____	
_____	_____	
_____	_____	

/TI/ QUESTIONS

1. How does the number of wave cycles seen on a screen for a fixed input frequency vary with the SWEEP TIME/DIV control setting? Why?

2. Explain why the 60-Hz Lissajous figure could be varied between a straight line and a circle.

EXPERIMENT 35

EXPERIMENT 36

Phase Measurements
and Resonance in ac Circuits

/TI/ *Advance Study Assignment*

Read the experiment and answer the following questions.

1. What is meant by the phase difference between the voltage and current in an ac circuit?

2. What is the effect of (a) capacitive reactance and (b) inductive reactance, and how does each affect the phase?

3. What is impedance, and is a circuit with impedance capacitive or inductive?

(continued)

4. State the condition for resonance in a series *RLC* circuit, and describe the circuit effect of resonance.

5. Explain how the phase difference between voltage and current can be measured with a single-beam oscilloscope.

CI *Advance Study Assignment*

Read the experiment and answer the following questions.

1. How many redians are there in one period of a sinusoidal function?

2. How is the phase difference (in degrees) between the voltage and the current measured from a graph that plots each quantity versus time?

516

Phase Measurements and Resonance in ac Circuits

OVERVIEW

Experiment examines phase measurements and resonance in ac circuits, using both TI and CI techniques. In the TI procedure, an oscilloscope and signal generator are used to examine *RL*, *RC*, and *RLC* circuits. The CI procedure offers an alternative to using a regular oscilloscope by employing voltage and current sensors, along with the Data Studio Scope. Resistive, capacitive, inductive, and *RLC* circuits are examined.

INTRODUCTION AND OBJECTIVES

In **ac (alternating current) circuits,** the current in the circuit alternates in direction or polarity. For an applied sinusoidal voltage, the current varies sinusoidally with time. However, the current is not always exactly in phase with the voltage. The current may "lead" or "lag" the voltage, depending on the circuit elements. That is, there is a *phase difference* between the voltage and the current.

In this experiment, the phase properties of particular circuit elements will be studied, together with the important condition of resonance. In this condition, the current and voltage are in phase and the current in the circuit is a maximum. This means that there is maximum power transfer from the voltage source to the circuit.

After performing this experiment and analyzing the data, you should be able to:

/TI/ OBJECTIVES

1. Describe phase differences in ac circuits.
2. Clearly distinguish among capacitive reactance, inductive reactance, and impedance.
3. Explain what is meant by *RLC* series circuit resonance and its circuit effects.

CI OBJECTIVES

1. Investigate phase differences between voltage and current for three different ac circuits: a resistive, a capacitive, and an inductive circuit.
2. Analyze a series *RLC*, circuit to determine whether the circuit is resistive, capacitive, or inductive.

Phase Measurements and Resonance in ac Circuits

/TI/ EQUIPMENT NEEDED

- Audio signal generator (sine-wave)
- Oscilloscope (with horizontal and vertical amplifiers in phase or completely out of phase)
- Standard decade resistance box (10 to 9990 Ω)

- Two capacitors (0.10 μF and 0.50 μF, or capacitor substitution box)
- Air-core inductor (250 mH)
- Connecting wires

/TI/ THEORY

When a dc (direct current) voltage source, such as a battery, is applied to an electrical circuit, the resulting current flows only in one direction, because the polarity (+ and −) of the voltage remains constant. For an ac (alternating current) voltage source, however, the polarity of the voltage alternates with time, and the direction of the current flow in the circuit alternates with the same frequency as the voltage source.

One of the most commonly used ac voltages is one that varies sinusoidally with time. This may be generally described by the equation

$$V(t) = V_m \sin(2\pi f t - \phi_v) \qquad \text{(TI 1)}$$

where V_m is the maximum voltage amplitude and f is the frequency of the source. The angle ϕ_v is called the **phase angle**. If $\phi_v = 0$, then

$$V(t) = V_m \sin 2\pi f t \qquad \text{(TI 2)}$$

which means that the voltage is zero at $t = 0$.

The current in the ac circuit may or may not be in phase with the voltage, depending on the nature of the components in the circuit. In any case, if the applied voltage is sinusoidal, the current I is also sinusoidal and, as a function of time, may be expressed as

$$I(t) = I_m \sin(2\pi f t - \phi) \qquad \text{(TI 3)}$$

where I_m is the maximum amplitude of the current and ϕ is a phase angle. In TI Eq. 2 the phase angle of the voltage has been assumed to be zero. In this case, ϕ is also the *phase difference* between the applied voltage and the resulting current. (More generally, $\phi = \phi_I - \phi_v$ and is the angle between I_m and V_m, as illustrated in ● TI Fig. 1, where time is expressed in terms of angular frequency $\omega = 2\pi f$.)

The phase angle ϕ in TI Eq. 3 can either be positive or negative. When it is positive, the maximum value of the current I_m is reached at a later time than the maximum value of the voltage V_m (TI Fig. 1). In this case, we say that the voltage *leads* the current or the current *lags* behind

the voltage by a phase difference ϕ. Similarly, if ϕ is negative, the current leads and the voltage lags.

When there is a capacitive element in the circuit (an *RC* circuit, ● TI Fig. 2a), the alternating charging and discharging of the capacitor opposes the current flow. This opposition is expressed as **capacitive reactance** X_C, and

$$X_C = \frac{1}{2\pi f C} \qquad \text{(TI 4)}$$

where C is the value of capacitance (in farads, F). The unit of X_C is ohms (Ω).

Similarly, when there is an inductive element in an ac circuit (an *RL* circuit, TI Fig. 2b), the self-induced counter emf in the induction coil opposes the current. The **inductive reactance** X_L is given by

$$X_L = 2\pi f L \qquad \text{(TI 5)}$$

where L is the inductance of the coil (in henrys, H). The unit of X_L is ohms.

Many ac circuits have both capacitive and inductive reactance elements. The combined opposition to the current flow of resistive and reactive elements in a series

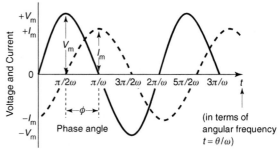

TI Figure 1   ac Voltage and current. The voltage and current are out of phase, and in this case, the voltage leads the current by a constant phase difference or angle ϕ.

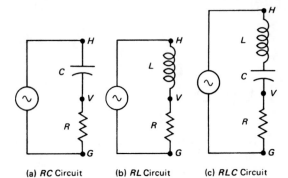

(a) *RC* Circuit   (b) *RL* Circuit   (c) *RLC* Circuit

TI Figure 2 Basic circuits. Circuit diagrams for (a) *RC*, (b) *RL*, and (c) *RLC* circuits.

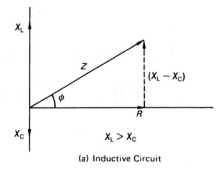

(a) Inductive Circuit

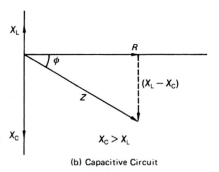

(b) Capacitive Circuit

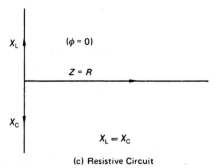

(c) Resistive Circuit

circuit as shown in TI Fig. 2c, a series *RLC* circuit, is expressed in terms of the **impedance** *Z* of the circuit, which is given by

$$Z = [R^2 + (X_L - X_C)^2]^{1/2}$$

$$= \left[R^2 + \left(2\pi f L - \frac{1}{2\pi f C} \right)^2 \right]^{1/2} \quad \textbf{(TI 6)}$$

The unit of *Z* is also ohms. (*R* is the total resistance of the circuit. In general, it is assumed that the resistance of the induction coil is negligible compared with the resistor element.)

For the series *RLC* circuit, the phase relation of the voltage and current is given by

$$\tan \phi = \frac{X_L - X_C}{R} \quad \textbf{(TI 7)}$$

This relationship is often represented in a *phasor diagram,* in which the resistance and reactances are added like vectors (● TI Fig. 3). Note that the angle ϕ is either positive or negative, depending on whether the inductive or capacitive reactance is greater. If X_L is greater than X_C, then ϕ is positive and the current lags behind the applied voltage. The circuit is then said to be *inductive.*

Similarly, if X_C is greater than X_L, ϕ is negative and the current leads the voltage. In this case, the circuit is said to be *capacitive.* (TI Eq. 7 can be applied to single-reactance circuits, as in TI Fig. 2a and b, by letting the appropriate reactance be zero.)

A common way to remember the phase relationship in inductive and capacitive circuits is by the phrase

ELI the *ICE* man

Here *E* represents the voltage (*V*). In an inductive circuit (*L*), the voltage leads the current as indicated by *E* "leading" *I* in *ELI* (see TI Fig. 1). Similarly, in a capacitive circuit (*C*), the current leads the voltage, *ICE.*

Since the voltage and current are continually changing in an ac circuit, it is convenient to consider effective or

TI Figure 3 Phasor diagrams. Phasor diagrams for (a) an inductive circuit, (b) a capacitive circuit, and (c) a resistive circuit. In a phasor diagram, resistances and reactances are added like vectors. See text for description.

time-average values of the voltage and current. These root-mean-square (rms) values are given by $V = V_m/\sqrt{2}$ and $I = I_m/\sqrt{2}$, where V_m and I_m are the maximum or "peak" voltage and current, respectively. The rms values of *V* and *I* are the values read on most ac voltmeters and ammeters.

An Ohm's-law type relationship holds between these alues and the impedance *Z*:

$$V = IZ \quad \textbf{(TI 8)}$$

and it follows that

$$V_m = I_m Z$$

Thus, for a given applied voltage, the smaller the impedance of a circuit, the greater the current in the circuit ($I = V/Z$).

Notice that in the reactance term in TI Eq. 6, there is a minus sign and the individual reactances are reciprocally frequency-dependent. As a result, for given L and C values, the total reactance can be zero for a particular frequency at which $X_L = X_C$. That is,

$$2\pi f L - \frac{1}{2\pi f C} = 0 \qquad \text{(TI 9)}$$

and solving for f yields

$$f_r = \frac{1}{2\pi \sqrt{LC}} \qquad \text{(TI 10)}$$

where f_r is called the **resonance frequency.** Driven at this frequency, the circuit is said to be **in resonance.** The impedance is then equal to the resistance in the circuit, $Z = R$, and the circuit is resistive (TI Fig. 3c).

Since the impedance is a minimum at the resonance condition, the current from the voltage source is maximum in the circuit, and maximum power $P = I^2 R$ (in this special case, $Z = R$). For fixed values of L and C, resonance occurs at the particular resonance frequency f_r given by TI Eq. 10. However, notice that for a given source frequency, resonance can also be obtained by varying L and/or C in the circuit.

TI Example 1   A 10 (rms)-V, 1000-Hz voltage is applied to a series RLC circuit with $R = 1.0$ kΩ and $L = 250$ mH. (a) What value of capacitance would put the circuit in resonance? (b) How much current would flow in the circuit in this condition?

Solution   Given: $V = 10$ (rms) V, $f = 10^3$ Hz, $R = 1.0$ k$\Omega = 1000\ \Omega$, and $L = 250$ mH $= 0.250$ H. (a) To have f_r be 10^3 Hz, the required capacitance is given by

$$f_r = \frac{1}{2\pi \sqrt{LC}}$$

or

$$C = \frac{1}{L4\pi^2 f_r^2} = \frac{1}{(0.250\ \text{H})(4\pi^2)(10^3\ \text{Hz})^2}$$

$$= 0.10 \times 10^{-6}\ \text{F} = 0.10\ \mu\text{F}$$

(b) Since $Z = R$, for resonance, and with $V = 10$ V and $R = 1000\ \Omega = 10^3\ \Omega$,

$$I = \frac{V}{Z} = \frac{V}{R} = \frac{10\ \text{V}}{10^3\ \Omega} = 0.010\ \text{A (rms)}$$

In this experiment, we want to measure the phase difference between the applied voltage and the current in ac circuits and to investigate resonance condition in a series (or parallel) RLC circuit using an oscilloscope. Notice from TI Eq. 7 that for the resonance condition, $X_L = X_C$, and the voltage and current are in phase ($\phi = 0$), since $\tan \phi = 0$ (and $\tan 0 = 0$).

To obtain a double-trace graph of V and I as in TI Fig. 1 requires a double-beam or dual-trace oscilloscope or special electronics, which are often not available. However, phase measurements can be made with an ordinary oscilloscope by the following method.

If the ratio of the horizontal and vertical input frequencies is an integer or ratio of integers, then a stationary pattern is observed—a Lissajous pattern. If the horizontal and vertical input frequencies are equal, a stationary pattern such as the one in ● TI Fig. 4 is observed. Assume that the applied voltages have the forms

$$x = A \sin 2\pi f t$$
$$y = B \sin (2\pi f t - \phi) \qquad \text{(TI 11)}$$

where A and B are the amplitudes. [Compare with TI Eqs. 1 and 3, which give the voltage $V(t)$ and circuit current $I(t)$, respectively.] Note that the y intercepts, $+b$ and $-b$, occur when $x = 0$, or when $2\pi f t = 0$ (since $\sin 0 = 0$).

Then, from the second equation, $b = B \sin (-\phi)$. Hence

$$\sin \phi = \pm \frac{b}{B}$$

and similarly,

$$\sin \phi = \pm \frac{a}{A} \qquad \text{(TI 12)}$$

You should be able to prove and understand that if $A = B$ and $\phi = 90°$, the trace is a circle. Also, if $\phi = 0°$, the trace is a straight line.

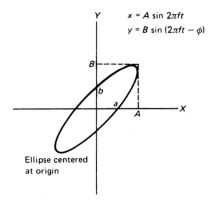

TI Figure 4   **Phase measurements.**   An elliptical pattern is used to measure the phase-angle difference. See text for description.

To measure the phase-angle difference of the voltage and current in a circuit, we apply the voltage signal from the signal generator to the horizontal input. For example, for the circuits in TI Fig. 2, connections are made to points H and G, where G is to the horizontal ground terminal. The voltage input signal has the form of TI Eq. 1.

The current signal is applied to the vertical input with connections made to points V and G, where G is to ground. This is actually a voltage input from across the resistor in the circuit.

However, it is proportional to and in phase with the current through the resistor (and therefore through the entire series circuit). Hence, the phase-angle difference between the voltage and the current can be determined from the shape of the resulting elliptical oscilloscope pattern, as described above.

TI / EXPERIMENTAL PROCEDURE

1. Turn on the oscilloscope and the signal generator. It is necessary for the horizontal and vertical amplifiers of the oscilloscope to be in phase (or completely out of phase, $\phi = 180°$) to measure the phase differences. See whether this is the case by connecting the signal generator output to both the horizontal and the vertical inputs of the oscilloscope. (Be sure the ground terminals of both inputs are connected to the same terminal of the signal generator, the grounded terminal if it has one.)

   Set the generator frequency at 1000 Hz with the wave amplitude near maximum. If the horizontal and vertical amplifiers are in phase, the resulting trace is a diagonal straight line through the first and third quadrants. If the amplifiers are completely out of phase, the diagonal line will be through the second and fourth quadrants.

   In either case, phase measurements can be made. Vary the frequency of the generator to see if the phase difference of the amplifiers remains reasonably constant, in particular over the range 100 to 1000 Hz.

A. Capacitive Circuit

2. Set up a circuit as in TI Fig. 2a. Use the standard decade resistance box as R set at 500 Ω and a capacitance of about 0.50 μF. If the signal generator has a grounded terminal, connect it to point G. Set the signal generator at a frequency of 1000 Hz. Connect point G to one of the oscilloscope ground terminals and points V and H to the vertical and horizontal inputs, respectively. The SWEEP (HOR) SELECTOR should be on EXT.

   Obtain on the screen an elliptical pattern of appropriate size for measurement by adjusting the HORIZONTAL and VERTICAL GAINS. Adjust the INTENSITY and FOCUS controls to obtain a sharp

pattern. (Use the position controls to center the pattern if necessary.)

3. Then vary (a) the generator frequency, (b) R, and (c) C (if a substitution box is used) one at a time and observe the effects.

4. Using the parameter values suggested in procedure 2, make measurements on the shape of the elliptical trace required to determine the phase-difference angle ϕ. (For convenience, measure the total distances from the left side to the right side of the pattern, $2b$ and $2B$; or the total distances from the bottom to the top of the pattern, $2a$ and $2A$: $\sin b/B = \sin 2b/2B$. Record the measurements in the Laboratory Report.

5. Compute the phase-difference angle (TI Eq. 12) and the value of R (TI Eq. 7). Compare the experimental and known values of R by finding the percent error.

B. Inductive Circuit

6. Replace the capacitor with the inductor (TI Fig. 2b) and make $R = 2000$ Ω. Repeat part A for this arrangement. Alternatively, if the value of L is unknown, use the known values of R and f to compute L.

C. RLC Circuit Resonance

7. Set up a circuit as in TI Fig. 2c with $R = 500$ Ω and $C = 0.50$ μF. Vary the frequency of the signal generator, and observe what happens. Determine, as accurately as possible, the frequency at which the circuit is in resonance with the trace a diagonal straight line. Record the resonance frequency in the Laboratory Report.

   Using the known values of L and C, compute the resonance frequency and compare to the measured value by finding the percent difference.

8. Repeat procedure 7 with the 0.50-μF capacitance replaced by a 0.10-μF capacitance.

9. Change the scope SWEEP TIME SELECTOR from EXT to a suitable SWEEP rate. Observe the magnitude of the current $I(t)$ as the generator frequency is varied around the resonance frequency f_r, with the circuit as shown in TI Fig. 2c. (What happens to I at f_r? How sharp is the effect?) Calculate and record

$$\frac{I(\text{at } f_r)}{I(\text{at } 2f_r \text{ or } 0.5f_r)}$$

(Use the number of scale divisions for I. This is proportional to the actual value.)

   Occasionally check the voltage of the frequency generator by moving the vertical connection from point V to point H in the circuit in TI Fig. 2c to verify that the voltage of the frequency generator remains unchanged. Adjust if necessary.

10. Rearrange the circuit for the parallel LC arrangement, as shown in ● TI Fig. 5. (This is sometimes called an *anti-resonance* circuit.) Repeat procedure 9. (Now what happens to I at f_r? How sharp is the effect?) Calculate and record

$$\frac{I(\text{at } 2f_r \text{ or } 0.5f_r)}{I(\text{at } f_r)}$$

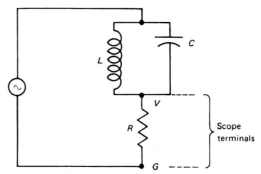

TI Figure 5   Anti-resonance.   The circuit diagram of a parallel LC circuit. See text for description.

EXPERIMENT 36

Phase Measurements and Resonance in ac Circuits

TI *Laboratory Report*

A. Capacitive Circuit

R _____

C _____

*Calculations
(show work)*

Generator frequency _____

2a or 2b _____

2A or 2B _____

Computed ϕ _____

Computed R _____

Percent error _____

B. Inductive Circuit

R _____

L _____

*Calculations
(show work)*

Generator frequency _____

2a or 2b _____

2A or 2B _____

Computed ϕ _____

Computed R _____

Percent error _____

Don't forget units

(continued)

C. RLC *Circuit Resonance*

1. R _____   Measured resonance frequency _____

    L _____   Computed resonance frequency _____

    C _____   Percent error _____

2. R _____   Measured resonance frequency _____

    L _____   Computed resonance frequency _____

    C _____   Percent error _____

Describe current behavior at $f = f_r$

Divisions at $f = f_r$ _____   Sharpness ratio _____

Divisions at $f = 2f_r$ or $0.5f_r$ _____

3. Parallel (anti-resonance) *LC* circuit

    R _____   Measured resonance frequency _____

    L _____   Computed resonance frequency _____

    C _____   Percent error _____

Describe current behavior at $f = f_r$.

Divisions at $f = f_r$ _____   Sharpness ratio _____

Divisions at $f = 2f_r$, or $0.5f_r$ _____

EXPERIMENT 36   *Laboratory Report*

/TI/ QUESTIONS

1. Speculate on the physical causes of (a) capacitive reactance and (b) inductive reactance.

2. Prove mathematically that the oscilloscope pattern for horizontal and vertical input signals of the same frequency and amplitude (a) is a circle when the phase difference of the signals is 90°, (b) is a straight line for the resonance condition.

3. A series RLC circuit is driven by a voltage signal of a given frequency other than f_r. For a fixed R, how should L and C be varied so that there is maximum current in the circuit?

4. Suppose that the resistance of the inductor coil were not negligible compared with the resistance in an RL or RLC circuit. How would this affect the impedence?

5. A parallel LC circuit is sometimes called an "*anti-resonance*" circuit. Why is this?

527

EXPERIMENT 36

C I **EXPERIMENT 36**

EQUIPMENT NEEDED

This activity is designed for the Science Workshop 750 interface, which has a built-in function generator.

- A resistor, a capacitor, and an inductor of suggested values: $R = 100\,\Omega$, $C = 100\,\mu F$ (nonpolar), and $L = 20$ mH

- 3 voltage sensors (PASCO CI-6503)
- Science Workshop 750 interface
- Multimeter with capacitance and inductance scales
- Optional: Color pencils

CI THEORY

This activity provides an alternative to using a standard oscilloscope when analyzing ac circuits. The suggested values for R, C, L, and the frequencies can be varied according to availability of equipment.

The voltage and current in a resistive, a capacitive, and an inductive ac circuit will be studied using voltage and current sensors and the Data Studio Scope, instead of the regular oscilloscope. The voltage and the current will appear plotted against time in the scope display, simulating an actual oscilloscope.

The phase difference between the voltage and the current will be determined directly from the scope display by using the time axis. The period of a sinusoidal function is 2π radians, or 360°. When plotted against time, the period (in time units) must be equivalent to a phase angle of exactly 360°. This is illustrated in ● CI Fig. 1 and ● CI Fig. 2.

Consider a sinusoidal voltage of $V_{max} = 3.0$ V, as illustrated in CI Fig. 1. The period T of the function is the peak-to-peak time difference. In this example, the period is

found to be approximately 3.3 s, read directly from the graph. If the plot were traced as voltage versus angle in degrees, then the 3.3 s would correspond to exactly a 360° angular distance between the peaks. Thus the angular separation between any two points in the plot can be found as follows:

$$\phi = \frac{\Delta t}{T} \cdot 360° \qquad \textbf{(CI 1)}$$

where Δt is the time shift between the two points and T is the period of the function.

CI Fig. 2 shows the same voltage trace as CI Fig. 1, but this time the current in the circuit is plotted also. Note that this is a case where the voltage leads the current. The phase difference between voltage and current is measured as the distance from a peak in the voltage graph to the nearest peak in the current graph. In this case, the time difference is found to be approximately $\Delta t = 0.6$ s. Thus the phase angle between the peaks is

$$\phi = \frac{\Delta t}{T} \cdot 360° = \frac{0.6\ \text{s}}{3.3\ \text{s}} \cdot 360° = 65°$$

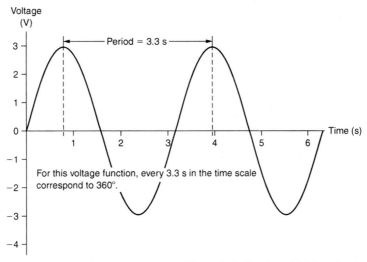

CI Figure 1   Sinusoidal function.   The period of a sinusoidal function is the peak-to-peak time difference.

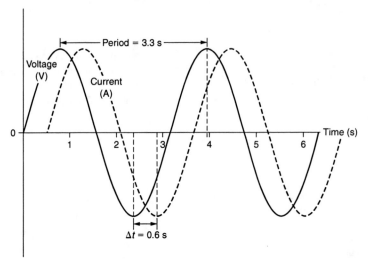

CI Figure 2 **Out-of-phase sinusoidal functions.** The phase angle is easily found by first measuring the period of the function.

SETTING UP DATA STUDIO

1. Open Data Studio and choose "Create Experiment."
2. From the sensor list, choose a voltage sensor. Connect the voltage sensor to the interface, as shown on the computer screen.
3. Double-click on the "Voltage Sensor" icon. The sensor properties window will open. Set the sample rate frequency to 200 Hz. Click OK to exit the sensor properties window.
4. Directly below the sensor list there is an icon for the signal output. Double-click on this icon. The signal generator window will open.
5. The default form of the signal generator function is a sine wave. Set the amplitude to 3.00 V. This will be the maximum output voltage, V_{max}.
6. The frequency of the sine function will need adjusting to a different value in each part of the experiment. To begin with the resistive circuit, set the frequency to 1.0 Hz.*
7. Click on the "Measurements And Sample Rate" button on the signal generator window. A list of measurements will open. Choose to measure both the output voltage and the output current.
8. Click the AUTO button of the signal generator. This makes the signal generator work only when the START button is pressed.
9. Drag the "Voltage, ChA" icon from the data list and drop it on the "Scope" icon in the displays list. The scope screen will open. Note that the volts/div (vertical axis) and the ms/div (horizontal time axis) can be controlled separately

*alue is appropriate for the suggested 100-Ω resistor.

10. Drag the "Output Current" icon and drop it in the middle of the scope window. A separate vertical control for the current will appear on the scope window, below the voltage control. The colors of the control boxes are also the colors that the voltage and current plots will have on the screen. ● CI Figure 3 shows what the screen should look like after the setup is complete.

◼CI EXPERIMENTAL PROCEDURE

A. *Resistive Circuit*

1. Measure the value of the resistor using a multimeter. Report the value in CI Data Table 1 in the Laboratory Report.
2. Connect the resistor to the output source of the 750 interface.
3. Put alligator clips on the probes of the voltage sensor, and connect the voltage sensor across the resistor. (Points *A–B*, as illustrated in ● CI Figure 4.) Check the polarity: The positive (red) lead of the voltage sensor must be connected to the end of the resistor that is connected to the positive output of the source (labeled with a wave symbol on the 750 interface.)
4. Press the START button. Data will be monitored in the scope, without being stored. Adjust the vertical and horizontal controls until both plots are clearly visible.
5. Press the STOP button. The process of pressing START/STOP can be repeated as many times as needed to get a nice plot on the screen.
6. Print the graph and label it "Resisti
printer is not available, use different color pencils to

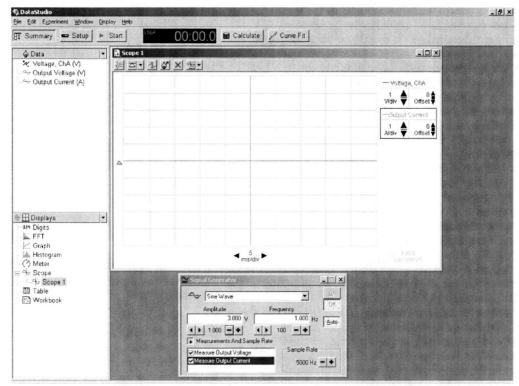

CI Figure 3 Data Studio setup. The scope display will monitor data for voltage and current in the circuit. The vertical and horizontal scales can be controlled independently, as in a regular oscilloscope. (Data displayed using DataStudio Software. Reprinted courtesy of PASCO scientific.)

trace the voltage and current plots in the graph, to make it easy to distinguish between them. Attach the graph to the Laboratory Report.

7. Looking directly at the graph, determine the phase difference, if any, between the voltage and the current for the resistive circuit. Is there a time difference between the peak of the voltage and the peak of the current? Report the values in CI Data Table 1.

8. Complete all other blanks in the table that have to do with the resistive circuit.

B. *Capacitive Circuit*

1. Measure the value of the capacitor using a multimeter. If a multimeter that measures capacitance is not available, use the value marked on the capacitor. Report the value in CI Data Table 1.

2. Introduce the capacitor into the previous circuit, connecting it in series with the resistor and the output voltage source. ● CI Fig. 5 is the circuit diagram for this part.

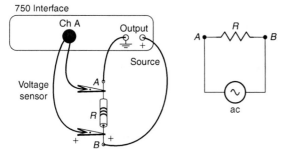

CI Figure 4 Equipment setup for resistive circuit. A resistor is connected to the output generator of the Science Workshop 750 interface. A voltage sensor measures the voltage across the resistor.

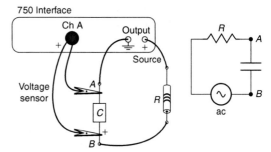

CI Figure 5 Equipment setup for capacitive circuit. The voltage sensor monitors the voltage across a capacitor connected in series with a resistor

3. Change the voltage sensor to measure the voltage across the capacitor. Some capacitors may be sensitive to polarity. Check before starting: Polar capacitors usually have one end or terminal marked negative or "−." These capacitors can be destroyed by reversing the voltage. Check with the instructor if you are not sure about the polarity of the capacitor.

4. Open the signal generator window and change the sine function frequency to 20.0 Hz.

5. Press the START button to begin monitoring data. Adjust the vertical and horizontal controls until both plots are clearly visible.

6. Press the STOP button.

7. Print the graph and label it "Capacitive Circuit." If a color printer is not available, use different color pencils to trace the voltage and current plots in the graph, to make it easy to distinguish between them. Attach the graph to the Laboratory Report.

8. It should be obvious that in this case, the voltage and the current are not in phase. To measure the phase angle, follow the steps below. The same steps will be used later on for the inductive circuit.

   a. Press the Smart Tool button on the scope toolbar. (It is the leftmost button in the toolbar.) A set of cross-hairs will appear. The cross-hairs can be dragged across the plot by grabbing them by the central square, when the mouse pointer becomes a hand. Drag the cross-hairs to a peak of the voltage graph.

   b. If the mouse is slightly moved to the bottom right corner of the central square of the cross-hairs, a small triangle appears in that corner. When the triangle is there, click and drag the cross-hairs to the next peak of the voltage graph. Note that the time difference between the two points is automatically presented on screen. This is the period of the voltage function. Report the period in CI Data Table 1. (● CI Fig. 6 shows an example of what this technique looks like on the screen.)

   c. With the same technique, measure the time difference Δt between a voltage peak and the nearest current peak. Report the time difference in CI Data Table 1.

9. Calculate the phase difference, in degrees, between the voltage and the current, using CI Eq. 1.

10. Complete the rest of the Capacitive column by carefully studying the graph.

C. Inductive Circuit

1. Measure the actual value of the inductor using a multimeter. If a multimeter that measures inductance is not available, use the value marked on the inductor. Report the value in CI Data Table 1 in the Laboratory Report.

2. Remove the capacitor from the circuit and put the inductor in its place. ● CI Fig. 47.7 is the circuit diagram for this part.

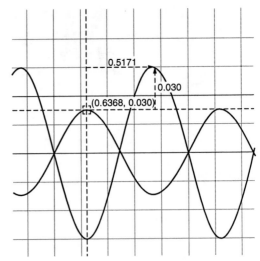

CI Figure 6 Sample measurement of Δt. This diagram is an example of what the technique for measuring interval looks like on the screen.

3. Connect the voltage sensor to measure the voltage across the inductor.

4. Open the signal generator window and change the sine function frequency to 390 Hz.*

5. Repeat the data collection process as done before for the capacitive circuit. This time, label the graph "Inductive Circuit."

D. Series RLC Circuit

In this part of the experiment, the resistor, capacitor, and inductor used in parts A, B, and C will all be connected in series with the voltage source. The behavior of the circuit will be studied at two different voltage frequencies, one "low" (100 Hz) and the other "high" (1000 Hz).[†] The phase

CI Figure 7 The inductive circuit. An inductor replaces the capacitor in the circuit. This time, the voltage sensor will measure the voltage across the inductor.

---

* Again, remember that the suggested values for the frequency are specific for the suggested 20-mH inductor. If the scope trace looks choppy, adjust the frequency until a smooth trace is obtained. The frequency can be changed with the scope working.

† The "low" and "high" frequencies are chosen to be well belo
well above the resonance frequency of the circuit. Adjust accordingly if using different than suggested values for R, C, and L.

differences between the voltage and the current, and the observation of which leads and which lags in each case, will enable us to classify each circuit as resistive, capacitive, or inductive.

1. Using the multimeter values of R, C, and L, fill in CI Data Table 2. These are the theoretical values for all quantities concerning the two circuits in this part of the experiment. Do not forget units!

2. Connect the resistor, the capacitor, and the inductor in series with the voltage source.

3. From the sensor list, select two other voltage sensors and connect them to the interface. There should be a voltage sensor in each of the channels A, B, and C of the interface.

4. Connect one voltage sensor across the resistor (this will measure V_R), a second across the capacitor (to measure V_C), and the third across the inductor (to measure V_L). Keep track of which is which! Remember that the red lead of the voltage sensor is the positive lead. It is very important that all sensors are connected with the correct polarity.

5. From the data list, drag all voltages (V_{out}, ChA, ChB and ChC) to the scope window.

6. Remove the output current from the scope display by clicking once in the vertical control for the current and hitting "delete." The current will not be directly monitored in this case, but its behavior, in comparison with the voltages, will be indirectly monitored by looking at V_R. Why? (*Hint:* See the results for the resistive circuit. How do V and I behave with each other in the resistor?)

7. **The Low-Frequency Circuit.** In the signal generator window, set the sine function frequency to 100 Hz, and the voltage output to 3.0 V. Press the START button to monitor the data. (If the data look choppy, increase the sample rate in the setup window.) Adjust all the vertical axis to the same number of volts/div, and adjust the horizontal scale until a clear picture of all the voltages is acquired. Press STOP and print the graph. Label this graph "Low-Frequency *RLC*."

8. Use the Smart Tool to measure the maximum voltage across the resistor, across the capacitor, and across the inductor. Report these values in CI Data Table 3.

9. Determine the phase-angle difference, ϕ, between the output voltage (V_{out}) and the voltage across the resistor (V_R). This is also the phase difference between the voltage and current in the circuit. (Why?)

10. How well does the measured angle compare to the expected, theoretical angle? Compare them by taking a percent error.

11. **The High-Frequency Circuit.** Open the signal generator window and change the sine function frequency to 1000 Hz. Repeat the procedure and calculations, as done for the low-frequency circuit. It may be necessary to increase the sampling rate of the voltage sensors.

EXPERIMENT 36

Phase Measurements and Resonance in ac Circuits

CI *Laboratory Report*

Parts A, B, and C

CI DATA TABLE 1

Purpose: To compare phase differences among the three types of ac circuits.

	Resistive R _____	Capacitive C _____	Inductive L _____
Period, T			
V-to-I time shift, Δt			
Phase angle, ϕ			
Are V and I in phase or out of phase?			
Which leads, V or I?			

Do not forget to include the graphs at the end of the report.

(continued)

Part D

CI **DATA TABLE 2: Theoretical Calculations for *RLC* Circuit**

Purpose: To compare theoretical values with experimental results.

	Low-frequency circuit $f = 100$ Hz	High-frequency circuit $f = 1000$ Hz
$\omega = 2\pi f$		
$X_C = 1/\omega C$		
$X_L = \omega L$		
$Z = \sqrt{R^2 + (X_L - X_C)^2}$		
$\phi = \tan^{-1}\dfrac{X_L - X_C}{R}$		

CI **DATA TABLE 3: Experimental Measurements for the *RLC* Circuit**

Purpose: To determine whether the circuits are resistive, capacitive or inductive, and to compare measured values with the theoretical values of ϕ.

		Low-frequency circuit 100 Hz	High-frequency circuit 1000 Hz
Maximum values	V_{out}		
	V_R		
	V_C		
	V_L		
Period, T			
Time shift, Δt (V_{out}-to-V_R)			
Phase angle, ϕ			
Percent error			

Do not forget to include the graph at the end of the report.

EXPERIMENT 36   *Laboratory Report*

CI QUESTIONS

Parts A, B, and C. Resistive, Capacitive, and Inductive Circuits

1. What can be concluded about the phase difference between voltage and current for each of the following:
   a. A resistive circuit

   b. A capacitive circuit

   c. An inductive circuit

2. Does the voltage across a resistor lead or lag the current, or are they in phase?

3. Does the voltage across a capacitor lead or lag the current, or are they in phase?

4. Does the voltage across an inductor lead or lag the current, or are they in phase?

Part D. Series *RLC* Circuit

1. Are the answers to Questions 2, 3, and 4 above the same for the *RLC* circuit?

(continued)

2. Consider the low-frequency circuit.
   a. Which leads, V_{out} or V_R?

   b. Which is higher, V_C or V_L?

   c. Therefore, this circuit is _____.
      (resistive, capacitive, or inductive?)

3. Consider the high-frequency circuit.
   a. Which leads, V_{out} or V_R?

   b. Which is higher, V_C or V_L?

   c. Therefore, this circuit is _____.
      (resistive, capacitive, or inductive?)

Electromagnetic Induction

/TI/ *Advance Study Assignment*

Read the experiment and answer the following questions.

1. Consider a straight wire in the plane of this paper and parallel to this printed line. If the wire carried a conventional current from left to right, what would be the directions of the magnetic field on each side of the wire? (Make a sketch.)

2. What does magnetic flux measure?

3. Explain Faraday's law of induction.

(continued)

4. Explain how Lenz's law gives the direction of the induced current in a wire when the change of flux is known.

5. What is magnetic permeability?

CI *Advance Study Assignment*

Read the experiment and answer the following questions.

1. What is represented by the area under the curve of a voltage (induced)-versus-time graph?

2. Why is it necessary that the coil used in this experiment be tightly wound (closed packed)?

Electromagnetic Induction

OVERVIEW

Experiment examines electromagnetic induction with TI and/or CI procedures. This is done with magnets and coils. The TI procedure uses a galvanometer to detect current in a coil.

In the CI procedure, a voltage sensor is used to measure the induced voltage. A graph of $\mathscr{E}$ versus t enables us to find the magnetic flux from the area under the curve.

INTRODUCTION AND OBJECTIVES

Electromagnetic induction is one of the most important and most importantly applied relationships between electricity and magnetism. A classic discovery was made by the Danish physicist Hans Oersted in 1820 when he noticed the deflection of the needle of a compass near a current-carrying wire. This led to the conclusion that a magnetic field is produced by an electrical current.

A reasonable question to ask is whether the reverse is possible. That is, can a magnetic field produce an electrical current, or potential difference, in a conductor? The English scientist Michael Faraday investigated this question and found that the answer was yes, under certain conditions. This phenomenon is expressed in Faraday's famous law of induction, which is the basis for generation of electrical power.

TI OBJECTIVES

After performing this experiment and analyzing the data, you should be able to:

1. Apply the right-hand rule to a current to determine the direction of the associated magnetic field.
2. Explain Faraday's law of induction and how a voltage can be generated.
3. Describe Lenz's law and its implications.
4. Explain what is meant by magnetic permeability.

CI OBJECTIVES

1. Experimentally measure the magnetic flux through a coil of wire when a magnet travels through it.
2. Experimentally observe the generation of an induced emf when a magnet travels through a coil of wire.

Electromagnetic Induction

EQUIPMENT NEEDED

- Pair of insulated cylindrical coils (many turns on secondary relative to primary)*
- Iron and brass or aluminum-core rods
- Two bar magnets of different pole strengths
- Low-voltage dc power supply or dry cell
- Magnetic compass

- Knife switch
- Portable galvanometer
- Metric ruler or meterstick
- Connecting wires
- 1 sheet of Cartesian graph paper

* The primary coil is to be inserted into the secondary coil.

THEORY

Oersted discovered that a magnetic field is associated with an electrical current. He further established a relationship between the direction of the current and the direction of the magnetic field. The magnetic field lines are found to be closed circles around the current-carrying wire, as may be illustrated with a compass.

The **direction of the magnetic field** is given by a **right-hand rule** (● TI Fig. 1):

If a current-carrying wire is grasped with the right hand with the thumb extended in the direction of the conventional current, the curled fingers indicate the circular sense of the magnetic field lines.

The direction of the magnetic field at any point is then tangential to the circle. (Recall that the direction of the con-

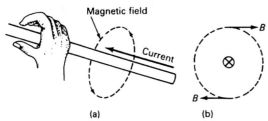

Magnetic field

Current

B

B

(a)   (b)

TI Figure 1   Right-hand rule for determining the direction of a magnetic field. (a) When the thumb of the right hand points in the direction of the conventional current, the fingers curl around the conductor in the circular sense of the magnetic field. (b) The magnetic field vector is tangent to the circle at any point on the circle. (The cross ⊗ symbol indicates that the current direction is into the page—that is, as viewed along the conductor in the direction of the current. A dot ⊙ symbol would indicate a current coming out of the page. Think of viewing an arrow. The dot corresponds to the tip that we would view if the arrow were coming out of the page, whereas the cross corresponds to the bowstring slot that we would view if the arrow were moving away.)

ventional current is the direction in which *positive* charges would flow in a circuit.)

Faraday investigated the possible reverse effect, a current being produced by a magnetic field in the vicinity of a wire. No effect was found with a stationary magnet or magnetic field. However, it was discovered that a current is induced when there is relative *motion* or a *changing* magnetic field.

Thus, electromagnetic induction involves a time-varying magnetic field—for example, when a magnet is moved toward or away from a loop of wire. Since magnetic induction depends on *relative* motion, the same effect could be produced by moving the coil toward or away from a stationary magnet.

Investigation led Faraday to the conclusion that the important factor in electromagnetic induction was the time rate of change of the magnetic field **B** through a loop. The "total" magnetic field through a fixed loop of wire can be characterized by what is called **magnetic flux Φ**:

$$\Phi = \mathbf{B} \cdot \mathbf{A} = BA \cos\theta \qquad \textbf{(TI 1)}$$

where A is the cross-sectional area of the loop and θ is the angle between a normal to the plane of the loop and the magnetic field.[†]

The result of this work was **Faraday's law of induction**, which relates the induced voltage, or emf, in a wire to the time rate of change of flux:

$$\mathscr{E} = -\frac{\Delta\Phi}{\Delta t} \qquad \textbf{(TI 2)}$$

where $\mathscr{E}$ is the average value of the induced emf over the time interval Δt. (By Ohm's law, the average induced current is $I = \mathscr{E}/R$, where R is the resistance of the circuit.)

[†] The SI unit for the magnetic field B is the tesla (T). Hence the units of the magnetic flux Φ are tesla-m² (T-m²). An older unit of magnetic field is weber/m² (Wb/m²), and the flux has the unit of weber (Wb) in this case.

Substituting the magnetic flux (TI Eq. 1) into TI Eq. 2 shows that a change in flux can result from changes other than a change in magnetic field. In general,

$$\frac{\Delta \Phi}{\Delta t} = -\left[\left(\frac{\Delta B}{\Delta t} \right)(A \cos \theta) + B \left(\frac{\Delta A}{\Delta t} \right)(\cos \theta) \right.$$
$$\left. + BA \left(\frac{\Delta (\cos \theta)}{\Delta t} \right) \right] \qquad \textbf{(TI 3)}$$

The first term within the brackets represents a flux change due to a time-varying magnetic field (with area and orientation of the loop constant). A time-varying magnetic field is easily obtained by varying the current producing it, as in the case of an alternating current.

The second term represents flux change due to a time-varying loop area. This could occur if a circular loop were stretched in a place or if it had an adjustable circumference. The third term represents a change in the orientation of the loop with time. This occurs when a loop is rotated in a magnetic field.

In all cases, the number, or density, of field lines through a loop changes. In this experiment, we are concerned only with the effect of the first term in TI Eq. 3—a flux change due to a time-varying magnetic field.

The negative sign in TI Eq. 2 expresses another important law of electromagnetic induction, **Lenz's law,*** which gives the direction of the induced current:

An induced current flows in such a direction that its effects oppose the change that produces it.

* First stated by Heinrich Lenz (1804–1865), a Russian physicist.

Essentially, this means the induced current gives rise to a magnetic field that opposes the change in the original magnetic field. If this were not the case and the magnetic field arising from the induced current augmented the original field (i.e., was in the same direction as the original field), the induced field would increase the flux, which would increase the current, which would increase the flux, and so on. This would give rise to a something-for-nothing situation that would violate the conservation of energy.

To investigate this phenomenon, consider producing a time-varying magnetic field through a stationary loop, as illustrated in ● TI Fig. 2. When the switch in the battery circuit is closed, the current in the loop goes from zero to a constant value in a short time. During this time, the magnetic field associated with the current also increases or changes with time.

The magnetic flux through the adjacent loop then changes with time, and an induced current momentarily flows in the loop, as indicated by the galvanometer deflection. The induced current goes to zero when the current in the battery circuit has a constant value. (Why?) Similarly, when the switch is opened, the magnetic field decreases and a current flows momentarily in the detector loop in the direction opposite to that of the first induced current. (Why?)

If there are a number of loops N in the detector circuit, the flux change through each loop contributes to the induced current, or emf, and Faraday's law becomes

$$\mathcal{E} = -N \frac{\Delta \Phi}{\Delta t} \qquad \textbf{(TI 4)}$$

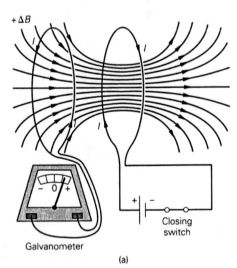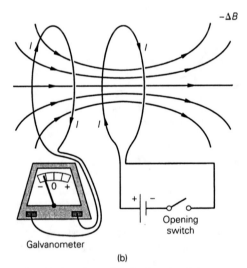

TI Figure 2 Electromagnetic induction. (a) A time-varying field is produced when the switch is closed and the current goes from zero to some steady value. As the current increases, the field builds up, and the number of field lines through the nearby loop changes, including the current in the loop, as indicated by the galvanometer deflection. (b) When the switch is opened, the current decays to zero, the field lines decrease, and a current is induced in the opposite direction in the nearby loop. Both processes take place in a very short time.

Similarly, the magnitude of the magnetic field (number of field lines) would be increased if there were a number of loops in the battery circuit. (Loops of wire wound in a tight helix so as to form a coil are called a **solenoid**.)

It can be shown that the magnetic field near and along the axis of a current-carrying solenoid is given by

$$B = \mu_0 n I \qquad \textbf{(TI 5)}$$

where n is the linear turn density of the coil (i.e., the number of turns per unit length N/L, where L is the length of the coil).

The constant μ_0 is called the **permeability of free space** (vacuum or, approximately, air) and indicates that the solenoid has an air core. If a material with a magnetic permeability μ is used as a solenoid core, μ_0 is replaced by μ in TI Eq. 5. The **permeability** μ expresses the magnetic characteristics of a material. Some materials with large permeabilities can be used to increase greatly the magnetic field in a current-carrying solenoid.

/TI/ EXPERIMENTAL PROCEDURE

1. In this experiment, it is important to know the direction of the induced current in a circuit. This is related to the positive and negative deflections of the galvanometer.

   To establish the direction of the galvanometer deflection due to a known current direction, connect one terminal of the dry cell (or dc power supply at 1.5 to 3.0 V) to one terminal of the galvanometer and the other source terminal *through a large resistance* to the other galvanometer terminal (● TI Fig. 3). Use yourself as the large resistance. (Really get into the circuit, so to speak.)

   From the known polarity of the source, relate the galvanometer deflection to the direction of current flow. For conventional current (assuming positive charge carriers), the current flows *from* the positive

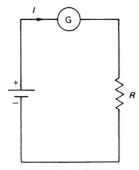

TI Figure 3   Galvanometer deflection.   A circuit diagram for determining the direction of the galvanometer deflection with respect to the current direction. See text for description.

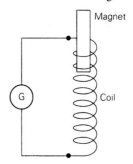

TI Figure 4   Induced current.   The experimental setup for determining the induced current due to a moving magnet. See text for description.

source terminal. Galvanometer deflections to the right are usually labeled as positive and deflections to the left as negative. It is convenient to have a conventional current entering the positive galvanometer terminal to give a positive deflection.

A. Induced Current

2. Connect the galvanometer to the terminals of the secondary coil (the larger coil with the greater number of turns) as shown in ● TI Fig. 4. Use the compass to determine the relative strengths of the bar magnets.

   Then move the stronger magnet into and out of the coil, noting and recording the effects (relative magnitude and direction of the galvanometer deflection) with (a) the speed at which the magnet is moved and (b) the change of the magnet's polarity. Also note the effect when the magnet is stationary in the coil. Record your results in the Laboratory Report.

3. Move magnet 2 in and out of the coil, noting and recording the effects as in Procedure 2. Draw conclusions based on the experimental data.

B. Induced Magnetic Field

4. Set up the primary coil circuit as shown in ● TI Fig. 5, with the knife switch S open. (A secondary coil circuit is not needed in this section and Part C below.) Close the switch and, with the compass, investigate the magnetic field around the coil. Make a sketch of the field pattern in the Laboratory Report.

C. Lenz's Law

5. Open the switch and insert the stronger bar magnet into the primary coil almost the full length of the magnet. Close the switch and slowly remove the magnet from the coil. Note, record, and explain any observed effects.

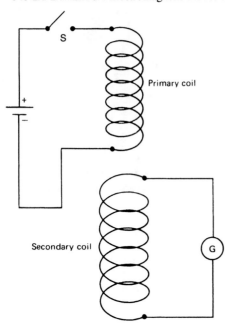

TI Figure 5 Coil effects.   A diagram illustrating the experimental setup for investigating electromagnetic induction. The primary coil is inserted into the secondary coil, as shown in the photos of the actual coil apparatus. [Courtesy of (top) PASCO Scientific and (bottom) Sargent-Welch.]

D. Permeability

6. Open the switch and insert the primary coil into the secondary coil, which is connected to the galvanometer (TI Fig. 5). Close and open the switch, noting and recording the magnitude and direction of the galvanometer deflections in each case.

7. Repeat procedure 6 with each of the two metal cores. Draw conclusions based on the experimental data.

E. Length of Coil

8. Measure and record the length of the primary coil. With the switch open, insert the primary coil with the iron core completely into the secondary coil. Make a series of observations of the magnitudes of the deflections as the switch is opened and closed, withdrawing the primary coil 1 cm between the observations. Record the length of the primary coil still inside the secondary coil in each case.

9. Find the average magnitude of the plus and minus deflections for each observation, and plot a graph of the average deflection magnitude versus the length of the primary inside the secondary. Use an abscissa scale of decreasing length. Interpret the results.

EXPERIMENT 37

Electromagnetic Induction

/TI/ *Laboratory Report*

A. Induced Current

	Maximum deflection (scale divisions)	Current direction (+ or −)

Magnet 1

(North end of magnet toward coil) Motion toward coil

v_1 _____ _____

$v_2 > v_1$ _____ _____

Motion away from coil

v_1 _____ _____

$v_2 > v_1$ _____ _____

(South end of magnet toward coil) Motion toward coil

v_1 _____ _____

Magnet 2

(North end of magnet toward coil) Motion toward coil

v_1 _____ _____

$v_2 > v_1$ _____ _____

Motion away from coil

v_1 _____ _____

$v_2 > v_1$ _____ _____

(South end of magnet toward coil) Motion toward coil

v_1 _____ _____

Conclusions

Don't forget units

(continued)

B. Induced Magnetic Field

(Sketch)

C. Lenz's Law

Observation and explanation

D. Permeability

	Galvanometer deflection (scale divisions)	Secondary current direction (+ or −)
Air core		
Switch closed	_____	_____
Switch open	_____	_____
Iron core		
Switch closed	_____	_____
Switch open	_____	_____
_____ **core**		
Switch closed	_____	_____
Switch open	_____	_____

EXPERIMENT 37

Conclusions

E. Length of Coil

⊤I DATA TABLE   Primary coil length L _____

Length of primary in secondary	Switch-closed deflection	Switch-open deflection	Average deflection

Interpretation of data

(continued)

QUESTIONS

1. A metal rod on a horizontal metal frame forms a conducting loop, as illustrated in
   ● TI Fig. 6. The rod is moved along the frame with a constant velocity v.

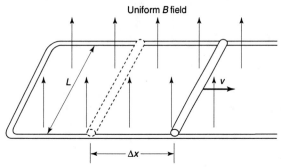

Uniform B field

TI Figure 6

(a) By which component of TI Eq. 3 is a voltage induced in the loop?

(b) Show that the induced voltage is given by $V = -BLv$.

2. Suppose that a bar magnet is dropped through a horizontal loop of wire connected to a
   galvanometer. Explain what would be observed on the galvanometer as the magnet enters,
   as it is in the middle, and as it leaves the loop. Why?

3. Describe the change of flux through, and the induced current in, a loop of wire rotated in a
   uniform magnetic field (rotational axis of loop perpendicular to field).

EXPERIMENT 37

Electromagnetic Induction

 EQUIPMENT NEEDED

- 1 short coil of many turns of copper wire, large enough in diameter to allow the magnet to drop through it. Suggested coils: The PASCO EM-8656 AC/DC Electronics Laboratory includes a small-diameter coil mounted on a board. Rolling a piece of paper into a tube and placing it in the hole of the coil makes it easy to drop a thin magnet from well above the coil. The PASCO EM-6711 is a larger-diameter coil. With a strong magnet it gives very good voltage readings.
- 1 bar magnet, of a size that permits it to be dropped through the coil.
- foam pads
- 1 voltage sensor (PASCO CI-6503)

THEORY

According to Faraday's law, if there is a changing magnetic flux through a coil, then an induced current appears in the coil. Faraday's law is more formally written as

$$\mathscr{E} = -N\frac{d\Phi_B}{dt} \qquad \textbf{(CI 1)}$$

where $\mathscr{E}$ is the voltage induced in the coil, which is sometimes also called the induced emf. N is the number of turns of wire in the coil, which is assumed to be tightly wound so that the same magnetic flux Φ_B passes through all turns.

In this experiment, a voltage sensor will be used to measure the induced voltage in a coil when a magnet is dropped through it. Variations in the polarity of the induced voltage can be easily observed as the voltage changes from positive to negative in a graph of voltage (induced) versus time. The magnetic flux can also be determined from the graph as follows. From equation CI 1, notice that

$$\mathscr{E}\,dt = N\,d\Phi_B$$

Integrating both sides of the equation, we have

$$N\Phi_B = \int \mathscr{E}\,dt \qquad \textbf{(CI 2)}$$

This integral is the area under the curve of $\mathscr{E}$ versus t. Since the number of turns of the coil is a constant, the area under the graph will be proportional to the magnetic flux through the coil.

SETTING UP DATA STUDIO

1. Open Data Studio and choose "Create Experiment."
2. From the sensor list, choose a voltage sensor. Connect the voltage sensor to the interface, as shown in the computer screen.
3. Double-click on the "Voltage Sensor" icon. The sensor properties window will open. Set the sampling rate to 1000 Hz, and then click OK to close the window.
4. In the main experiment window, click the "Options" button. The sampling options window will open. Set Automatic Stop to 0.5 s, and then click OK. This will set an automatic stop after 0.5 s of data recording.
5. Create a graph by dragging the "Voltage" icon from the data list and dropping it on the "Graph" icon in the displays list. A graph of voltage versus time will open in a window called Graph 1.

EXPERIMENTAL PROCEDURE

1. Connect the voltage sensor to the coil.
2. Position the coil flat against a table, but with its center beyond the edge of the table. In this way, a magnet can fall freely through it.
3. Place foam pads on the floor under the coil to catch the magnet. Permanent magnets tend to be brittle, so do not let the magnet strike the floor. Crumpled newspaper in a waste basket also works fine to catch the magnet.
4. Hold the magnet north end up so that the south pole is about 3 cm above the center of the coil. Click the START button and let the magnet drop vertically through the center of the coil. Data recording will stop automatically. Repeat the process several times until you get an idea of what the plot looks like.
5. The only important part of the plot is the two peaks, so zoom in to display only this part; then print a copy of the plot and attach it to the Laboratory Report. If no printer is available, make a careful drawing of the

graph, paying special attention to the peaks and their voltage and time values.

6. From the graph, determine the maximum positive voltage and the maximum negative voltage. Record the values in the Laboratory Report.

7. The statistics button is a drop menu marked with a Greek letter sigma, Σ, on the graph toolbar. From this menu, select "Area." This will calculate the area under the curve. Use the cursor to select only the portions of the graph for which you want the area. Measure and record in the report:

    a. The area of the first peak (incoming flux).

    b. The area of the second peak (outgoing flux).

Don't forget units!

C I   EXPERIMENT 37

Electromagnetic Induction

CI *Laboratory Report*

Induced emf:

    Maximum positive voltage _____

    Maximum negative voltage _____

Magnetic flux:

    Incoming peak area _____

    Outgoing peak area _____

Do not forget to include the graph at the end of the report.

CI QUESTIONS

1. Is the incoming flux (area) equal to the outgoing flux? Explain why or why not.

2. Why are the peaks in opposite direction?

3. Why is the outgoing-voltage peak higher than the incoming-voltage peak?

(continued)

Optional (requires having studied Lenz's law):

4. The diagrams in ● CI Fig. 1 represent top views of a coil laid flat. The magnetic field
   B is represented as coming out of the plane of the paper. The left diagram represents a
   magnet approaching the coil from above (getting closer to the coil), with the south end
   going into the coil first. The right diagram is the magnet below the coil, moving away from
   it toward the floor.
   a. In each case, fill in the blank at the bottom to indicate whether the flux through the coil
      is increasing or decreasing?
   b. Apply Lenz's law to the coil and draw, at the center of each coil, the direction of the
      induced magnetic field.
   c. Draw the direction of the induced current in the coil.

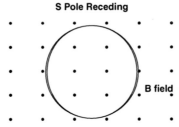

Magnet approaching the coil from above,   Magnet below coil, moving away
            S-pole entering first   from the coil

      S Pole Approaching   **S Pole Receding**

The magnetic flux is _____.   The magnetic flux is _____.

CI Figure 1   Top views of a laid-flat coil.

Name _____ Section _____ Date _____

Lab Partner(s) _____

EXPERIMENT 38

The Mass of an Electron:
e/m Measurement
/TI/ *Advance Study Assignment*

Read the experiment and answer the following questions.

1. Why does an electron traveling perpendicular to a uniform magnetic field describe a circular path?

2. What causes the fan-shaped pattern in the vacuum tube?

3. How is the radius of curvature of the electron path measured in the experiment?

4. How does changing the current in the solenoid change the curvature of the electron path?

EXPERIMENT 38

The Mass of an Electron: *e/m* Measurement

INTRODUCTION AND OBJECTIVES

The mass of an electron, 9.1×10^{-31} kg, is much too small to measure directly. However, using electromagnetic methods involving the charge of the electron in electric and magnetic fields, we can determine the electron mass indirectly.

In several situations, the ratio of the electron's charge to its mass (e/m) can be determined in terms of directly measurable parameters. Then, knowing the value of the electron's charge ($e = 1.6 \times 10^{-19}$ C)—for example, from

Millikan's oil-drop experiment—we can compute the mass of an electron. This is the procedure we will use in this experiment.

After performing this experiment and analyzing the data, you should be able to:

1. Explain on what the velocity of an electron moving in a magnetic field depends.
2. Explain how the charge/mass (e/m) ratio of an electron can be measured experimentally.

EQUIPMENT NEEDED

- Tuning-eye vacuum tube (6AF6)
- Air-core solenoid
- Variable dc power supply (250 V dc)
- ac power supply (6.3 V ac)
- dc power supply (12 V dc)

- Rheostat
- dc ammeter (0 to 5 A)
- Connecting wires
- Vernier calipers and metric ruler
- Wooden dowels of different diameters

THEORY

An electron traveling with a speed v perpendicular to a uniform magnetic field **B** will experience a force **F** with a magnitude of

$$F = evB \tag{1}$$

where e is the charge of the electron.

The direction of the force ($\mathbf{F} = e\mathbf{v} \times \mathbf{B}$) is perpendicular to the plane defined by **v** and **B**. Thus, the force is always at right angles to the direction of the electron's motion. This causes the electron to describe a circular path of radius r, with the magnetic force supplying the required centripetal force of magnitude F_c. That is,

$$F = evB = F_c = \frac{mv^2}{r} \tag{2}$$

where m is the mass of the electron. Solving for v, we have

$$v = \frac{eBr}{m} \tag{3}$$

The initial electron speed v is normally attained by accelerating the electron through an electric potential V, and by the work-energy theorem,

$$W = eV = \tfrac{1}{2}mv^2 \tag{4}$$

Then, squaring Eq. 3 and substituting for v^2, we have, after rearranging,

$$\frac{e}{m} = \frac{2V}{B^2 r^2} \tag{5}$$

Hence, by knowing or measuring e, V, B, and r, we can compute the mass of an electron.

In this experiment, the electron(s) are produced, accelerated, and deflected in a commercial "tuning-eye" vacuum tube (● Fig. 1). Such tubes were once commonly used in tuning radios. The magnetic field is supplied by a solenoid. In the tube, the electrons emitted by the cathode are accelerated by the potential difference applied between the center cathode and the outer conical anode (● Fig. 2).

The electrons move radially outward from the cathode coil, reaching nearly their maximum speed by the time they

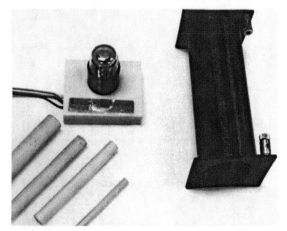

Figure 1   Experimental equipment.   The "tuning-eye" vacuum tube, air core solenoid, and wooden dowels used in measuring *e/m*.

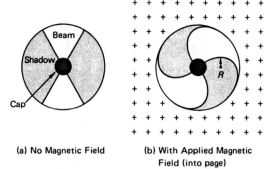

(a) No Magnetic Field   (b) With Applied Magnetic
                                   Field (into page)

Figure 3   "Tuning eye."   Illustrations of the observed patterns of the "tuning eye" in the tube.

emerge from beneath the black metal cap covering the center of the tube. The conical anode is coated with a fluorescent material (e.g., zinc sulfide), which emits light when the electrons strike it. Hence, we can visually trace the beam (● Fig. 3).

The beam pattern in the tube appears fan-shaped owing to two deflecting electrodes in the tube (Fig. 2). These electrodes are connected to the cathode; therefore, they are negatively charged and repel electrons moving toward them from the cathode, forming wedge-shaped shadows.

When a uniform magnetic field is applied parallel to the axis of the tube, the electrons are deflected in an almost circular path. This is indicated by the edge of the shadow, from which the radius of curvature can be measured (Fig. 3).

In the experiment, the magnetic field is produced by the solenoid, which conveniently fits over the tube. The magnitude of the magnetic field at the center of the solenoid coil carrying a current *I* is given by

$$B = \mu_o n I \qquad (6)$$

where $n = N/L$ is the turn density (number of turns N per coil length L in meters).

EXPERIMENTAL PROCEDURE

1. Set up the circuits as shown in the schematic diagrams in ● Fig. 4, and place the coil over the tube. Connect the tube leads to the ac and dc power sources and adjust the anode potential to about 150 V dc. Note the fan-shaped tuning-eye pattern.

   Then connect the solenoid to the 12-V dc power supply, and observe how the electron beam is deflected into a curved path as the coil current is increased from zero to a maximum of about 5.0 A.

2. The radius of curvature is measured by matching it to that of a wooden dowel, a dime, or some other appropriate circular object. The radius of the circular reference object is determined using vernier calipers.

3. With the anode voltage at 150 V dc, adjust the coil current until the curvature of the shadow matches the reference object. Record the ammeter reading in the Laboratory Report. Determine the current value for two more trials.

   Reverse the connections to the coil so the electron beam bends the other way. Make two more current determinations, and compute the average of all the measurements. Do not watch the ammeter while matching the radii.

4. Repeat procedure 3 for anode voltages of 200 and 250 V.

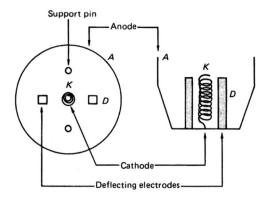

Support pin

Anode

(a) Top View   (b) Cutaway Side View

Figure 2   Elements of a "tuning-eye" tube.   See text for description.

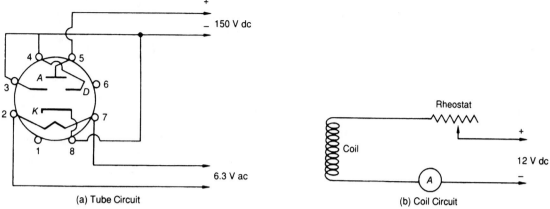

Figure 4   Experimental circuits.   Circuit diagrams for the *e/m* apparatus. See text for description.

5. Disconnect the apparatus from the power supplies. Measure the length L of the solenoid and count the number of turns N, so as to determine the turn density n if this is not given by the instructor. (The coil usually has 540 turns.)

6. Using the average current in each voltage case, compute the magnitude of the magnetic field B of the solenoid for each set of data.

7. Compute the charge/mass (e/m) ratio of the electron for each set of data, and find the average.

8. From the average value of e/m, compute the mass of the electron using $e = 1.6 \times 10^{-19}$ C. Compare this value to the accepted mass value of the electron.

EXPERIMENT 38

The Mass of an Electron: e/m Measurement

/TI/ *Laboratory Report*

DATA TABLE

Purpose: To determine the magnitude of B for computation of m_e.

Anode voltage ( )				
Radius r ( )				
Current ( )	Trial 1			
	2			
	3			
	4			
	5			
Average current				
Magnitude of field B ( )				
Calculated e/m				

Turn density

N _____

L _____

n _____

Average e/m _____

Computed m_e _____

Accepted m_e _____

Percent error _____

Calculations
(show work)

Don't forget units

(continued)

/TI/ QUESTIONS

1. What are the major sources of error in the experiment? Did you expect the percent error that you obtained? Explain.

2. Suppose that protons were emitted in the vacuum tube instead of electrons. How would this affect the experiment?

3. Suppose that only a constant solenoid current were available. Could the mass of an electron still be determined? Explain.

Material Properties

TABLE A1   Densities of Materials

Substance	(g/cm^3)	(kg/m^3)
Solids		
Aluminum	2.7	2.7 × 10^3
Brass	8.4	8.4 × 10^3
Copper	8.9	8.9 × 10^3
Glass		
crown	2.5–2.7	2.5–2.7 × 10^3
flint	3.0–3.6	3.0–3.6 × 10^3
Gold	19.3	19.3 × 10^3
Iron and steel		
(general)	7.88	7.88 × 10^3
Lead	11.3	11.3 × 10^3
Nickel	8.8	8.8 × 10^3
Silver	10.5	10.5 × 10^3
Wood		
oak	0.60–0.90	0.60–0.90 × 10^3
pine	0.35–0.50	0.35–0.50 × 10^3
Zinc	7.1	7.1 × 10^3
Liquids		
Alcohol		
ethyl	0.79	0.79 × 10^3
methyl	0.81	0.81 × 10^3
Carbon tetra-chloride	1.60	1.60 × 10^3
Gasoline	0.68–0.75	0.68–0.75 × 10^3
Glycerine	1.26	1.26 × 10^3
Mercury	13.6	13.6 × 10^3
Turpentine	0.87	0.87 × 10^3
Water	1.00	1.00 × 10^3
Gases (at STP):		
Air	0.001293	0.001293 × 10^3
Carbon dioxide	0.001975	0.001975 × 10^3
Helium	0.000179	0.000179 × 10^3
Hydrogen	0.000089	0.000089 × 10^3
Nitrogen	0.000125	0.000125 × 10^3
Oxygen	0.00143	0.00143 × 10^3

TABLE A2   Young's Modulus for Some Metals

Metals	(N/m^2)
Aluminum	6.5 × 10^{10}
Brass	9.0 × 10^{10}
Copper	12.0 × 10^{10}
Iron	
cast	9.0 × 10^{10}
wrought	19.0 × 10^{10}
Steel	19.2 × 10^{10}

TABLE A3   Coefficients of Linear Thermal Expansion

Substance	(1/C°)
Aluminum	24.0 × 10^{-6}
Brass	18.8 × 10^{-6}
Copper	16.8 × 10^{-6}
Glass	
window	8.5 × 10^{-6}
Pyrex	3.3 × 10^{-6}
Iron	11.4 × 10^{-6}
Lead	29.4 × 10^{-6}
Nickel	12.8 × 10^{-6}
Silver	18.8 × 10^{-6}
Steel	13.4 × 10^{-6}
Tin	26.9 × 10^{-6}
Zinc	26.4 × 10^{-6}

TABLE A4   Specific Heats

Substance	kcal/(kg-C°) or cal/(g-C°)	J/(kg-C°)
Aluminum	0.22	921
Brass	0.092	385
Copper	0.093	389
Glass	0.16	670
Iron	0.11	460
Lead	0.031	130
Mercury	0.033	138
Nickel	0.11	460
Silver	0.056	234
Steel	0.11	460
Tin	0.054	226
Water	1.00	4186
Zinc	0.093	389

TABLE A5   Color Code for Resistors (Composition Type)

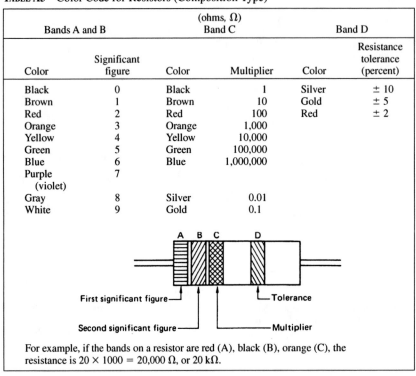

Bands A and B		Band C		Band D	
Color	Significant figure	Color	Multiplier	Color	Resistance tolerance (percent)
Black	0	Black	1	Silver	± 10
Brown	1	Brown	10	Gold	± 5
Red	2	Red	100	Red	± 2
Orange	3	Orange	1,000		
Yellow	4	Yellow	10,000		
Green	5	Green	100,000		
Blue	6	Blue	1,000,000		
Purple (violet)	7				
Gray	8	Silver	0.01		
White	9	Gold	0.1		

First significant figure —

Second significant figure —

— Tolerance

— Multiplier

For example, if the bands on a resistor are red (A), black (B), orange (C), the resistance is $20 \times 1000 = 20{,}000 \ \Omega$, or 20 k$\Omega$.

TABLE A6   Resistivities and Temperature
Coefficients

Substance	Resistivity ρ (Ω-cm)	Temperature coefficient (1/C°)
Aluminum	2.8×10^{-6}	0.0039
Brass	7×10^{-6}	0.002
Constantan	49×10^{-6}	0.00001
Copper	1.72×10^{-6}	0.00393
German silver (18% Ni)	33×10^{-6}	0.0004
Iron	10×10^{-6}	0.005
Manganin	44×10^{-6}	0.00001
Mercury	95.8×10^{-6}	0.00089
Nichrome	100×10^{-6}	0.0004
Nickel	7.8×10^{-6}	0.006
Silver	1.6×10^{-6}	0.0038
Tin	11.5×10^{-6}	0.0042

TABLE A7   Wire Sizes [American Wire Gauge (AWG)]

Gauge No.	Diameter	
	in.	cm
0000	0.4600	1.168
000	0.4096	1.040
00	0.3648	0.9266
0	0.3249	0.8252
1	0.2893	0.7348
2	0.2576	0.6543
3	0.2294	0.5827
4	0.2043	0.5189
5	0.1819	0.4620
6	0.1620	0.4115
7	0.1443	0.3665
8	0.1285	0.3264
9	0.1144	0.2906
10	0.1019	0.2588
11	0.09074	0.2305
12	0.08081	0.2053
13	0.07196	0.1828
14	0.06408	0.1628
15	0.05707	0.1450
16	0.05082	0.1291
17	0.04526	0.1150
18	0.04030	0.1024
19	0.03589	0.09116
20	0.03196	0.08118
21	0.02846	0.07229
22	0.02535	0.06439
23	0.02257	0.05733
24	0.02010	0.05105
25	0.01790	0.04547
26	0.01594	0.04049
27	0.01419	0.03604
28	0.01264	0.03211
29	0.01126	0.02860
30	0.01003	0.02548
31	0.008928	0.02268
32	0.007950	0.02019
33	0.007080	0.01798
34	0.006304	0.01601
35	0.005614	0.01426
36	0.005000	0.01270
37	0.004453	0.01131
38	0.003965	0.01007
39	0.003531	0.008969
40	0.003145	0.007988

TABLE A8 Major Visible Spectral Lines of Some Elements

Element	Wavelength (nm)	Color	Relative intensity
Helium	388.9	Violet	1000
	396.5 (near)	Violet	50
	402.6 (near)	Violet	70
	438.8	Blue-violet	30
	447.1	Dark blue	100
	471.3	Blue	40
	492.2	Blue-green	50
	501.5	Green	100
	587.6	Yellow	1000
	667.8	Red	100
	706.5	Red	70
Mercury	404.7	Violet	300
	407.8	Violet	150
	435.8	Blue	500
	491.6	Blue-green	50
	546.1	Green	2000
	577.0	Yellow	200
	579.0	Yellow	1000
	690.7	Red	125
Sodium	449.4	Blue	60
	449.8	Blue	70
	466.5	Blue	80
	466.9	Blue	200
	498.3	Green	200
	514.9	Green	400
	515.3	Green	600
	567.0	Green	100
	567.5	Green	150
	568.3	Green	80
	568.8	Green	300
	589.0	Yellow-orange	9000
	589.6	Yellow-orange	5000
	615.4	Orange	500
	616.1	Orange	500

Wavelengths of various colors

Color	Representative (nm)	General ranges (nm)
Red	650.0	647.0–700.0
Orange	600.0	584.0–647.0
Yellow	580.0	575.0–585.0
Green	520.0	491.2–575.0
Blue	470.0	424.0–491.2
Violet	410.0	400.0–420.0

Visible spectrum $\approx$ 400.0–700.0 nm

TABLE A9   Radioisotopes

Isotope	Half-life	Principal Radiations (MeV)		
		Alpha	Beta	Gamma
Barium-133	10.4 years			0.356
Bismuth-210	5.01 days	4.654, 4.691	1.161	
Carbon-14	5730 years		0.156	
Cesium-137	30.1 years		0.512, 1.173	
Barium-137m	2.6 min			0.662
Cobalt-60	5.26 years		0.315	
Iodine-131	8.07 days		0.606	
Lead-210	22.3 years		0.017, 0.061	0.0465
Manganese-54	312.5 days			0.835
Phosphorus-32	14.3 days		1.710	
Polonium-210	138.4 days	5.305		
Potassium-42	12.4 hours		3.52	
			1.97	
Radium-226	1600 years	4.781		0.186
		4.598		
Sodium-22	2.60 years	0.545		1.275
		1.82		
Strontium-90	28.1 years		0.546	
Thallium-204	3.78 years		0.763	
Uranium-238	4.5×10^6 years	4.195		0.48
Yttrium-90	64.0 hours		2.27	
Zinc-65	243.6 days		0.329	1.116

TABLE A10   Elements: Atomic Numbers and Atomic Weights

The atomic weights are based on $^{12}C = 12.0000$. If the element does not occur naturally, the mass number of the most stable isotope is given in parentheses.

	Symbol	Atomic number	Atomic weight		Symbol	Atomic number	Atomic weight
Actinium	Ac	89	(227)	Mercury	Hg	80	200.59
Aluminum	Al	13	26.9815	Molybdenum	Mo	42	95.94
Americium	Am	95	(243)	Neodymium	Nd	60	144.24
Antimony	Sb	51	121.75	Neon	Ne	10	20.179
Argon	Ar	18	39.948	Neptunium	Np	93	(237)
Arsenic	As	33	74.9216	Nickel	Ni	28	58.71
Astatine	At	85	(210)	Niobium	Nb	41	92.9064
Barium	Ba	56	137.34	Nitrogen	N	7	14.0067
Berkelium	Bk	97	(247)	Nobelium	No	102	(253)
Beryllium	Be	4	9.01218	Osmium	Os	76	190.2
Bismuth	Bi	83	208.9806	Oxygen	O	8	15.9994
Boron	B	5	10.81	Palladium	Pd	46	106.4
Bromine	Br	35	79.90	Phosphorus	P	15	30.9738
Cadmium	Cd	48	112.40	Platinum	Pt	78	195.09
Calcium	Ca	20	40.08	Plutonium	Pu	94	(224)
Californium	Cf	98	(251)	Polonium	Po	84	(209)
Carbon	C	6	12.011	Potassium	K	19	39.102
Cerium	Ce	58	140.12	Praseodymium	Pr	59	140.9077
Cesium	Cs	55	132.9055	Promethium	Pm	61	(145)
Chlorine	Cl	17	35.453	Protactinium	Pa	91	(231)
Chromium	Cr	24	51.996	Radium	Ra	88	(226)
Cobalt	Co	27	58.9332	Radon	Rn	86	(222)
Copper	Cu	29	63.545	Rhenium	Re	75	186.2
Curium	Cm	96	(247)	Rhodium	Rh	45	102.9055
Dysprosium	Dy	66	162.50	Rubidium	Rb	37	85.4678
Einsteinium	Es	99	(254)	Ruthenium	Ru	44	101.07
Erbium	Er	68	167.26	Rutherfordium	Rf	104	(257)
Europium	Eu	63	151.96	Samarium	Sm	62	150.4
Fermium	Fm	100	(253)	Scandium	Sc	21	44.9559
Fluorine	F	9	18.9984	Selenium	Se	34	78.96
Francium	Fr	87	(223)	Silicon	Si	14	28.086
Gadolinium	Gd	64	157.25	Silver	Ag	47	107.868
Gallium	Ga	31	69.72	Sodium	Na	11	22.9898
Germanium	Ge	32	72.59	Strontium	Sr	38	87.62
Gold	Au	79	196.967	Sulfur	S	16	32.06
Hafnium	Hf	72	178.49	Tantalum	Ta	73	180.9479
Hahnium	Ha	105	(260)	Technetium	Tc	43	(99)
Helium	He	2	4.00260	Tellerium	Te	52	127.60
Holmium	Ho	67	164.9303	Terbium	Tb	65	158.9254
Hydrogen	H	1	1.0080	Thallium	Tl	81	204.37
Indium	In	49	114.82	Thorium	Th	90	232.0381
Iodine	I	53	126.9045	Thulium	Tm	69	168.9342
Iridium	Ir	77	192.22	Tin	Sn	50	118.69
Iron	Fe	26	55.847	Titanium	Ti	22	47.90
Krypton	Kr	36	83.80	Tungsten	W	74	183.85
Lanthanium	La	57	138.9055	Uranium	U	92	238.029
Lawrencium	Lr	103	(257)	Vanadium	Vy	23	50.9414
Lead	Pb	82	207.12	Xenon	Xe	54	131.30
Lithium	Li	3	6.941	Ytterbium	Yb	70	173.04
Lutetium	Lu	71	174.97	Yttrium	Y	39	88.9059
Magnesium	Mg	12	24.305	Zinc	Zn	30	65.37
Manganese	Mn	25	54.9380	Zirconium	Zr	40	91.22
Mendelevium	Md	101	(256)				

Mathematical and Physical Constants

TABLE B1   Metric Prefixes

Multiple		Name	Abbreviation
1,000,000,000,000,000,000	10^{18}	exa	E
1,000,000,000,000,000	10^{15}	peta	P
1,000,000,000,000	10^{12}	tera	T
1,000,000,000	10^{9}	giga	G
1,000,000	10^{6}	mega	M
1,000	10^{3}	kilo	k
100	10^{2}	hecto	h
10	10^{1}	deka	da
1	1	—	—
0.1	10^{-1}	deci	d
0.01	10^{-2}	centi	c
0.001	10^{-3}	milli	m
0.000001	10^{-6}	micro	μ
0.000000001	10^{-9}	nano	n
0.000000000001	10^{-12}	pico	p
0.000000000000001	10^{-15}	femto	f
0.000000000000000001	10^{-18}	atto	a

TABLE B2   Physical Constants

Acceleration due to gravity	g	$9.8 \text{ m/s}^2 = 980 \text{ cm/s}^2 = 32.2 \text{ ft/s}^2$
Universal gravitational constant	G	$6.67 \times 10^{-11} \dfrac{\text{N-m}^2}{\text{kg}^2}$
Electron charge	e	$1.60 \times 10^{-19} \text{ C}$
Speed of light	c	$3.0 \times 10^{8} \text{ m/s} = 3.0 \times 10^{10} \text{ cm/s}$
		$= 1.86 \times 10^{5} \text{ mi/s}$
Boltzmann's constant	k	$1.38 \times 10^{-23} \text{ J/K}$
Planck's constant	h	$6.63 \times 10^{-34} \text{ J-s} = 4.14 \times 10^{-15} \text{ eV-s}$
	$\hbar$	$h/2\pi = 1.05 \times 10^{-34} \text{ J-s} = 6.58 \times 10^{-16} \text{ eV-s}$
Electron rest mass	m_e	$9.11 \times 10^{-31} \text{ kg} = 5.49 \times 10^{-4} \text{ u} \leftrightarrow 0.511 \text{ MeV}$
Proton rest mass	m_p	$1.672 \times 10^{-27} \text{ kg} = 1.00783 \text{ u} \leftrightarrow 938.3 \text{ MeV}$
Neutron rest mass	m_n	$1.674 \times 10^{-27} \text{ kg} = 1.00867 \text{ u} \leftrightarrow 939.1 \text{ MeV}$
Coulomb's law constant	k	$1/4\pi\varepsilon_0 = 9.0 \times 10^{9} \text{ N-m}^2/\text{C}^2$
Permittivity of free space	ε_0	$8.85 \times 10^{-12} \text{ C}^2/\text{N-m}^2$
Permeability of free space	μ_0	$4\pi \times 10^{-7} = 1.26 \times 10^{-6} \text{ Wb/A-m (T-M/A)}$
Astronomical and Earth data		
Radius of Earth		
equatorial		$3963 \text{ mi} = 6.378 \times 10^{6} \text{ m}$
polar		$3950 \text{ mi} = 6.357 \times 10^{6} \text{ m}$
Mass of Earth		$6.0 \times 10^{24} \text{ kg}$
Mass of Moon		$7.4 \times 10^{22} \text{ kg} = \frac{1}{81} \text{ mass of Earth}$
Mass of Sun		$2.0 \times 10^{30} \text{ kg}$
Average distance of Earth		
from Sun		$93 \times 10^{6} \text{ mi} = 1.5 \times 10^{8} \text{ km}$
Average distance of Moon		
from Earth		$2.4 \times 10^{5} \text{ mi} = 3.8 \times 10^{5} \text{ km}$
Diameter of Moon		$2160 \text{ mi} \approx 3500 \text{ km}$
Diameter of Sun		$864,000 \text{ mi} \approx 1.4 \times 10^{6} \text{ km}$

TABLE B3 Conversion Factors

Mass	$1 \text{ g} = 10^{-3} \text{ kg}$ 1 metric ton = 1000 kg $1 \text{ kg} = 10^3 \text{ g}$ $1 \text{ u} = 1.66 \times 10^{-24} \text{ g} = 1.66 \times 10^{-27} \text{ kg}$
Length	$1 \text{ cm} = 10^{-2} \text{ m} = 0.394 \text{ in.}$ $1 \text{ m} = 10^{-3} \text{ km} = 3.28 \text{ ft} = 39.4 \text{ in.}$ $1 \text{ km} = 10^3 \text{ m} = 0.621 \text{ mi}$ $1 \text{ in.} = 2.54 \text{ cm} = 2.54 \times 10^{-2} \text{ m}$ $1 \text{ ft} = 12 \text{ in.} = 30.48 \text{ cm} = 0.3048 \text{ m}$ $1 \text{ mi} = 5280 \text{ ft} = 609 \text{ m} = 1.609 \text{ km}$
Area	$1 \text{ cm}^2 = 10^{-4} \text{ m}^2 = 0.1550 \text{ in}^2 = 1.08 \times 10^{-3} \text{ ft}^2$ $1 \text{ m}^2 = 10^4 \text{ cm}^2 = 10.76 \text{ ft}^2 = 1550 \text{ in}^2$ $1 \text{ in}^2 = 6.94 \times 10^{-3} \text{ ft}^2 = 6.45 \text{ cm}^2 = 6.45 \times 10^{-4} \text{ m}^2$ $1 \text{ ft}^2 = 144 \text{ in}^2 = 9.29 \times 10^{-2} \text{ m}^2 = 929 \text{ cm}^2$
Volume	$1 \text{ cm}^3 = 10^{-6} \text{ m}^3 = 3.53 \times 10^{-5} \text{ ft}^3 = 6.10 \times 10^{-2} \text{ in}^3$ $1 \text{ m}^3 = 10^6 \text{ cm}^3 = 10^3 \text{ liters} = 35.3 \text{ ft}^3 = 6.10 \times 10^4 \text{ in}^3 = 264 \text{ gal}$ $1 \text{ liter} = 10^3 \text{ cm}^3 = 10^{-3} \text{ m}^3 = 1.056 \text{ qt} = 0.264 \text{ gal}$ $1 \text{ in}^3 = 5.79 \times 10^{-4} \text{ ft}^3 = 16.4 \text{ cm}^3 = 1.64 \times 10^{-5} \text{ m}^3$ $1 \text{ ft}^3 = 1728 \text{ in}^3 = 7.48 \text{ gal} = 0.0283 \text{ m}^3 = 28.3 \text{ liters}$ $1 \text{ qt} = 2 \text{ pt} = 946.5 \text{ cm}^3 = 0.946 \text{ liter}$ $1 \text{ gal} = 4 \text{ qt} = 231 \text{ in}^3 = 3.785 \text{ liters}$
Time	$1 \text{ h} = 60 \text{ min} = 3600 \text{ s}$ $1 \text{ day} = 24 \text{ h} = 1440 \text{ min} = 8.64 \times 10^4 \text{ s}$ $1 \text{ year} = 365 \text{ days} = 8.76 \times 10^3 \text{ h} = 5.26 \times 10^5 \text{ min} = 3.16 \times 10^7 \text{ s}$
Angle	$360° = 2\pi \text{ rad}$ $180° = \pi \text{ rad}$ $1 \text{ rad} = 57.3°$ $90° = \pi/2 \text{ rad}$ $60° = \pi/3 \text{ rad}$ $1° = 0.0175 \text{ rad}$ $45° = \pi/4 \text{ rad}$ $30° = \pi/6 \text{ rad}$
Speed	$1 \text{ m/s} = 3.6 \text{ km/h} = 3.28 \text{ ft/s} = 2.24 \text{ mi/h}$ $1 \text{ km/h} = 0.278 \text{ m/s} = 0.621 \text{ mi/h} = 0.911 \text{ ft/s}$ $1 \text{ ft/s} = 0.682 \text{ mi/h} = 0.305 \text{ m/s} = 1.10 \text{ km/h}$ $1 \text{ mi/h} = 1.467 \text{ ft/s} = 1.609 \text{ km/h} = 0.447 \text{ m/s}$ $60 \text{ mi/h} = 88 \text{ ft/s}$
Force	$1 \text{ newton} = 10^5 \text{ dynes} = 0.225 \text{ lb}$ $1 \text{ lb} = 4.45 \text{ N}$ Equivalent weight of 1-kg mass on Earth's surface $= 2.2 \text{ lb} = 9.8 \text{ N}$
Pressure	$1 \text{ Pa (N/m}^2) = 1.45 \times 10^{-4} \text{ lb/in}^2 = 7.4 \times 10^{-3} \text{ torr (mm Hg)}$ $1 \text{ tor (mm Hg)} = 133 \text{ Pa (N/m}^2) = 0.02 \text{ lb/in}^2$ $1 \text{ atm} = 14.7 \text{ lb/in}^2 = 1.013 \times 10^5 \text{ N/m}^2 \text{ (Pa)}$ $= 30 \text{ in. Hg} = 76 \text{ cm Hg}$ $1 \text{ bar} = 10^5 \text{ N/m}^2 \text{ (Pa)}$ $1 \text{ millibar} = 10^2 \text{ N/m}^2 \text{ (Pa)}$
Energy	$1 \text{ J} = 10^7 \text{ ergs} = 0.738 \text{ ft-lb} = 0.239 \text{ cal} = 9.48 \times 10^{-4} \text{ Btu} = 6.24 \times 10^{18} \text{ eV}$ $1 \text{ kcal} = 4186 \text{ J} = 3.968 \text{ Btu}$ $1 \text{ Btu} = 1055 \text{ J} = 778 \text{ ft-lb} = 0.252 \text{ kcal}$ $1 \text{ cal} = 4.186 \text{ J} = 3.97 \times 10^{-3} \text{ Btu} = 3.09 \text{ ft-lb}$ $1 \text{ ft-lb} = 1.356 \text{ J} = 1.29 \times 10^{-3} \text{ Btu}$ $1 \text{ eV} = 1.60 \times 10^{-19} \text{ J}$
Power	$1 \text{ W} = 0.738 \text{ ft-lb/s} = 1.34 \times 10^{-3} \text{ hp} = 3.41 \text{ Btu/h}$ $1 \text{ ft-lb/s} = 1.36 \text{ W} = 1.82 \times 10^{-3} \text{ hp}$ $1 \text{ hp} = 550 \text{ ft-lb/s} = 745.7 \text{ W} = 2545 \text{ Btu/h}$
Rest mass- energy equivalents	$1 \text{ u} = 1.66 \times 10^{-27} \text{ kg} \leftrightarrow 931 \text{ MeV}$ $1 \text{ electron mass} = 9.11 \times 10^{-31} \text{ kg} = 5.49 \times 10^{-4} \text{ u} \leftrightarrow 0.511 \text{ MeV}$ $1 \text{ proton mass} = 1.672 \times 10^{-27} \text{ kg} = 1.00728 \text{ u} \leftrightarrow 938.3 \text{ MeV}$ $1 \text{ neutron mass} = 1.674 \times 10^{-27} \text{ kg} = 1.00867 \text{ u} \leftrightarrow 939.1 \text{ MeV}$

TABLE B4   Trigonometric Relationships

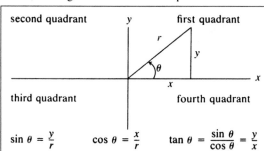

second quadrant $\qquad$ y $\qquad$ first quadrant

third quadrant $\qquad$ fourth quadrant

$$\sin \theta = \frac{y}{r} \qquad \cos \theta = \frac{x}{r} \qquad \tan \theta = \frac{\sin \theta}{\cos \theta} = \frac{y}{x}$$

$\theta °$ (rad)	$\sin \theta$	$\cos \theta$	$\tan \theta$
0° (0)	0	1	0
30° ($\pi/6$)	0.500	0.866	0.577
45° ($\pi/4$)	0.707	0.707	1.00
60° ($\pi/3$)	0.866	0.500	1.73
90° ($\pi/2$)	1	0	→ ∞

The sign of trigonometric functions depends on the quadrant, or sign of x and y, e.g., in the second quadrant $(-x, y)$, $-x/r = \cos \theta$ and $x/r = \sin \theta$, or by:

Reduction Formulas

	(θ in second quadrant)		(θ in third quadrant)		(θ in fourth quadrant)
$\sin \theta =$	$\cos (\theta - 90°)$	$=$	$-\sin (\theta - 180°)$	$=$	$-\cos (\theta - 270°)$
$\cos \theta =$	$-\sin (\theta - 90°)$	$=$	$-\cos (\theta - 180°)$	$=$	$\sin (\theta - 270°)$

Fundamental Identities
$$\sin^2\theta + \cos^2\theta = 1$$
$$\sin 2\theta = 2 \sin \theta \cos \theta$$
$$\cos 2\theta = \cos^2\theta - \sin^2\theta = 2 \cos^2\theta - 1 = 1 - 2 \sin^2\theta$$
$$\sin^2\theta = \tfrac{1}{2}(1 - \cos 2\theta)$$
$$\cos^2\theta = \tfrac{1}{2}(1 + \cos 2\theta)$$

For half-angle ($\theta/2$) identities, replace θ with $\theta/2$, e.g.,
$$\sin^2\theta/2 = \tfrac{1}{2}(1 - \cos \theta) \qquad \cos^2 \theta/2 = \tfrac{1}{2}(1 + \cos \theta)$$

For very small angles:
$$\cos \theta \approx 1$$

$$\sin (\alpha \pm \beta) = \sin \alpha \cos \beta \pm \cos \alpha \sin \beta$$

$$\sin \theta \approx \theta \text{ (radians)} \qquad \tan \theta = \frac{\sin \theta}{\cos \theta} \approx \theta$$

$$\cos (\alpha \pm \beta) = \cos \alpha \cos \beta \mp \sin \alpha \sin \beta$$

Law of sines:
$$\frac{a}{\sin \alpha} = \frac{b}{\sin \beta} = \frac{c}{\sin \gamma}$$

Law of cosines:
$$a^2 = b^2 + c^2 - 2bc \cos \alpha$$
$$b^2 = a^2 + c^2 - 2ac \cos \beta$$
$$c^2 = a^2 + b^2 - 2ab \cos \gamma$$

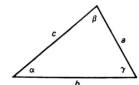

Standard Deviation and Method of Least Squares

STANDARD DEVIATION

To avoid the problem of negative deviations and absolute values, it is statistically convenient to use the square of the deviation.

The **variance** σ^2 of a set of measurements is the average of the squares of the deviations:

$$\sigma^2 = \frac{(x_1 - \bar{x})^2 + (x_2 - \bar{x})^2 + (x_3 - \bar{x})^2 + \cdots + (x_N - \bar{x})^2}{N}$$

$$= d_1^2 + d_2^2 + d_3^2 + \cdots d_N^2$$

$$= \frac{1}{N} \sum_{i=1}^{N} (x_i - \bar{x})^2 = \frac{1}{N} \sum_{i=1}^{N} d_i^2 \qquad \text{(C.1)}$$

The square root of the variance σ is called the **standard deviation***:

$$\sigma = \sqrt{\frac{1}{N} \sum_{i=1}^{N} (x_i - \bar{x})^2} = \sqrt{\frac{1}{N} \sum_{i=1}^{N} d_i^2} \qquad \text{(C.2)}$$

Because we take the average of the squares of the deviations and then the square root, the standard deviation is sometimes called the **root-mean-square deviation,** or simply the **root mean square.** Notice that σ always has the same units as x_i and that it is always positive.

Example C.1   What is the standard deviation of the set of numbers given in Example 1.6 in Experiment 1?

Solution   First find the square of the deviation of each of the numbers.

$$d_1^2 = (5.42 - 5.93)^2 = 0.26$$
$$d_2^2 = (6.18 - 5.93)^2 = 0.06$$
$$d_3^2 = (5.70 - 5.93)^2 = 0.05$$
$$d_4^2 = (6.01 - 5.93)^2 = 0.01$$
$$d_5^2 = (6.32 - 5.93)^2 = 0.15$$

Then

$$\sigma = \sqrt{\frac{1}{N} \sum_{i=1}^{5} d_i^2}$$

---

* For a small number of measurements, it can be statistically shown that a better value of the standard deviation is given by $\sigma = \sqrt{[1/(N-1)]\Sigma d_i^2}$, where N is replaced by $N - 1$. Your instructor may want you to use this form of the standard deviation.

$$= \left(\frac{0.26 + 0.06 + 0.05 + 0.01 + 0.15}{5} \right)^{1/2}$$

$$= 0.33$$

The experimental value E is then commonly reported as

$$E = \bar{x} \pm \sigma = 5.93 \pm 0.33$$

The standard deviation is used to describe the precision of the mean of a set of measurements. For a normal distribution of random errors,[†] it can be statistically shown that the probability that an individual measurement will fall within one standard deviation of the mean, which is assumed to be the true value, is 68% (● Fig. C.1). The probability of a measurement falling within two standard deviations is 95%.

METHOD OF LEAST SQUARES

Let $y' = m'x + b'$ be the predicted equation of the best-fitting straight line for a set of data. The vertical deviation of the ith data point from this line is then $(y_i - y_i')$.

The principle of least squares may be stated as follows: The "best-fitting" straight line is the one that minimizes the

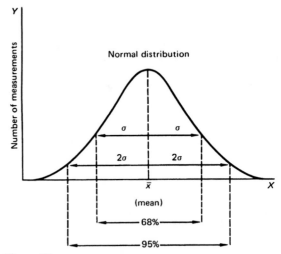

Figure C.1

[†] This *normal*, or *Gaussian, distribution* is represented by a "bell-shaped" curve (Fig. C.1). That is, the scatter, or dispersion, of the measurements is assumed to be symmetric about the true value of a quantity.

sum of the squares of the deviations of the measured y values from those of the predicted equation $y' = m'x + b'$.

The numerical values of the slope m' and intercept b' that minimize the sum of the squares of the deviations, $\sum_{i=1}^{N} (y_i - y_i')^2$, may be found using differential calculus. The results are as follows:

$$m' = \frac{M_{xy}}{M_{xx}}$$

and

$$b' = \bar{y} - m'\bar{x}$$

where $\bar{x}$ and $\bar{y}$ are the mean values, $\bar{x} = \sum_{i=1}^{N} x_i$ and $\bar{y} = \sum_{i=1}^{N} y_i$,

and

$$M_{xy} = \sum_{i=1}^{N} (x_i - \bar{x})(y_i - \bar{y})$$

$$= \sum_{i=1}^{N} x_i y_i - \frac{\left(\sum_{i=1}^{N} x_i\right)\left(\sum_{i=1}^{N} y_i\right)}{N}$$

$$M_{xx} = \sum_{i=1}^{N} (x_i - \bar{x})^2 = \sum_{i=1}^{N} x_i^2 - \frac{\left(\sum_{i=1}^{N} x_i\right)^2}{N}$$

where the sums of the deviations, e.g., $\sum_{i=1}^{N} (x_i - \bar{x})$, are zero.

Exercises

1. Plot the data given in Data Table 1 on a sheet of graph paper, and draw the straight line you judge to fit the data best.
2. Using the method of least squares, find the slope and intercept of the "best-fitting" straight line, and compare them with the slope and intercept of the line you drew in Exercise 1. Plot this "best-fitting" line on the graph. (Recall that the slope of a line is the change in y for a one-unit increase in x.)

DATA TABLE 1

	y_i	x_i	x_i^2	$x_i y_i$
	25	12		
	44	28		
	78	47		
	80	70		
	43	16		
	58	53		
	95	72		
	67	38		
Sums (Σ)				

Graphing Exponential Functions

In some cases, exponential functions of the form

$$N = N_0 e^{\lambda t} \qquad \text{(D.1)}$$

$$(\text{or } y = A e^{ax})$$

are plotted on Cartesian coordinates in linear form by first taking the natural, or Naperian, logarithm (base e) of both sides of the equation. For example, for $N = N_0 e^{\lambda t}$,

$$\ln N = \ln (N_0 e^{\lambda t}) = \ln N_0 + \ln e^{\lambda t} = \ln N_0 + \lambda t$$

or

$$\ln N = \lambda t + \ln N_0 \qquad \text{(D.2)}$$

Similarly, for $y = A e^{ax}$,

$$\ln y = \ln A + \ln e^{ax} = \ln A + ax$$

or

$$\ln y = ax + \ln A \qquad \text{(D.3)}$$

These equations have the general form of a straight line when plotted on a Cartesian graph ($y = mx + b$). For example, when we plot $\ln N$ versus t as Cartesian coordinates, the slope of the line is λ and the intercept is $\ln N_0$. The value of N_0 is obtained by taking the *antilog* of the intercept value $\ln N_0$. (For a decaying exponential, $N = N_0 e^{-\lambda t}$, the slope would be negative. Note that before plotting $\ln N$ versus t on Cartesian graph paper, we must find $\ln N$ for each value of N.)

Because logarithmic functions occur quite often in physics, special graph paper, called *semi-log graph paper,* is printed with graduations along the y or ordinate axis that are spaced logarithmically rather than linearly. The x or abscissa axis is graduated linearly. (Look at a sheet of semi-log graph paper.)

If a quantity is plotted on the ordinate axis of semi-log paper, the logarithmic graduated scale automatically takes the logarithm, so it is not necessary to look up the logarithm for each y value. However, commercial logarithmic graph paper is set up for common (base 10) logarithms rather than natural (base e) logarithms. Exponential functions may be treated as follows. Taking the (common) log of each side of $y = A e^{ax}$ yields

$$\log y = \log A + \log e^{ax}$$

$$= \log A + ax \log e$$

$$= \log A + (0.4343)ax \qquad \text{(D.4)}$$

where $\log e = 0.4343$.

Hence the slope of the resulting straight line is (0.4343)a rather than simply a.

The logarithmic ordinate scale is called "one-cycle," "two-cycle," and so on, depending on the number of powers of 10 covered on the axis. The beginnings of the cycles are consecutively labeled in multiples of 10 (e.g., 0.1, 1.0, 10, or 1.0, 10, 100, etc.), depending on the range (cycles) of the function. (Common logarithms can also be plotted on semi-log paper.)

Care must be taken in determining the slope of the line on a semi-log plot. On an ordinary Cartesian graph, the slope of a line is given by $\Delta y/\Delta x = (y_2 - y_1)/(x_2 - x_1)$. However, on a semi-log graph, the slope of a line is given by

$$\text{slope} = \frac{\Delta \log y}{\Delta x} \qquad \left(\text{or } \frac{\Delta \log N}{\Delta t} \right) \qquad \text{(D.5)}$$

On a semi-log plot, the listed ordinate values are y, not $\ln y$. Hence, one must explicitly take the logs of the ordinate values of the endpoints of the slope interval, y_2 and y_1, or the log of their ratio:

$$\text{slope} = \frac{\Delta \log y}{\Delta x} = \frac{\log y_2 - \log y_1}{x_2 - x_1}$$

$$= \frac{\log y_2/y_1}{x_2 - x_1} \qquad \text{(D.6)}$$

The value of N_0 can be read directly from the y-intercept of the graph.

Another common equation form in physics is

$$y = ax^n \qquad \text{(D.7)}$$

For example, the electric field, $E = kq/r^2 = kqr^{-2}$, is of this form, with $a = kq$ and $n = -2$. By plotting y versus x^n on Cartesian graph paper, we obtain a straight line with a slope of a. However, in an experiment the measured values are usually y and x, so computation of the x^n's is required.

But in some instances the exponent n may not be known. This constant, along with the constant a, may be found by plotting y versus x on log graph paper. (This is commonly called *log-log graph paper* because of the logarithmic graduations on both axes. Look at a sheet of log-log graph paper.)

At logarithmic graduations on axes, we again automatically take the logarithms of x and y. Working with common logarithms (base 10) in this instance, we find that the log-log plot of y versus x yields a straight line, as can be seen by taking the (common) log of both sides of Eq. D.7.

$$\log y = \log (ax^n) = \log a + \log x^n$$
$$= \log a + n \log x$$

or

$$\log y = n \log x + \log a \qquad \textbf{(D.8)}$$

which has the general form of a straight line with a slope of n and an intercept of $\log a$. For the electric field example, this would be

$$E = \frac{kq}{r^2} = kqr^{-2}$$

$$\log E = -2 \log r + \log kq$$

Again, care must be taken in determining the slope of a straight line on a log-log graph. In this case,

$$\text{slope} = \frac{\Delta \log y}{\Delta \log x}$$

$$= \frac{\log y_2 - \log y_1}{\log x_2 - \log x_1} = \frac{\log y_2/y_1}{\log x_2/x_1} \qquad \textbf{(D.9)}$$

and the logs of the endpoints of the slope interval or their ratio must be found explicitly. (The ordinate and abscissa values on the log-log plot are y and x, *not* $\log y$ and $\log x$.)

As in the case of the semi-log plot, the value of a in $y = ax^n$ can be read directly from the y-intercept of the graph. However, in this case, the intercept is not at $x = 0$ *but* at $x = 1$, since the intercept $\log y = \log a$ requires that $\log x = 0$ and $\log 1 = 0$.